Year 3B

A Guide to Teaching for Mastery

Series Editor: Tony Staneff
Lead author: Josh Lury

Contents

Introduction to the author team	4
What is *Power Maths*?	5
What's different in the new edition?	6
Your *Power Maths* resources	7
The *Power Maths* teaching model	10
The *Power Maths* lesson sequence	12
Using the *Power Maths* Teacher Guide	15
Power Maths Year 3, yearly overview	16
Mindset: an introduction	22
The *Power Maths* characters	23
Mathematical language	24
The role of talk and discussion	25
Assessment strategies	26
Keeping the class together	28
Same-day intervention	29
The role of practice	30
Structures and representations	31
Variation helps visualisation	32
Practical aspects of *Power Maths*	33
Working with children below age-related expectation	35
Providing extra depth and challenge with *Power Maths*	37
Using *Power Maths* with mixed age classes	39
List of practical resources	40
Getting started with *Power Maths*	43

Unit 6 – Multiplication and division (3) — 44
Multiples of 10	46
Related calculations	50
Reasoning about multiplication	54
Multiply 2-digits by 1-digit – no exchange	58
Multiply 2-digits by 1-digit – exchange	62
Expanded written method	66
Link multiplication and division	70
Divide 2-digits by 1-digit – no exchange	74
Divide 2-digits by 1-digit – flexible partitioning	78
Divide 2-digits by 1-digit with remainders	82
How many ways?	86
Problem solving – mixed problems (1)	90
Problem solving – mixed problems (2)	94
End of unit check	98

Unit 7 – Length and perimeter — 100
Measure in m and cm	102
Measure in cm and mm	106
Metres, centimetres and millimetres	110
Equivalent lengths (m and cm)	114
Equivalent lengths (mm and cm)	118

Compare lengths   122
Add lengths   126
Subtract lengths   130
Measure perimeter   134
Calculate perimeter   138
Problem solving – length   142
End of unit check   146

Unit 8 – Fractions (1)   148
Understand the denominator of unit fractions   150
Compare and order unit fractions   154
Understand the numerator of non-unit fractions   158
Understand the whole   162
Compare and order non-unit fractions   166
Divisions on a number line   170
Count in fractions on a number line   174
Equivalent fractions as bar models   178
Equivalent fractions on a number line   182
Equivalent fractions   186
End of unit check   190

Unit 9 – Mass   192
Use scales   194
Measure mass   198
Measure mass in kilograms and grams   202
Equivalent masses   206
Compare mass   210
Add and subtract mass   214
Problem solving – mass   218
End of unit check   222

Unit 10 – Capacity   224
Measure capacity and volume in litres and millilitres   226
Measure in litres and millilitres   230
Equivalent capacities and volumes (litres and millilitres)   234
Compare capacity and volume   238
Add and subtract capacity and volume   242
Problem solving – capacity   246
End of unit check   250

Introduction to the author team

Power Maths arises from the work of maths mastery experts who are committed to proving that, given the right mastery mindset and approach, **everyone can do maths**. Based on robust research and best practice from around the world, *Power Maths* was developed in partnership with a group of UK teachers to make sure that it not only meets our children's wide-ranging needs but also aligns with the National Curriculum in England.

Power Maths – White Rose Maths edition

This edition of *Power Maths* has been developed and updated by:

Tony Staneff, Series Editor and Author

Vice Principal at Trinity Academy, Halifax, Tony also leads a team of mastery experts who help schools across the UK to develop teaching for mastery via nationally recognised CPD courses, problem-solving and reasoning resources, schemes of work, assessment materials and other tools.

Josh Lury, Lead Author

Josh is a specialist maths teacher, author and maths consultant with a passion for innovative and effective maths education.

The first edition of *Power Maths* was developed by a team of experienced authors, including:

- **Tony Staneff and Josh Lury**
- **Trinity Academy Halifax** (Michael Gosling CEO, Emily Fox, Kate Henshall, Rebecca Holland, Stephanie Kirk, Stephen Monaghan and Rachel Webster)
- **David Board, Belle Cottingham, Jonathan East, Tim Handley, Derek Huby, Neil Jarrett, Stephen Monaghan, Beth Smith, Tim Weal, Paul Wrangles** – skilled maths teachers and mastery experts
- **Cherri Moseley** – a maths author, former teacher and professional development provider
- **Professors Liu Jian and Zhang Dan**, Series Consultants and authors, and their team of mastery expert authors: **Wei Huinv, Huang Lihua, Zhu Dejiang, Zhu Yuhong, Hou Huiying, Yin Lili, Zhang Jing, Zhou Da and Liu Qimeng**

    Used by over 20 million children, Professor Liu Jian's textbook programme is one of the most popular in China. He and his author team are highly experienced in intelligent practice and in embedding key maths concepts using a C-P-A approach.

- **A group of 15 teachers and maths co-ordinators**

    We consulted our teacher group throughout the development of *Power Maths* to ensure we are meeting their real needs in the classroom.

What is *Power Maths*?

Created especially for UK primary schools, and aligned with the new National Curriculum, *Power Maths* is a whole-class, textbook-based mastery resource that empowers every child to understand and succeed. *Power Maths* rejects the notion that some people simply 'can't do' maths. Instead, it develops growth mindsets and encourages hard work, practice and a willingness to see mistakes as learning tools.

Best practice consistently shows that mastery of small, cumulative steps builds a solid foundation of deep mathematical understanding. *Power Maths* combines interactive teaching tools, high-quality textbooks and continuing professional development (CPD) to help you equip children with a deep and long-lasting understanding. Based on extensive evidence, and developed in partnership with practising teachers, *Power Maths* ensures that it meets the needs of children in the UK.

Power Maths and Mastery

Power Maths makes mastery practical and achievable by providing the structures, pathways, content, tools and support you need to make it happen in your classroom.

To develop mastery in maths, children must be enabled to acquire a deep understanding of maths concepts, structures and procedures, step by step. Complex mathematical concepts are built on simpler conceptual components and when children understand every step in the learning sequence, maths becomes transparent and makes logical sense. Interactive lessons establish deep understanding in small steps, as well as effortless fluency in key facts such as tables and number bonds. The whole class works on the same content and no child is left behind.

Power Maths

- Builds every concept in small, progressive steps
- Is built with interactive, whole-class teaching in mind
- Provides the tools you need to develop growth mindsets
- Helps you check understanding and ensure that every child is keeping up
- Establishes core elements such as intelligent practice and reflection

The *Power Maths* approach

Everyone can!
Founded on the conviction that every child can achieve, *Power Maths* enables children to build number fluency, confidence and understanding, step by step.

Child-centred learning
Children master concepts one step at a time in lessons that embrace a concrete-pictorial-abstract (C-P-A) approach, avoid overload, build on prior learning and help them see patterns and connections. Same-day intervention ensures sustained progress.

Continuing professional development
Embedded teacher support and development offer every teacher the opportunity to continually improve their subject knowledge and manage whole-class teaching for mastery.

Whole-class teaching
An interactive, whole-class teaching model encourages thinking and precise mathematical language and allows children to deepen their understanding as far as they can.

What's different in the new edition?

If you have previously used the first editions of *Power Maths*, you might be interested to know how this edition is different. All of the improvements described below are based on feedback from *Power Maths* customers.

Changes to units and the progression

- The order of units has been slightly adjusted, creating closer alignment between adjacent year groups, which will be useful for mixed age teaching.
- The flow of lessons has been improved within units to optimise the pace of the progression and build in more recap where needed. For key topics, the sequence of lessons gives more opportunities to build up a solid base of understanding. Other units have fewer lessons than before, where appropriate, making it possible to fit in all the content.
- Overall, the lessons put more focus on the most essential content for that year, with less time given to non-statutory content.
- The progression of lessons matches the steps in the new White Rose Maths schemes of learning.

Lesson resources

- There is a Quick recap for each lesson in the Teacher Guide, which offers an alternative lesson starter to the Power Up for cases where you feel it would be more beneficial to surface prerequisite learning than general number fluency.
- In the **Discover** and **Share** sections there is now more of a progression from 1 a) to 1 b). Whereas before, 1 b) was mainly designed as a separate question, now 1 a) leads directly into 1 b). This means that there is an improved whole-class flow, and also an opportunity to focus on the logic and skills in more detail. As a teacher, you will be using 1 a) to lead the class into the thinking, then 1 b) to mould that thinking into the core new learning of the lesson.
- In the **Share** section, for KS1 in particular, the number of different models and representations has been reduced, to support the clarity of thinking prompted by the flow from 1 a) into 1 b).
- More fluency questions have been built into the guided and independent practice.
- Pupil pages are as easy as possible for children to access independently. The pages are less full where this supports greater focus on key ideas and instructions. Also, more freedom is offered around answer format, with fewer boxes scaffolding children's responses; squared paper backgrounds are used in the Practice Books where appropriate. Artwork has also been revisited to ensure the highest standards of accessibility.

New components

480 Individual Practice Games are available in *ActiveLearn* for practising key facts and skills in Years 1 to 6. These are designed in an arcade style, to feel like fun games that children would choose to play outside school. They can be accessed via the Pupil World for homework or additional practice in school – and children can earn rewards. There are Support, Core and Extend levels to allocate, with Activity Reporting available for the teacher. There is a Quick Guide on *ActiveLearn* and you can use the Help area for support in setting up child accounts.

There is also a new set of lesson video resources on the Professional Development tile, designed for in-school training in 10- to 20-minute bursts. For each part of the *Power Maths* lesson sequence, there is a slide deck with embedded video, which will facilitate discussions about how you can take your *Power Maths* teaching to the next level.

Your *Power Maths* resources

Pupil Textbooks

Discover, **Share** and **Think together** sections promote discussion and introduce mathematical ideas logically, so that children understand more easily.

Using a Concrete-Pictorial-Abstract approach, clear mathematical models help children to make connections and grasp concepts.

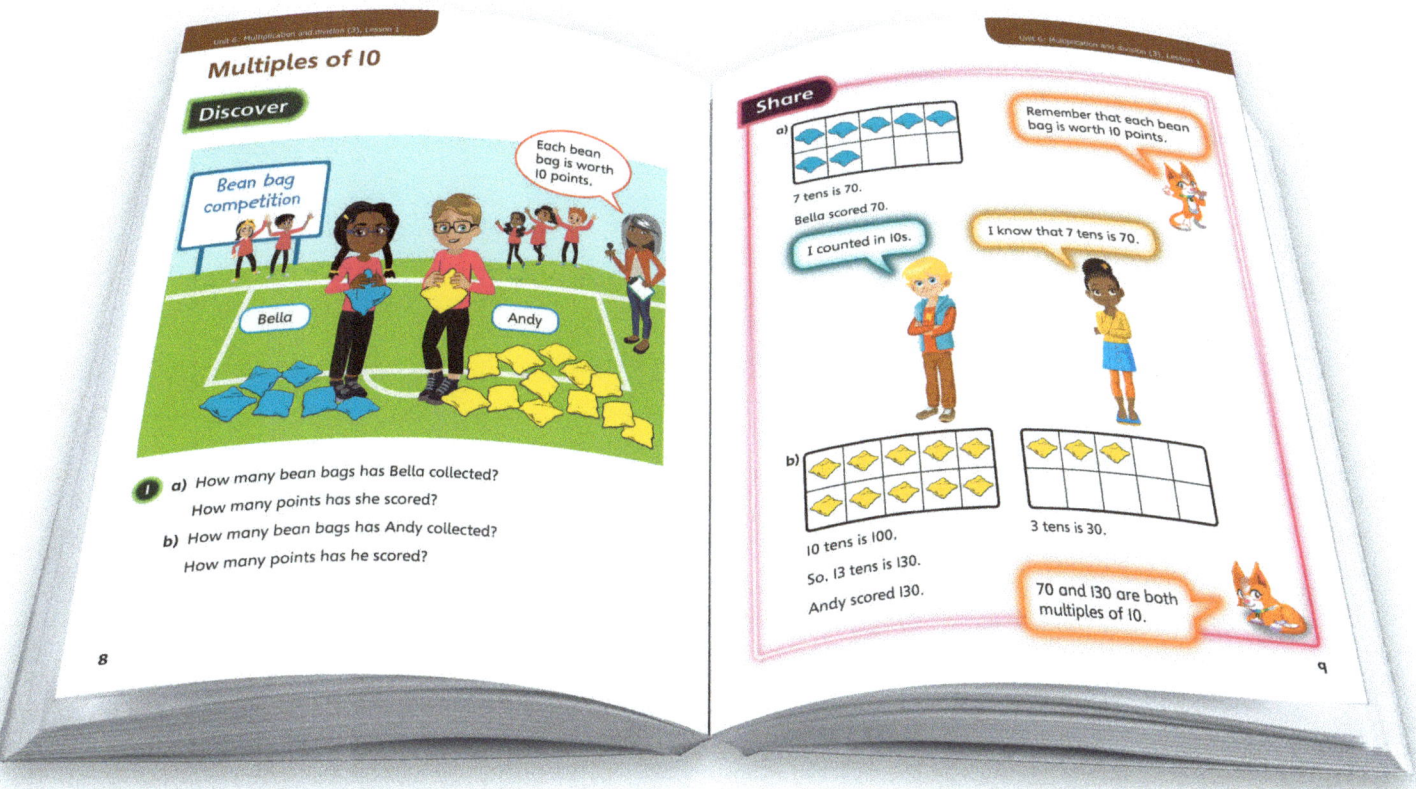

Appealing scenarios stimulate curiosity, helping children to identify the maths problem and discover patterns and relationships for themselves.

Friendly, supportive characters help children develop a growth mindset by prompting them to think, reason and reflect.

To help you teach for mastery, *Power Maths* comprises a variety of high-quality resources.

The coherent *Power Maths* lesson structure carries through into the vibrant, high-quality textbooks. Setting out the core learning objectives for each class, the lesson structure follows a carefully mapped journey through the curriculum and supports children on their journey to deeper understanding.

Pupil Practice Books

The Practice Books offer just the right amount of intelligent practice for children to complete independently in the final section of each lesson.

Practice questions are finely tuned to move children forward in their thinking and to reveal misconceptions.

The practice questions are for everyone – each question varies one small element to move children on in their thinking.

Calculations are connected so that children think about the underlying concept.

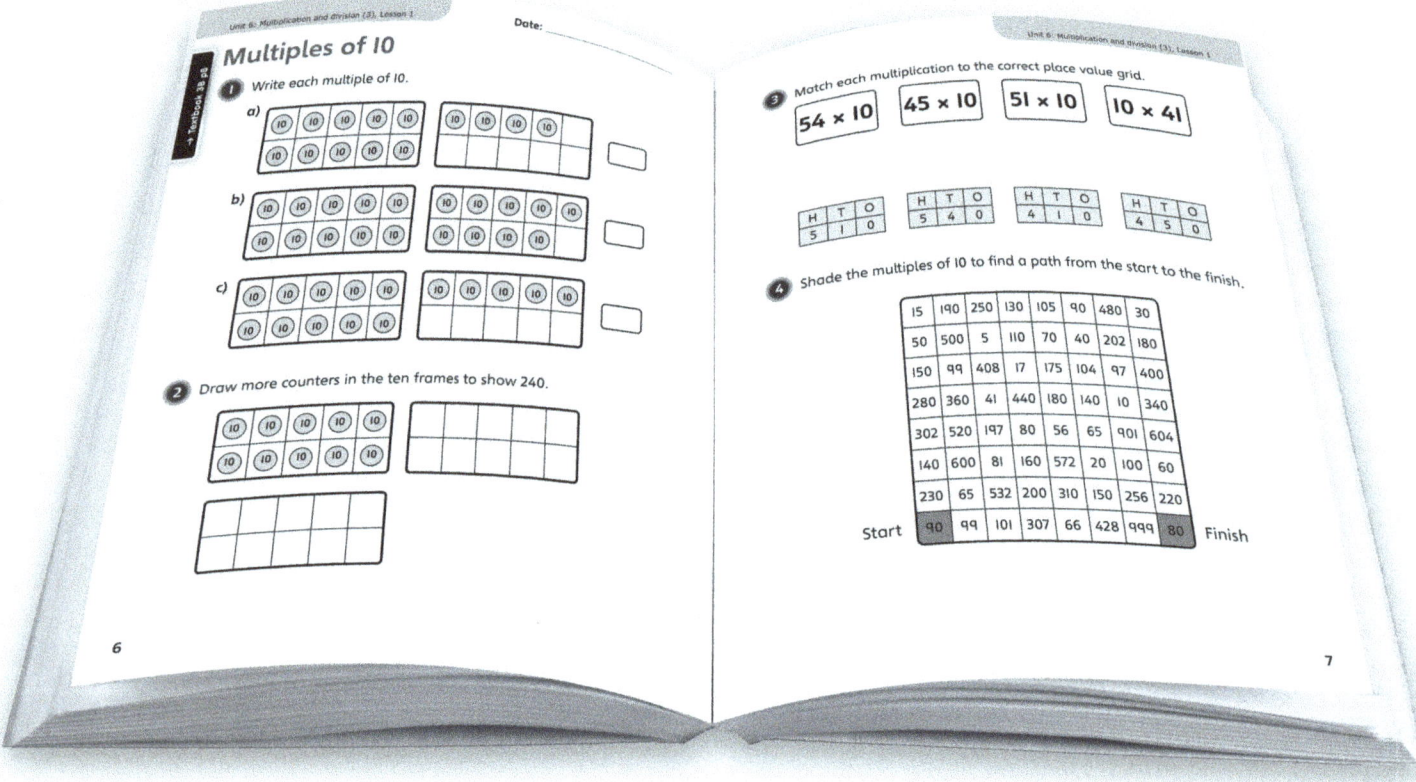

Challenge questions allow children to delve deeper into a concept.

The *Power Maths* characters support and encourage children to think and work in different ways.

Think differently questions encourage children to use reasoning as well as their mathematical knowledge to reach a solution.

Reflect questions reveal the depth of each child's understanding before they move on.

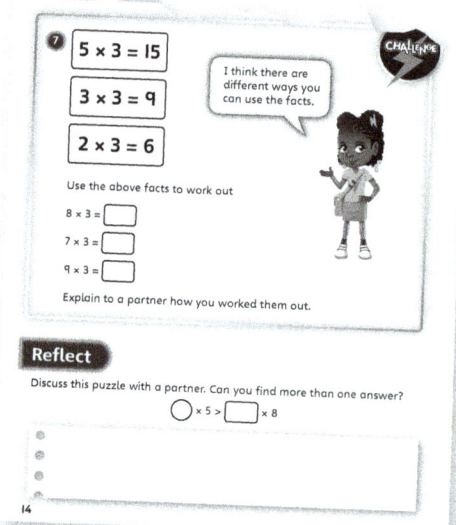

Online subscription

The online subscription will give you access to additional resources and answers from the Textbook and Practice Book.

eTextbooks

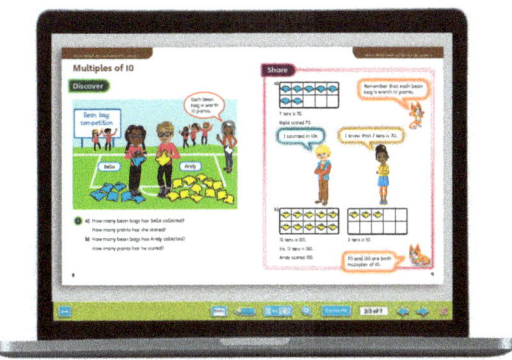

Digital versions of Power Maths Textbooks allow class groups to share and discuss questions, solutions and strategies. They allow you to project key structures and representations at the front of the class, to ensure all children are focusing on the same concept.

Teaching tools

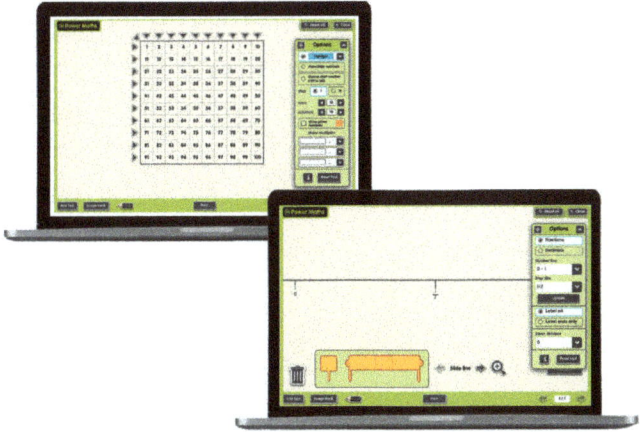

Here you will find interactive versions of key Power Maths structures and representations.

Power Ups

Use this series of daily activities to promote and check number fluency.

Online versions of Teacher Guide pages

PDF pages give support at both unit and lesson levels. You will also find help with key strategies and templates for tracking progress.

Unit videos

Watch the professional development videos at the start of each unit to help you teach with confidence. The videos explore common misconceptions in the unit, and include intervention suggestions as well as suggestions on what to look out for when assessing mastery in your students.

End of unit Strengthen and Deepen materials

The Strengthen activity at the end of every unit addresses a key misconception and can be used to support children who need it. The Deepen activities are designed to be low ceiling/high threshold and will challenge those children who can understand more deeply. These resources will help you ensure that every child understands and will help you keep the class moving forward together. These printable activities provide an optional resource bank for use after the assessment stage.

Individual Practice Games

These enjoyable games can be used at home or at school to embed key number skills (see page 6).

Professional Development videos and slides

These slides and videos of Power Maths lessons can be used for ongoing training in short bursts or to support new staff.

The *Power Maths* teaching model

At the heart of *Power Maths* is a clearly structured teaching and learning process that helps you make certain that every child masters each maths concept securely and deeply. For each year group, the curriculum is broken down into core concepts, taught in units. A unit divides into smaller learning steps – lessons. Step by step, strong foundations of cumulative knowledge and understanding are built.

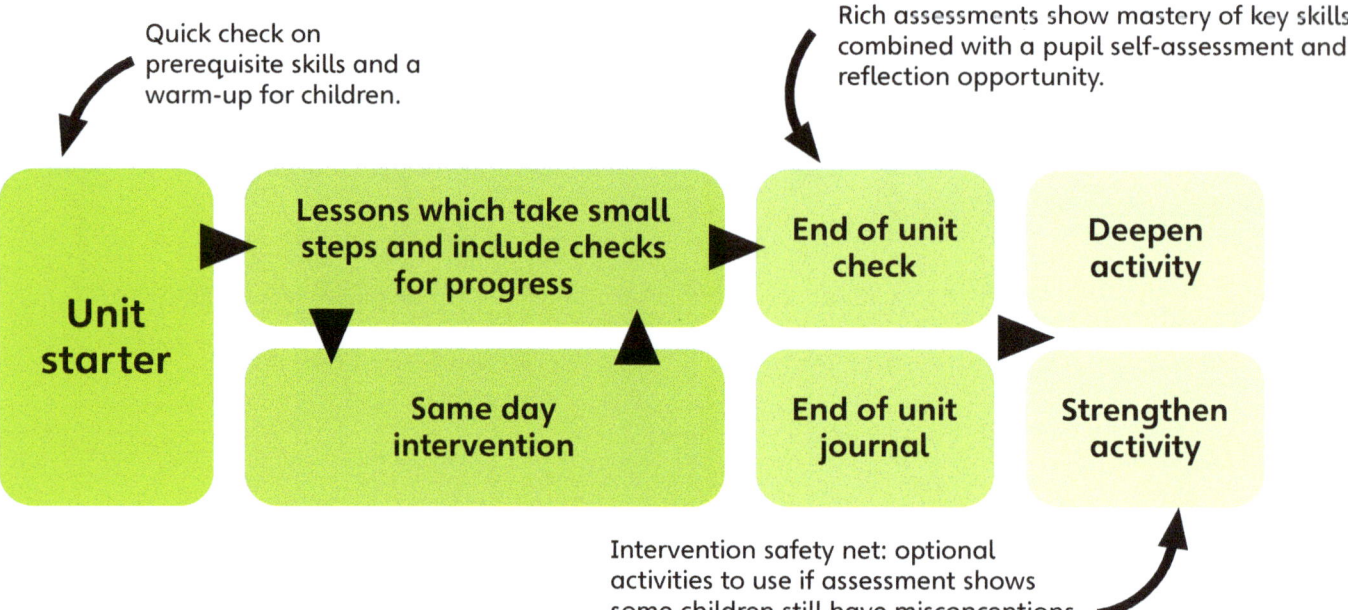

Unit starter

Each unit begins with a unit starter, which introduces the learning context along with key mathematical vocabulary and structures and representations.

- The Textbooks include a check on readiness and a warm-up task for children to complete.
- Your Teacher Guide gives support right from the start on important structures and representations, mathematical language, common misconceptions and intervention strategies.
- Unit-specific videos develop your subject knowledge and insights so you feel confident and fully equipped to teach each new unit. These are available via the online subscription.

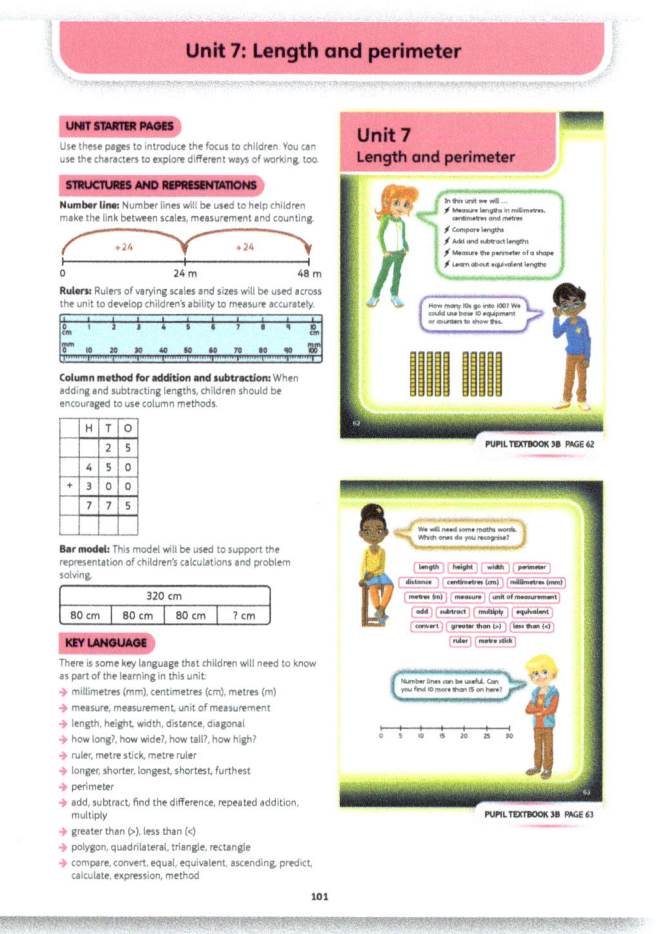

10

Lesson

Once a unit has been introduced, it is time to start teaching the series of lessons.

- Each lesson is scaffolded with Textbook and Practice Book activities and begins with a Power Up activity (available via online subscription) or the Quick recap activity in the Teacher Guide (see page 15).
- *Power Maths* identifies lesson by lesson what concepts are to be taught.
- Your Teacher Guide offers lots of support for you to get the most from every child in every lesson. As well as highlighting key points, tricky areas and how to handle them, you will also find question prompts to check on understanding and clarification on why particular activities and questions are used.

Same-day intervention

Same-day interventions are vital in order to keep the class progressing together. This can be during the lesson as well as afterwards (see page 29). Therefore, *Power Maths* provides plenty of support throughout the journey.

- Intervention is focused on keeping up now, not catching up later, so interventions should happen as soon as they are needed.
- Practice section questions are designed to bring misconceptions to the surface, allowing you to identify these easily as you circulate during independent practice time.
- Child-friendly assessment questions in the Teacher Guide help you identify easily which children need to strengthen their understanding.

End of unit check and journal

For each unit, the End of unit check in the Textbook lets you see which children have mastered the key concepts, which children have not and where their misconceptions lie. The Practice Books also include an End of unit journal in which children can reflect on what they have learned. Each unit also offers Strengthen and Deepen activities, available via the online subscription.

> The Teacher Guide offers different ways of managing the End of unit assessments as well as giving support with handling misconceptions.

> The End of unit check presents multiple-choice questions. Children think about their answer, decide on a solution and explain their choice.

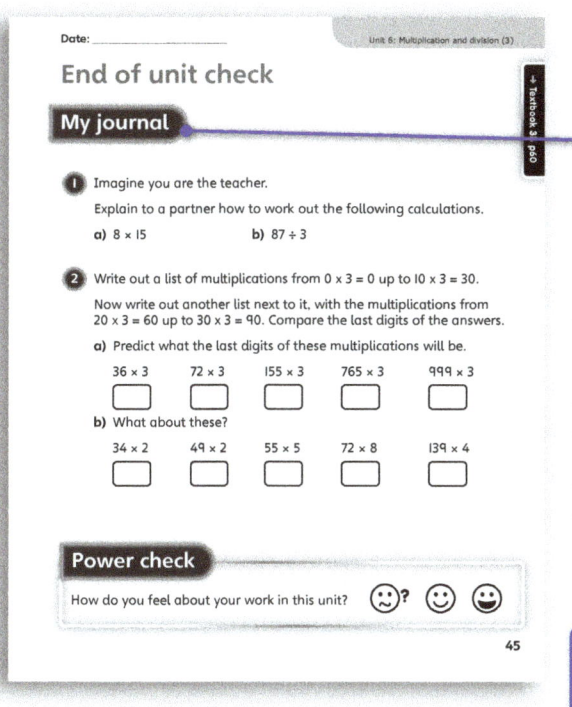

> The End of unit journal is an opportunity for children to test out their learning and reflect on how they feel about it. Tackling the 'journal' problem reveals whether a child understands the concept deeply enough to move on to the next unit.

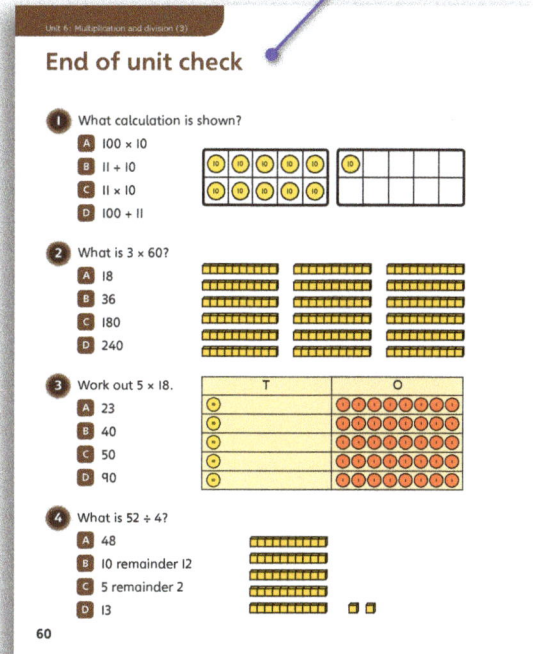

> In KS2, the End of unit assessment will also include at least one SATs-style question.

The *Power Maths* lesson sequence

At the heart of *Power Maths* is a unique lesson sequence designed to empower children to understand core concepts and grow in confidence. Embracing the National Centre for Excellence in the Teaching of Mathematics' (NCETM's) definition of mastery, the sequence guides and shapes every *Power Maths* lesson you teach.

Flexibility is built into the *Power Maths* programme so there is no one-to-one mapping of lessons and concepts and you can pace your teaching according to your class. While some children will need to spend longer on a particular concept (through interventions or additional lessons), others will reach deeper levels of understanding. However, it is important that the class moves forward together through the termly schedules.

Power Up  5 minutes

- Each lesson begins with a Power Up activity (available via the online subscription) which supports fluency in key number facts.

- The whole-class approach depends on fluency, so the Power Up is a powerful and essential activity.

- The Quick recap is an alternative starter, for when you think some or all children would benefit more from revisiting pre-requisite work (see page 15).

TOP TIP
If the class is struggling with the task, revisit it later and check understanding.

- Power Ups reinforce the two key things that are essential for success: times-tables and number bonds.

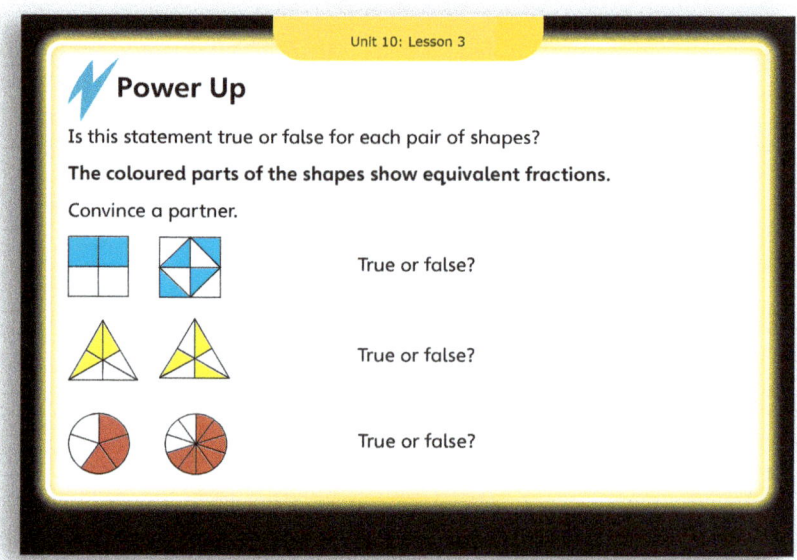

Discover 10 minutes

- A practical, real-life problem arouses curiosity. Children find the maths through story telling.

- A real-life scenario is provided for the **Discover** section but feel free to build upon these with your own examples that are more relevant to your class, or get creative with the context.

TOP TIP
Discover works best when run at tables, in pairs with concrete objects.

- Question ❶ a) tackles the key concept and question ❶ b) digs a little deeper. Children have time to explore, play and discuss possible strategies.

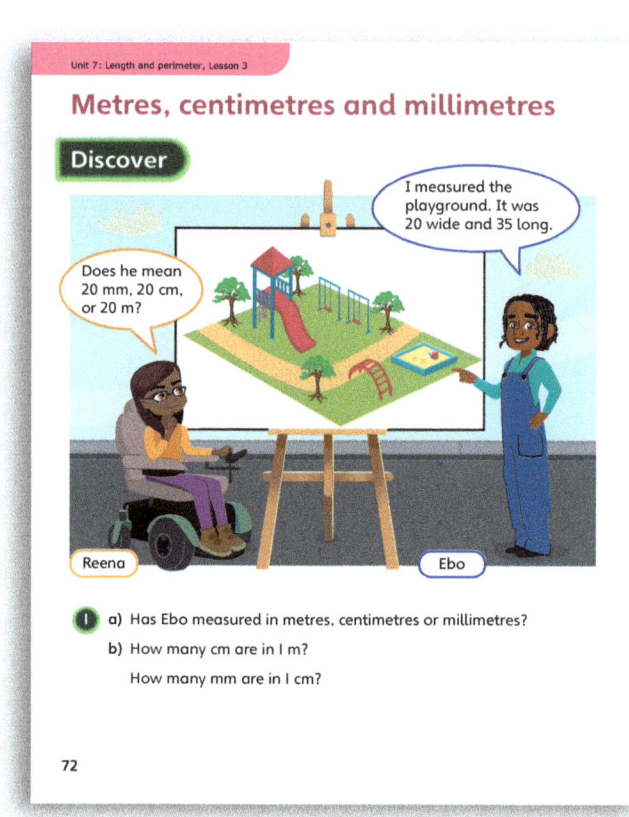

12

Share ⏱ 10 minutes

Teacher-led, this interactive section follows the **Discover** activity and highlights the variety of methods that can be used to solve a single problem.

TOP TIP
Pairs sharing a textbook is a great format for **Share**!

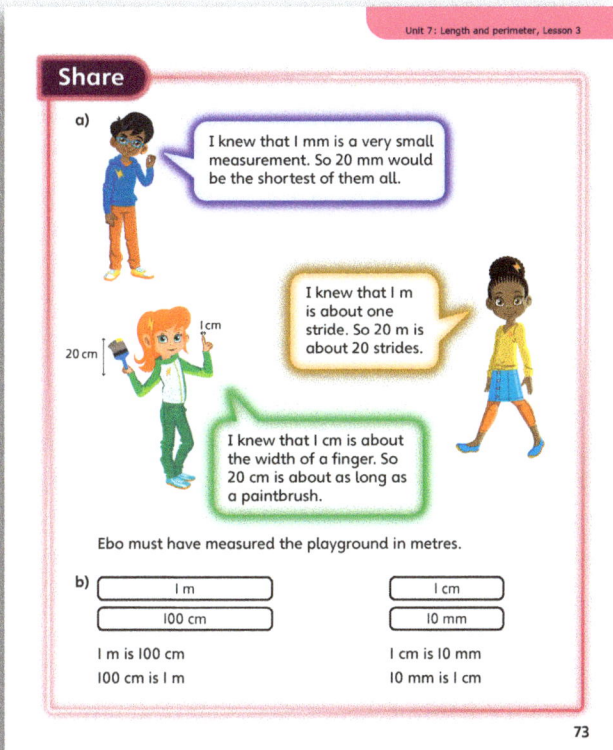

Your Teacher Guide gives target questions for children. The online toolkit provides interactive structures and representations to link concrete and pictorial to abstract concepts.

Bring children to the front to share and celebrate their solutions and strategies.

Think together

⏱ 10 minutes

Children work in groups on the carpet or at tables, using their textbooks or eBooks.

TOP TIP
Make sure children have mini whiteboards or pads to write on if they are not at their tables.

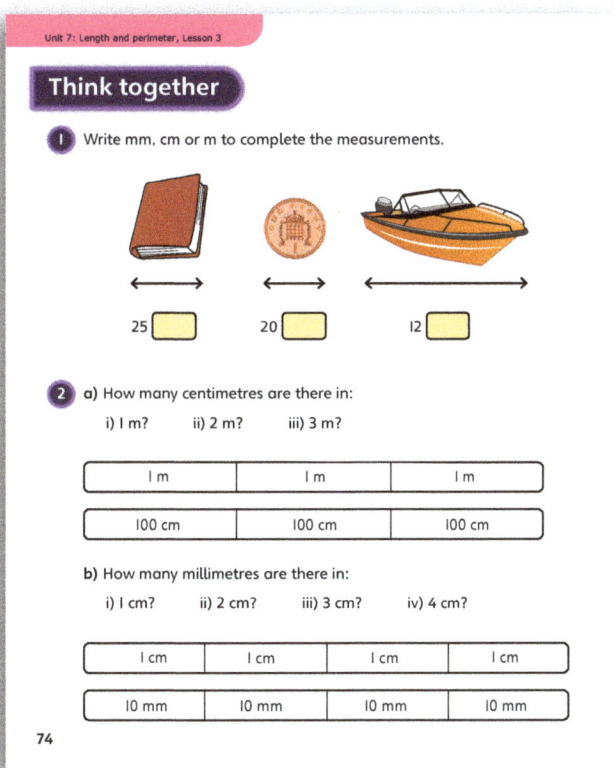

Using the Teacher Guide, model question ❶ for your class.

Question ❷ is less structured. Children will need to think together in their groups, then discuss their methods and solutions as a class.

In question ❸ children try working out the answer independently. The openness of the **Challenge** question helps to check depth of understanding.

Practice ⏱ 15 minutes

Using their Practice Books, children work independently while you circulate and check on progress.

Questions follow small steps of progression to deepen learning.

TOP TIP
Some children could work separately with a teacher or assistant.

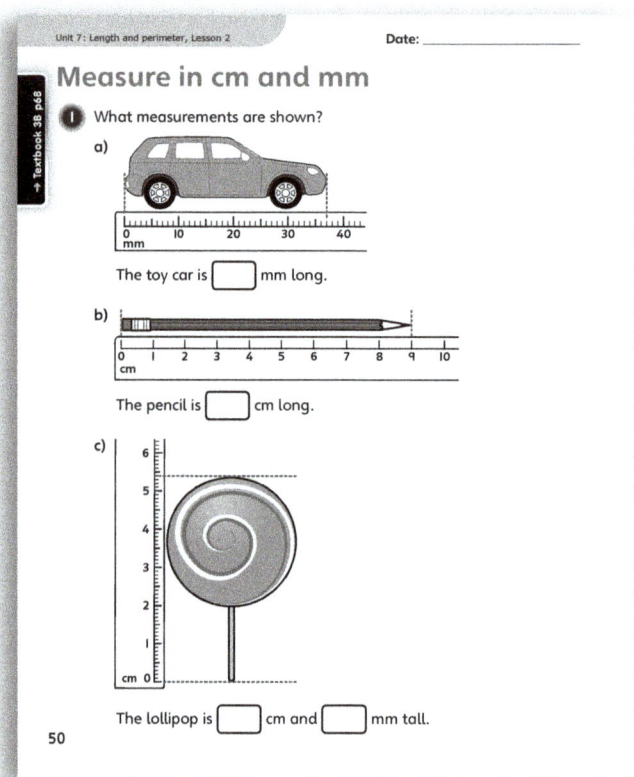

Are some children struggling? If so, work with them as a group, using mathematical structures and representations to support understanding as necessary.

There are no set routines: for real understanding, children need to think about the problem in different ways.

Reflect ⏱ 5 minutes

'Spot the mistake' questions are great for checking misconceptions.

The **Reflect** section is your opportunity to check how deeply children understand the target concept.

The Practice Books use various approaches to check that children have fully understood each concept.

Looking like they understand is not enough! It is essential that children can show they have grasped the concept.

Using the *Power Maths* Teacher Guide

Think of your Teacher Guides as *Power Maths* handbooks that will guide, support and inspire your day-to-day teaching. Clear and concise, and illustrated with helpful examples, your Teacher Guides will help you make the best possible use of every individual lesson. They also provide wrap-around professional development, enhancing your own subject knowledge and helping you to grow in confidence about moving your children forward together.

There is a Teacher Guide per year group for every term, with unit and lesson level guidance and support.

Never feel stuck! You will find ideas for introducing every unit and lesson and questions to encourage teacher reflection before and after each lesson.

Tips and advice on key elements such as C-P-A approaches, misconceptions, language, modelling growth mindsets and same day intervention.

Annotations for every Textbook and Practice Book page, providing prompts for key questions to ask to expose understanding and explanations as to why key questions have been chosen.

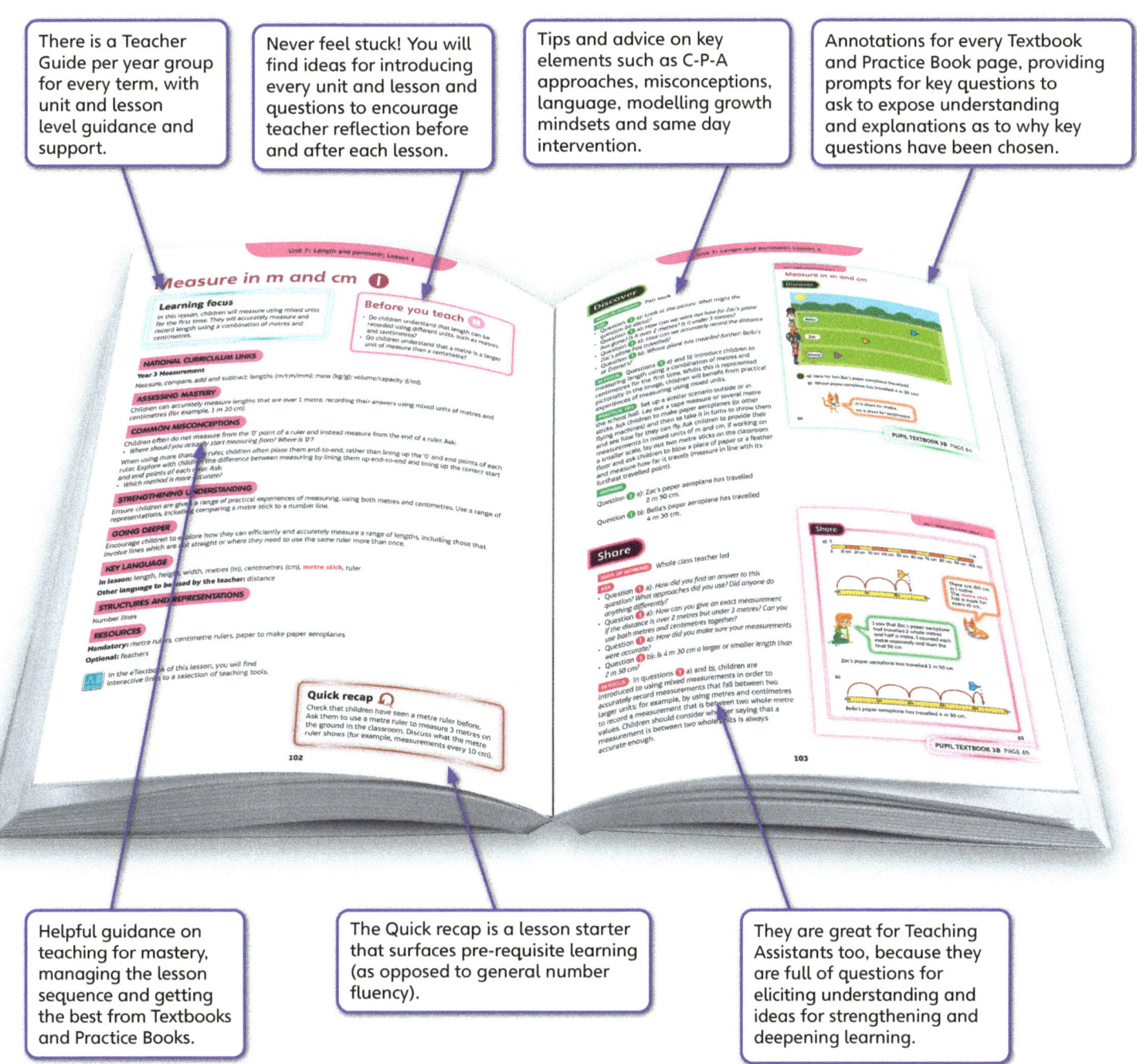

Helpful guidance on teaching for mastery, managing the lesson sequence and getting the best from Textbooks and Practice Books.

The Quick recap is a lesson starter that surfaces pre-requisite learning (as opposed to general number fluency).

They are great for Teaching Assistants too, because they are full of questions for eliciting understanding and ideas for strengthening and deepening learning.

At the end of each unit, your Teacher Guide helps you identify who has fully grasped the concept, who has not and how to move every child forward. This is covered later in the Assessment strategies section.

Power Maths Year 3, yearly overview

Textbook	Strand	Unit		Number of lessons
Textbook A / Practice Workbook A (Term 1)	Number – number and place value	1	Place value within 1,000	13
	Number – addition and subtraction	2	Addition and subtraction (1)	10
	Number – addition and subtraction	3	Addition and subtraction (2)	13
	Number – multiplication and division	4	Multiplication and division (1)	5
	Number – multiplication and division	5	Multiplication and division (2)	13
Textbook B / Practice Workbook B (Term 2)	Number – multiplication and division	6	Multiplication and division (3)	13
	Measurement	7	Length and perimeter	11
	Number – fractions	8	Fractions (1)	10
	Measurement	9	Mass	7
	Measurement	10	Capacity	6
Textbook C / Practice Workbook C (Term 3)	Number – fractions	11	Fractions (2)	8
	Measurement	12	Moneys	5
	Measurement	13	Time	12
	Geometry – properties of shapes	14	Angles and properties of shapes	9
	Statistics	15	Statistics	7

Power Maths Year 3, Textbook 3B (Term 2) overview

Strand	Unit	Unit title	Lesson number	Lesson title	NC Objective 1	NC Objective 2
Number – multiplication and division	6	Multiplication and division (3)	1	Multiples of 10	Write and calculate mathematical statements for multiplication and division using the multiplication tables that they know, including for two-digit numbers times one-digit numbers, using mental and progressing to formal written methods	
Number – multiplication and division	6	Multiplication and division (3)	2	Related calculations	Write and calculate mathematical statements for multiplication and division using the multiplication tables that they know, including for two-digit numbers times one-digit numbers, using mental and progressing to formal written methods	
Number – multiplication and division	6	Multiplication and division (3)	3	Reasoning about multiplication	Solve problems, including missing number problems, involving multiplication and division, including positive integer scaling problems and correspondence problems in which *n* objects are connected to *m* objects	
Number – multiplication and division	6	Multiplication and division (3)	4	Multiply 2-digits by 1-digit – no exchange	Write and calculate mathematical statements for multiplication and division using the multiplication tables that they know, including for two-digit numbers times one-digit numbers, using mental and progressing to formal written methods	
Number – multiplication and division	6	Multiplication and division (3)	5	Multiply 2-digits by 1-digit – exchange	Write and calculate mathematical statements for multiplication and division using the multiplication tables that they know, including for two-digit numbers times one-digit numbers, using mental and progressing to formal written methods	
Number – multiplication and division	6	Multiplication and division (3)	6	Expanded written method	Write and calculate mathematical statements for multiplication and division using the multiplication tables that they know, including for two-digit numbers times one-digit numbers, using mental and progressing to formal written methods	

Strand	Unit	Unit title	Lesson number	Lesson title	NC Objective 1	NC Objective 2
Number – multiplication and division	6	Multiplication and division (3)	7	Link multiplication and division	Solve problems, including missing number problems, involving multiplication and division, including positive integer scaling problems and correspondence problems in which n objects are connected to m objects	
Number – multiplication and division	6	Multiplication and division (3)	8	Divide 2-digits by 1-digit – no exchange	Write and calculate mathematical statements for multiplication and division using the multiplication tables that they know, including for two-digit numbers times one-digit numbers, using mental and progressing to formal written methods	
Number – multiplication and division	6	Multiplication and division (3)	9	Divide 2-digits by 1-digit –flexible partitioning	Write and calculate mathematical statements for multiplication and division using the multiplication tables that they know, including for two-digit numbers times one-digit numbers, using mental and progressing to formal written methods	
Number – multiplication and division	6	Multiplication and division (3)	10	Divide 2-digits by 1-digit with remainders	Write and calculate mathematical statements for multiplication and division using the multiplication tables that they know, including for two-digit numbers times one-digit numbers, using mental and progressing to formal written methods	
Number – multiplication and division	6	Multiplication and division (3)	11	How many ways?	Solve problems, including missing number problems, involving multiplication and division, including positive integer scaling problems and correspondence problems in which n objects are connected to m objects	
Number – multiplication and division	6	Multiplication and division (3)	12	Problem solving – mixed problems (1)	Solve problems, including missing number problems, involving multiplication and division, including positive integer scaling problems and correspondence problems in which n objects are connected to m objects	Write and calculate mathematical statements for multiplication and division using the multiplication tables that they know, including for two-digit numbers times one-digit numbers, using mental and progressing to formal written methods
Number – multiplication and division	6	Multiplication and division (3)	13	Problem solving – mixed problems (2)	Solve problems, including missing number problems, involving multiplication and division, including positive integer scaling problems and correspondence problems in which n objects are connected to m objects	Write and calculate mathematical statements for multiplication and division using the multiplication tables that they know, including for two-digit numbers times one-digit numbers, using mental and progressing to formal written methods

Strand	Unit	Unit title	Lesson number	Lesson title	NC Objective 1	NC Objective 2
Measurement	7	Length and perimeter	1	Measure in m and cm	Measure, compare, add and subtract: lengths (m/cm/mm); mass (kg/g); volume/capacity (l/ml)	
Measurement	7	Length and perimeter	2	Measure in cm and mm	Measure, compare, add and subtract: lengths (m/cm/mm); mass (kg/g); volume/capacity (l/ml)	
Measurement	7	Length and perimeter	3	Metres, centimetres and millimetres	Measure, compare, add and subtract: lengths (m/cm/mm); mass (kg/g); volume/capacity (l/ml)	
Measurement	7	Length and perimeter	4	Equivalent lengths (m and cm)	Measure, compare, add and subtract: lengths (m/cm/mm); mass (kg/g); volume/capacity (l/ml)	
Measurement	7	Length and perimeter	5	Equivalent lengths (mm and cm)	Measure, compare, add and subtract: lengths (m/cm/mm); mass (kg/g); volume/capacity (l/ml)	
Measurement	7	Length and perimeter	6	Compare lengths	Measure, compare, add and subtract: lengths (m/cm/mm); mass (kg/g); volume/capacity (l/ml)	
Measurement	7	Length and perimeter	7	Add lengths	Measure, compare, add and subtract: lengths (m/cm/mm); mass (kg/g); volume/capacity (l/ml)	
Measurement	7	Length and perimeter	8	Subtract lengths	Measure, compare, add and subtract: lengths (m/cm/mm); mass (kg/g); volume/capacity (l/ml)	
Measurement	7	Length and perimeter	9	Measure perimeter	Measure the perimeter of simple 2D shapes	
Measurement	7	Length and perimeter	10	Calculate perimeter	Measure the perimeter of simple 2D shapes	
Measurement	7	Length and perimeter	11	Problem solving – length	Measure the perimeter of simple 2D shapes	
Number – fractions	8	Fractions (1)	1	Understand the denominator of unit fractions	Recognise and use fractions as numbers: unit fractions and non-unit fractions with small denominators	
Number – fractions	8	Fractions (1)	2	Compare and order unit fractions	Recognise and use fractions as numbers: unit fractions and non-unit fractions with small denominators	
Number – fractions	8	Fractions (1)	3	Understand the numerator of non-unit fractions	Recognise and use fractions as numbers: unit fractions and non-unit fractions with small denominators	
Number – fractions	8	Fractions (1)	4	Understand the whole	Recognise and use fractions as numbers: unit fractions and non-unit fractions with small denominators	

Strand	Unit	Unit title	Lesson number	Lesson title	NC Objective 1	NC Objective 2
Number – fractions	8	Fractions (1)	5	Compare and order non-unit fractions	Compare and order unit fractions, and fractions with the same denominators	
Number – fractions	8	Fractions (1)	6	Divisions on a number line	Compare and order unit fractions, and fractions with the same denominators	
Number – fractions	8	Fractions (1)	7	Count in fractions on a number line	Compare and order unit fractions, and fractions with the same denominators	
Number – fractions	8	Fractions (1)	8	Equivalent fractions as bar models	Recognise and show, using diagrams, equivalent fractions with small denominators	
Number – fractions	8	Fractions (1)	9	Equivalent fractions on a number line	Recognise and show, using diagrams, equivalent fractions with small denominators	
Number – fractions	8	Fractions (1)	10	Equivalent fractions	Recognise and show, using diagrams, equivalent fractions with small denominators	
Measurement	9	Mass	1	Use scales	Measure, compare, add and subtract: lengths (m/cm/mm); mass (kg/g); volume/capacity (l/ml)	
Measurement	9	Mass	2	Measure mass	Measure, compare, add and subtract: lengths (m/cm/mm); mass (kg/g); volume/capacity (l/ml)	
Measurement	9	Mass	3	Measure mass in kilograms and grams	Measure, compare, add and subtract: lengths (m/cm/mm); mass (kg/g); volume/capacity (l/ml)	
Measurement	9	Mass	4	Equivalent masses	Measure, compare, add and subtract: lengths (m/cm/mm); mass (kg/g); volume/capacity (l/ml)	
Measurement	9	Mass	5	Compare mass	Measure, compare, add and subtract: lengths (m/cm/mm); mass (kg/g); volume/capacity (l/ml)	
Measurement	9	Mass	6	Add and subtract mass	Measure, compare, add and subtract: lengths (m/cm/mm); mass (kg/g); volume/capacity (l/ml)	
Measurement	9	Mass	7	Problem solving – mass	Measure, compare, add and subtract: lengths (m/cm/mm); mass (kg/g); volume/capacity (l/ml)	
Measurement	10	Capacity	1	Measure capacity and volume in litres and millilitres	Measure, compare, add and subtract: lengths (m/cm/mm); mass (kg/g); volume/capacity (l/ml)	
Measurement	10	Capacity	2	Measure in litres and millilitres	Measure, compare, add and subtract: lengths (m/cm/mm); mass (kg/g); volume/capacity (l/ml)	
Measurement	10	Capacity	3	Equivalent capacities and volumes (litres and millilitres)	Measure, compare, add and subtract: lengths (m/cm/mm); mass (kg/g); volume/capacity (l/ml)	

Strand	Unit	Unit title	Lesson number	Lesson title	NC Objective 1	NC Objective 2
Measurement	10	Capacity	4	Compare capacity and volume	Measure, compare, add and subtract: lengths (m/cm/mm); mass (kg/g); volume/capacity (l/ml)	
Measurement	10	Capacity	5	Add and subtract capacity and volume	Measure, compare, add and subtract: lengths (m/cm/mm); mass (kg/g); volume/capacity (l/ml)	
Measurement	10	Capacity	6	Problem solving – capacity	Measure, compare, add and subtract: lengths (m/cm/mm); mass (kg/g); volume/capacity (l/ml)	

Mindset: an introduction

Global research and best practice deliver the same message: learning is greatly affected by what learners perceive they can or cannot do. What is more, it is also shaped by what their parents, carers and teachers perceive they can do. Mindset – the thinking that determines our beliefs and behaviours – therefore has a fundamental impact on teaching and learning.

Everyone can!

Power Maths and mastery methods focus on the distinction between 'fixed' and 'growth' mindsets (Dweck, 2007).[1] Those with a fixed mindset believe that their basic qualities (for example, intelligence, talent and ability to learn) are pre-wired or fixed: 'If you have a talent for maths, you will succeed at it. If not, too bad!' By contrast, those with a growth mindset believe that hard work, effort and commitment drive success and that 'smart' is not something you are or are not, but something you become. In short, everyone can do maths!

Key mindset strategies

A growth mindset needs to be actively nurtured and developed. *Power Maths* offers some key strategies for fostering healthy growth mindsets in your classroom.

It is okay to get it wrong

Mistakes are valuable opportunities to re-think and understand more deeply. Learning is richer when children and teachers alike focus on spotting and sharing mistakes as well as solutions.

Praise hard work

Praise is a great motivator, and by focusing on praising effort and learning rather than success, children will be more willing to try harder, take risks and persist for longer.

Mind your language!

The language we use around learners has a profound effect on their mindsets. Make a habit of using growth phrases, such as, 'Everyone can!', 'Mistakes can help you learn' and 'Just try for a little longer'. The king of them all is one little word, 'yet'... I can't solve this...yet!' Encourage parents and carers to use the right language too.

Build in opportunities for success

The step-by-small-step approach enables children to enjoy the experience of success. In addition, avoid ability grouping and encourage every child to answer questions and explain or demonstrate their methods to others.

---

[1] Dweck, C (2007) *The New Psychology of Success*, Ballantine Books: New York

The *Power Maths* characters

The *Power Maths* characters model the traits of growth mindset learners and encourage resilience by prompting and questioning children as they work. Appearing frequently in the Textbooks and Practice Books, they are your allies in teaching and discussion, helping to model methods, alternatives and misconceptions, and to pose questions. They encourage and support your children, too: they are all hardworking, enthusiastic and unafraid of making and talking about mistakes.

Meet the team!

Creative Flo is open-minded and sometimes indecisive. She likes to think differently and come up with a variety of methods or ideas.

Determined Dexter is resolute, resilient and systematic. He concentrates hard, always tries his best and he'll never give up – even though he doesn't always choose the most efficient methods!

'Let's try again.'
'Mistakes are cool!'
'Have I found all of the solutions?'

'Let's try it this way…'
'Can we do it differently?'
'I've got another way of doing this!'

'I'm going to try this!'
'I know how to do that!'
'Want to share my ideas?'

Curious Ash is eager, interested and inquisitive, and he loves solving puzzles and problems. Ash asks lots of questions but sometimes gets distracted.

'What if we tried this…?'
'I wonder…'
'Is there a pattern here?'

Sparks the Cat

Miaow!

Brave Astrid is confident, willing to take risks and unafraid of failure. She's never scared to jump straight into a problem or question, and although she often makes simple mistakes, she's happy to talk them through with others.

Mathematical language

Traditionally, we in the UK have tended to try simplifying mathematical language to make it easier for young children to understand. By contrast, evidence and experience show that by diluting the correct language, we actually mask concepts and meanings for children. We then wonder why they are confused by new and different terminology later down the line! *Power Maths* is not afraid of 'hard' words and avoids placing any barriers between children and their understanding of mathematical concepts. As a result, we need to be deliberate, precise and thorough in building every child's understanding of the language of maths. Throughout the Teacher Guides you will find support and guidance on how to deliver this, as well as individual explanations throughout the pupil Textbooks.

Use the following key strategies to build children's mathematical vocabulary, understanding and confidence.

Precise and consistent

Everyone in the classroom should use the correct mathematical terms in full, every time. For example, refer to 'equal parts', not 'parts'. Used consistently, precise maths language will be a familiar and non-threatening part of children's everyday experience.

Full sentences

Teachers and children alike need to use full sentences to explain or respond. When children use complete sentences, it both reveals their understanding and embeds their knowledge.

Stem sentences

These important sentences help children express mathematical concepts accurately, and are used throughout the *Power Maths* books. Encourage children to repeat them frequently, whether working independently or with others. Examples of stem sentences are:

'4 is a part, 5 is a part, 9 is the whole.'

'There are …. groups. There are …. in each group.'

Key vocabulary

The unit starters highlight essential vocabulary for every lesson. In the pupil books, characters flag new terminology and the Teacher Guide lists important mathematical language for every unit and lesson. New terms are never introduced without a clear explanation.

Mathematical signs

Mathematical signs are used early on so that children quickly become familiar with them and their meaning. Often, the *Power Maths* characters will highlight the connection between language and particular signs.

The role of talk and discussion

When children learn to talk purposefully together about maths, barriers of fear and anxiety are broken down and they grow in confidence, skills and understanding. Building a healthy culture of 'maths talk' empowers their learning from day one.

Explanation and discussion are integral to the *Power Maths* structure, so by simply following the books your lessons will stimulate structured talk. The following key 'maths talk' strategies will help you strengthen that culture and ensure that every child is included.

Sentences, not words

Encourage children to use full sentences when reasoning, explaining or discussing maths. This helps both speaker and listeners to clarify their own understanding. It also reveals whether or not the speaker truly understands, enabling you to address misconceptions as they arise.

Working together

Working with others in pairs, groups or as a whole class is a great way to support maths talk and discussion. Use different group structures to add variety and challenge. For example, children could take timed turns for talking, work independently alongside a 'discussion buddy', or perhaps play different *Power Maths* character roles within their group.

Think first – then talk

Provide clear opportunities within each lesson for children to think and reflect, so that their talk is purposeful, relevant and focused.

Give every child a voice

Where the 'hands up' model allows only the more confident child to shine, *Power Maths* involves everyone. Make sure that no child dominates and that even the shyest child is encouraged to contribute – and praised when they do.

Assessment strategies

Teaching for mastery demands that you are confident about what each child knows and where their misconceptions lie; therefore, practical and effective assessment is vitally important.

Formative assessment within lessons

The **Think together** section will often reveal any confusions or insecurities; try ironing these out by doing the first **Think together** question as a class. For children who continue to struggle, you or your Teaching Assistant should provide support and enable them to move on.

▶ Performance in practice can be very revealing: check Practice Books and listen out both during and after practice to identify misconceptions.

▶ The **Reflect** section is designed to check on the all-important depth of understanding. Be sure to review how the children performed in this final stage before you teach the next lesson.

End of unit check – Textbook

Each unit concludes with a summative check to help you assess quickly and clearly each child's understanding, fluency, reasoning and problem solving skills. Your Teacher Guide will suggest ideal ways of organising a given activity and offer advice and commentary on what children's responses mean. For example, 'What misconception does this reveal?'; 'How can you reinforce this particular concept?'

For younger children, assess in small, teacher-led groups, giving each child time to think and respond while also consolidating correct mathematical language. Assessment with young children should always be an enjoyable activity, so avoid one-to-one individual assessments, which they may find threatening or scary. If you prefer, the End of unit check can be carried out as a whole-class group using whiteboards and Practice Books.

End of unit check – Practice Book

The Practice Book contains further opportunities for assessment, and can be completed by children independently whilst you are carrying out diagnostic assessment with small groups. Your Teacher Guide will advise you on what to do if children struggle to articulate an explanation – or perhaps encourage you to write down something they have explained well. It will also offer insights into children's answers and their implications for next learning steps. It is split into three main sections, outlined below.

My journal is designed to allow children to show their depth of understanding of the unit. It can also serve as a way of checking that children have grasped key mathematical vocabulary. The question children should answer is first presented in the Textbook in the Think! section. This provides an opportunity for you to discuss the question first as a class to ensure children have understood their task. Children should have some time to think about how they want to answer the question, and you could ask them to talk to a partner about their ideas. Then children should write their answer in their Practice Book, using the word bank provided to help them with vocabulary.

The **Power check** allows pupils to self-assess their level of confidence on the topic by colouring in different smiley faces. You may want to introduce the faces as follows:

Each unit ends with either a Power play or a Power puzzle. This is an activity, puzzle or game that allows children to use their new knowledge in a fun, informal way.

Progress Tests

There are *Power Maths* Progress Tests for each half term and at the end of the year, including an Arithmetic test and Reasoning test in each case. You can enter results in the online markbook to track and analyse results and see the average for all schools' results. The tests use a 6-step scale to show results against age-related expectation.

How to ask diagnostic questions

The diagnostic questions provided in children's Practice Books are carefully structured to identify both understanding and misconceptions (if children answer in a particular way, you will know why). The simple procedure below may be helpful:

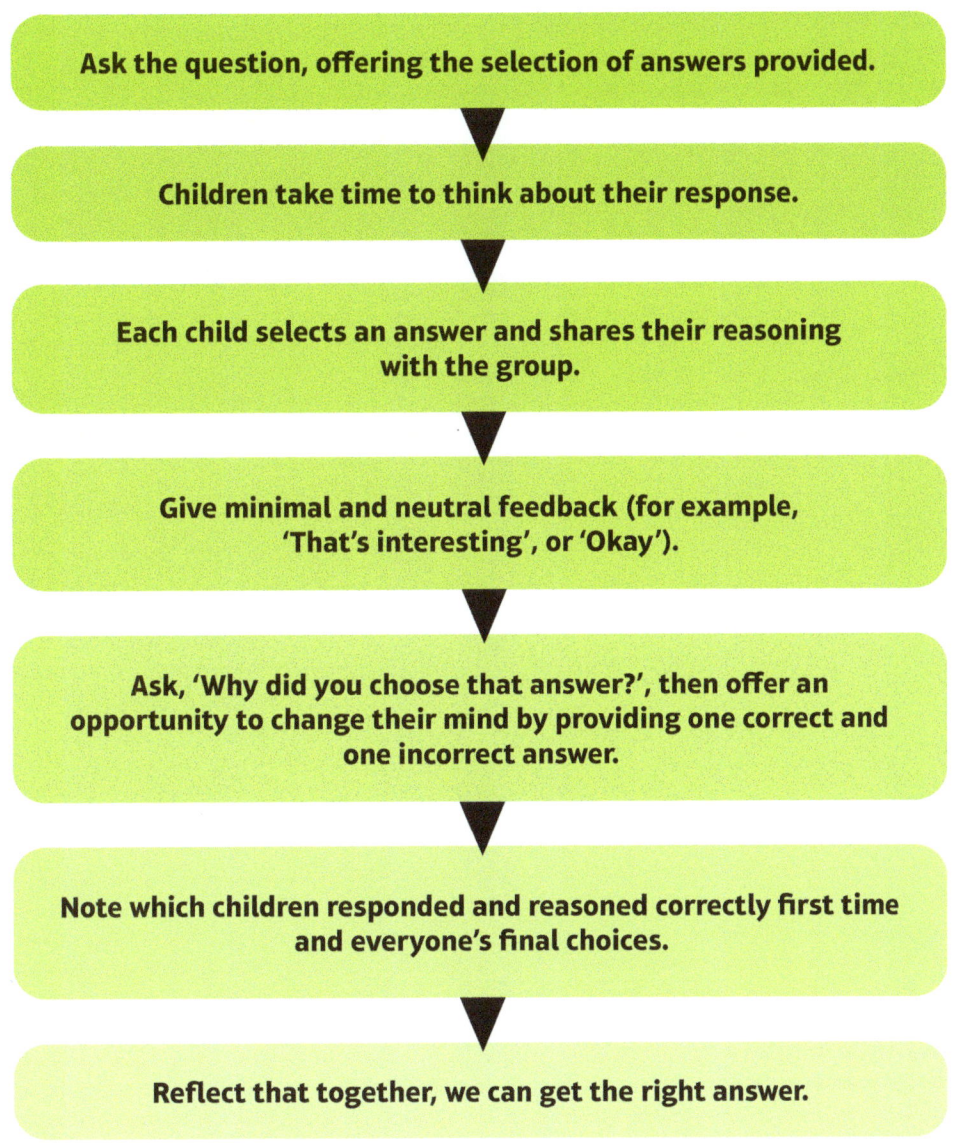

Keeping the class together

Traditionally, children who learn quickly have been accelerated through the curriculum. As a consequence, their learning may be superficial and will lack the many benefits of enabling children to learn with and from each other.

By contrast, *Power Maths'* mastery approach values real understanding and richer, deeper learning above speed. It sees all children learning the same concept in small, cumulative steps, each finding and mastering challenge at their own level. Remember that when you teach for mastery, EVERYONE can do maths! Those who grasp a concept easily have time to explore and understand that concept at a deeper level. The whole class therefore moves through the curriculum at broadly the same pace via individual learning journeys.

For some teachers, the idea that a whole class can move forward together is revolutionary and challenging. However, the evidence of global good practice clearly shows that this approach drives engagement, confidence, motivation and success for all learners, and not just the high flyers. The strategies below will help you keep your class together on their maths journey.

Mix it up

Do not stick to set groups at each table. Every child should be working on the same concept, and mixing up the groupings widens children's opportunities for exploring, discussing and sharing their understanding with others.

Recycling questions

Reuse the Textbook and Practice Book questions with concrete materials to allow children to explore concepts and relationships and deepen their understanding. This strategy is especially useful for reinforcing learning in same-day interventions.

Strengthen at every opportunity

The next lesson in a *Power Maths* sequence always revises and builds on the previous step to help embed learning. These activities provide golden opportunities for individual children to strengthen their learning with the support of Teaching Assistants.

Prepare to be surprised!

Children may grasp a concept quickly or more slowly. The 'fast graspers' won't always be the same individuals, nor does the speed at which a child understands a concept predict their success in maths. Are they struggling or just working more slowly?

Same-day intervention

Since maths competence depends on mastering concepts one by one in a logical progression, it is important that no gaps in understanding are ever left unfilled. Same-day interventions – either within or after a lesson – are a crucial safety net for any child who has not fully made the small step covered that day. In other words, intervention is always about keeping up, not catching up, so that every child has the skills and understanding they need to tackle the next lesson. That means presenting the same problems used in the lesson, with a variety of concrete materials to help children model their solutions.

We offer two intervention strategies below, but you should feel free to choose others if they work better for your class.

Within-lesson intervention

The **Think together** activity will reveal those who are struggling, so when it is time for practice, bring these children together to work with you on the first practice questions. Observe these children carefully, ask questions, encourage them to use concrete models and check that they reach and can demonstrate their understanding.

After-lesson intervention

You might like to use the **Think together** questions to recap the lesson with children who are working behind expectations during assembly time. Teaching Assistants could also work with these children at other convenient points in the school day. Some children may benefit from revisiting work from the same topic in the previous year group. Note also the suggestion for recycling questions from the Textbook and Practice Book with concrete materials on page 28.

The role of practice

Practice plays a pivotal role in the *Power Maths* approach. It takes place in class groups, smaller groups, pairs, and independently, so that children always have the opportunities for thinking as well as the models and support they need to practise meaningfully and with understanding.

Intelligent practice

In *Power Maths*, practice never equates to the simple repetition of a process. Instead we embrace the concept of intelligent practice, in which all children become fluent in maths through varied, frequent and thoughtful practice that deepens and embeds conceptual understanding in a logical, planned sequence. To see the difference, take a look at the following examples.

Traditional practice
- Repetition can be rote – no need for a child to think hard about what they are doing
- Praise may be misplaced
- Does this prove understanding?

Intelligent practice
- Varied methods – concrete, pictorial and abstract
- Equation expressed in different ways, requiring thought and understanding
- Constructive feedback

All practice questions are designed to move children on and reveal misconceptions.

Simple, logical steps build onto earlier learning.

C-P-A runs throughout – different ways of modelling and understanding the same concept.

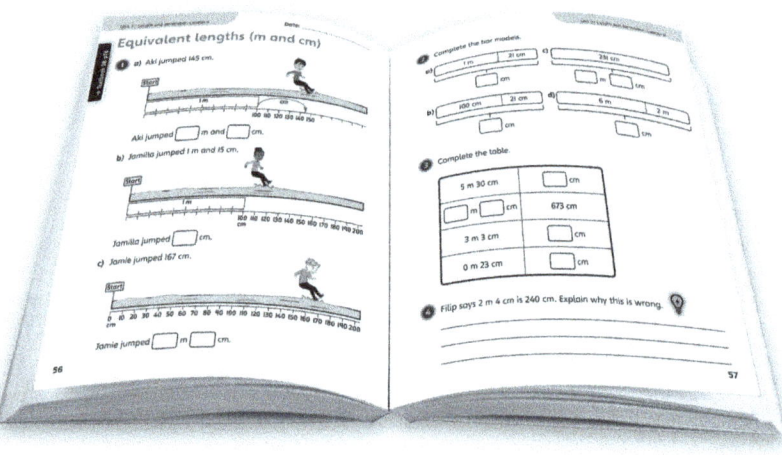

Conceptual variation – children work on different representations of the same maths concept.

Friendly characters offer support and encourage children to try different approaches.

A carefully designed progression

The Practice Books provide just the right amount of intelligent practice for children to complete independently in the final sections of each lesson. It is really important that all children are exposed to the practice questions, and that children are not directed to complete different sections. That is because each question is different and has been designed to challenge children to think about the maths they are doing. The questions become more challenging so children grasping concepts more quickly will start to slow down as they progress. Meanwhile, you have the chance to circulate and spot any misconceptions before they become barriers to further learning.

Homework and the role of parents and carers

While *Power Maths* does not prescribe any particular homework structure, we acknowledge the potential value of practice at home. For example, practising fluency in key facts, such as number bonds and times-tables, is an ideal homework task. You can share the Individual Practice Games for homework (see page 6), or parents and carers could work through uncompleted Practice Book questions with children at either primary stage.

However, it is important to recognise that many parents and carers may themselves lack confidence in maths, and few, if any, will be familiar with mastery methods. A Parents' and Carers' evening that helps them understand the basics of mindsets, mastery and mathematical language is a great way to ensure that children benefit from their homework. It could be a fun opportunity for children to teach their families that everyone can do maths!

Structures and representations

Unlike most other subjects, maths comprises a wide array of abstract concepts – and that is why children and adults so often find it difficult. By taking a concrete-pictorial-abstract (C-P-A) approach, *Power Maths* allows children to tackle concepts in a tangible and more comfortable way.

Non-linear stages

Concrete

Replacing the traditional approach of a teacher working through a problem in front of the class, the concrete stage introduces real objects that children can use to 'do' the maths – any familiar object that a child can manipulate and move to help bring the maths to life. It is important to appreciate, however, that children must always understand the link between models and the objects they represent. For example, children need to first understand that three cakes could be represented by three pretend cakes, and then by three counters or bricks. Frequent practice helps consolidate this essential insight. Although they can be used at any time, good concrete models are an essential first step in understanding.

Pictorial

This stage uses pictorial representations of objects to let children 'see' what particular maths problems look like. It helps them make connections between the concrete and pictorial representations and the abstract maths concept. Children can also create or view a pictorial representation together, enabling discussion and comparisons. The *Power Maths* teaching tools are fantastic for this learning stage, and bar modelling is invaluable for problem solving throughout the primary curriculum.

Abstract

Our ultimate goal is for children to understand abstract mathematical concepts, symbols and notation and of course, some children will reach this stage far more quickly than others. To work with abstract concepts, a child must be comfortable with the meaning of and relationships between concrete, pictorial and abstract models and representations. The C-P-A approach is not linear, and children may need different types of models at different times. However, when a child demonstrates with concrete models and pictorial representations that they have grasped a concept, we can be confident that they are ready to explore or model it with abstract symbols such as numbers and notation.

Use at any time and with any age to support understanding

Variation helps visualisation

Children find it much easier to visualise and grasp concepts if they see them presented in a number of ways, so be prepared to offer and encourage many different representations.

For example, the number six could be represented in various ways:

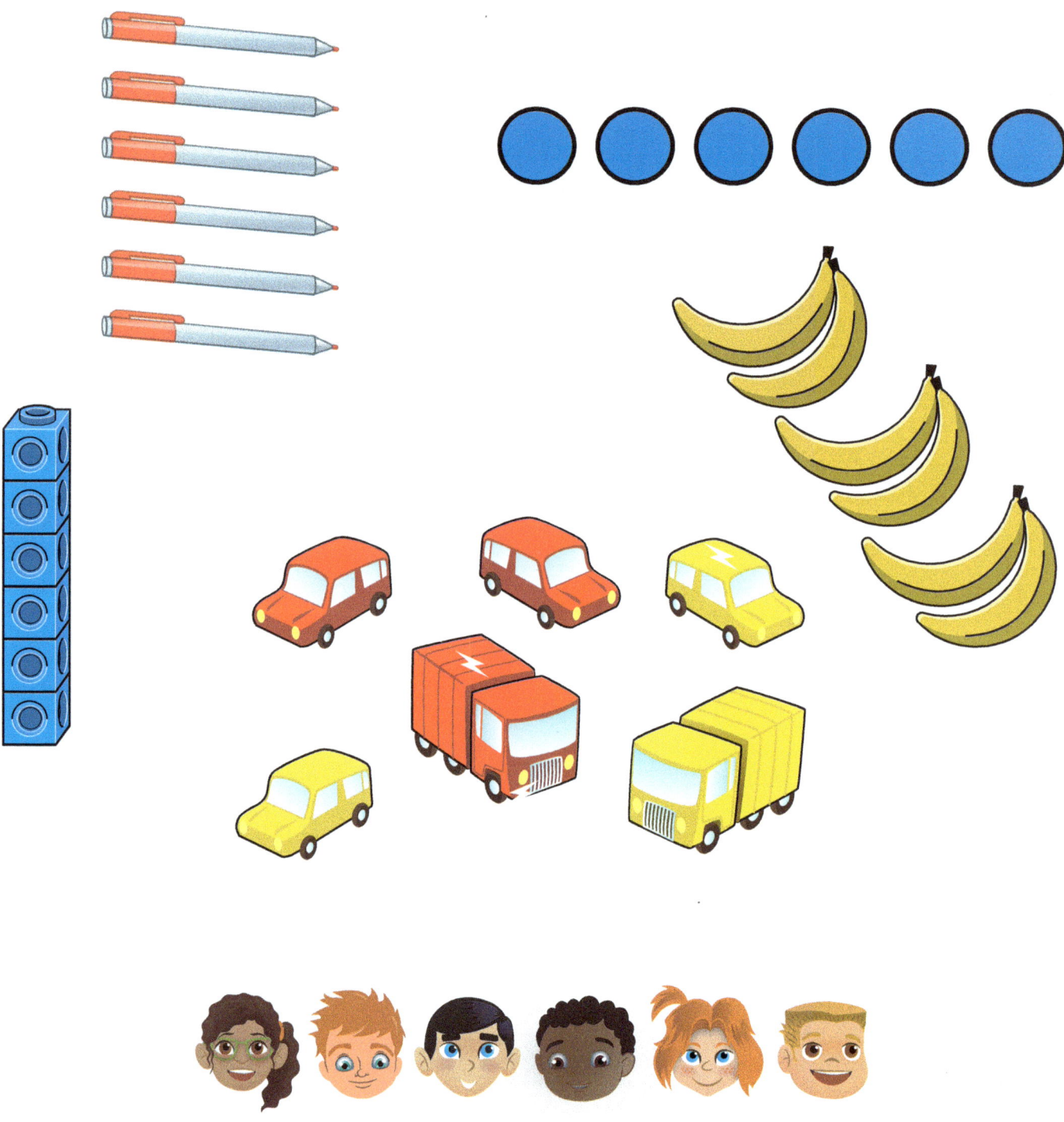

Practical aspects of *Power Maths*

One of the key underlying elements of *Power Maths* is its practical approach, allowing you to make maths real and relevant to your children, no matter their age.

Manipulatives are essential resources for both key stages and *Power Maths* encourages teachers to use these at every opportunity, and to continue the Concrete-Pictorial-Abstract approach right through to Year 6.

The Textbooks and Teacher Guides include lots of opportunities for teaching in a practical way to show children what maths means in real life.

Discover and Share

The **Discover** and **Share** sections of the Textbook give you scope to turn a real-life scenario into a practical and hands-on section of the lesson. Use these sections as inspiration to get active in the classroom. Where appropriate, use the **Discover** contexts as a springboard for your own examples that have particular resonance for your children – and allow them to get their hands dirty trying out the mathematics for themselves.

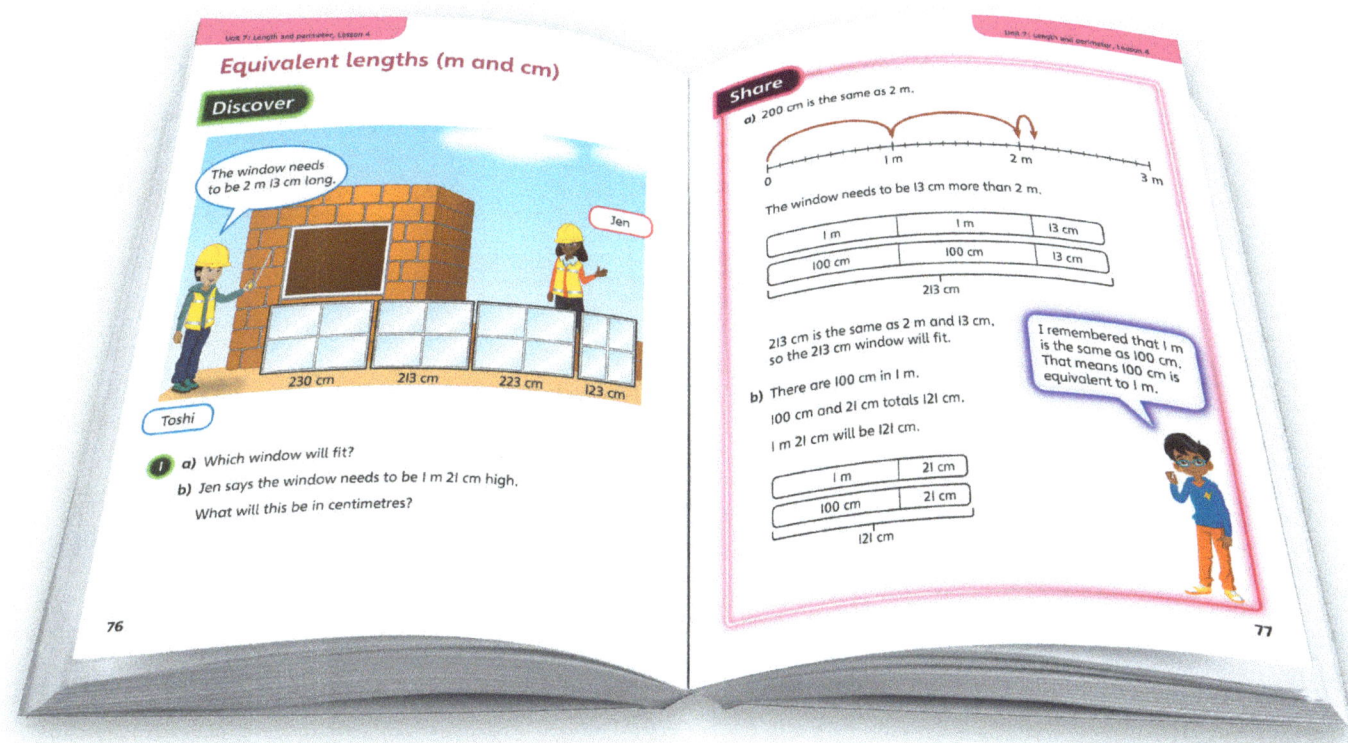

Unit videos

Every term has one unit video which incorporates real-life classroom sequences.

These videos show you how the reasoning behind mathematics can be carried out in a practical manner by showing real children using various concrete and pictorial methods to come to the solution. You can see how using these practical models, such as part-whole and bar models, helps them to find and articulate their answer.

Mastery tips

Mastery Experts give anecdotal advice on where they have used hands-on and real-life elements to inspire their children.

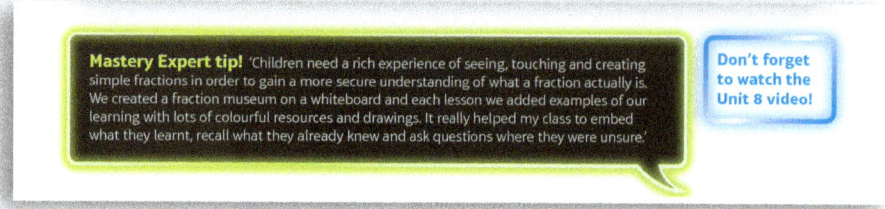

Concrete-Pictorial-Abstract (C-P-A) approach

Each **Share** section uses various methods to explain an answer, helping children to access abstract concepts by using concrete tools, such as counters. Remember, this isn't a linear process, so even children who appear confident using the more abstract method can deepen their knowledge by exploring the concrete representations. Encourage children to use all three methods to really solidify their understanding of a concept.

Pictorial representation – drawing the problem in a logical way that helps children visualise the maths

Concrete representation – using manipulatives to represent the problem. Encourage children to physically use resources to explore the maths.

Abstract representation – using words and calculations to represent the problem.

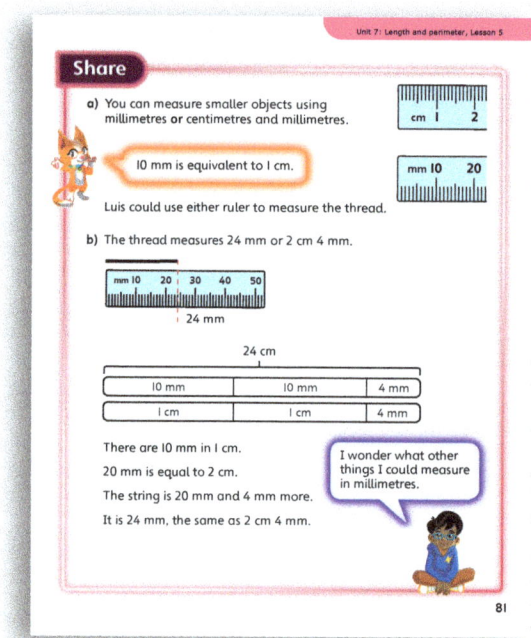

Practical tips

Every lesson suggests how to draw out the practical side of the **Discover** context.

You'll find these in the **Discover** section of the Teacher Guide for each lesson.

> **PRACTICAL TIPS** Ask children to count aloud in 100s when finding the total number of bricks in each part. For support, children could represent the number of bricks using base 10 equipment.

Resources

Every lesson lists the practical resources you will need or might want to use. There is also a summary of all of the resources used throughout the term on page 40 to help you be prepared.

> **RESOURCES**
> **Mandatory:** base 10 equipment
> **Optional:** place value counters, place value grids

Working with children below age-related expectation

This section offers advice on using *Power Maths* with children who are significantly behind age-related expectation. Teacher judgement will be crucial in terms of where and why children are struggling, and in choosing the right approach. The suggestions can of course be adapted for children with special educational needs, depending on the specific details of those needs.

General approaches to support children who are struggling

Keeping the pace manageable
Remember, you have more teaching days than *Power Maths* lessons so you can cover a lesson over more than one day, and revisit key learning, to ensure all children are ready to move on. You can use the + and – buttons to adjust the time for each unit in the online planning. The NCETM's Ready-to-Progress criteria can be used to help determine what should be highest priority.

Same-day intervention
You could go over the Textbook pages or revisit the previous year's work if necessary (see Addressing gaps). Remember that same-day intervention can be within the lesson, as well as afterwards (see page 29). As children start their independent practice, you can work with those who found the first part of the lesson difficult, checking understanding using manipulatives.

Fluency sessions
Fit in as much practice as you can for number bonds and times-tables, etc., at other times of the day. If you can, plan a short 'maths meeting' for this in the afternoon. You might choose to use a Power Up you haven't used already.

Addressing gaps
Use material from the same topic in the previous year to consolidate or address gaps in learning, e.g. Textbook pages and Strengthen activities. The End of unit check will help gauge children's understanding.

Pre-teaching
Find a 5- to 10-minute slot before the lesson to work with the children you feel would benefit. The afternoon before the lesson can work well, because it gives children time to think in between. Recap previous work on the topic (addressing any gaps you're aware of) and do some fluency practice, targeting number facts etc. that will help children access the learning.

Focusing on the key concepts
If children are a long way behind, it can be helpful to take a step back and think about the key concepts for children to engage with, not just the fine detail of the objective for that year group (e.g. addition with a specific number of columns). Bearing that in mind, how could children advance their understanding of the topic?

Providing extra support within the lesson

Support in the Teacher Guide
First of all, use the Strengthen support in the Teacher Guide for guided and independent work in each lesson, and share this with Teaching Assistants, where relevant. As you read through the lesson content and corresponding Teacher Guide pages before the lesson, ask yourself what key idea or nugget of understanding is at the heart of the lesson. If children are struggling, this should help you decide what's essential for all children before they move on.

Annotating pages
You can annotate questions to provide extra scaffolding or hints if you need to, but aim to build up children's ability to access questions independently wherever you can. Children tend to get used to the style of the *Power Maths* questions over time.

Quick recap as lesson starter
The Quick recap for each lesson in the Teacher Guide is an alternative starter activity to the Power Up. You might choose to use this with some or all children if you feel they will need support accessing the main lesson.

Consolidation questions
If you think some children would benefit from additional questions at the same level before moving on, write one or two similar questions on the board. (This shouldn't be at the expense of reasoning and problem-solving opportunities: take longer over the lesson if you need to.)

Hard copy Textbooks
The Textbooks help children focus in more easily on the mathematical representations, read the text more comfortably, and revisit work from a previous lesson that you are building on, as well as giving children ownership of their learning journey. In main lessons, it can work well to use the e-Textbook for **Discover** and give out the books when discussing the methods in the **Share** section.

Reading support
It's important that all children are exposed to problem solving and reasoning questions, which often involve reading. For whole-class work you can read questions together. For independent practice you could consider annotating pages to help children see what the question is asking, and stem sentences to help structure their answer. A general focus on specific mathematical language and vocabulary will help children access the questions. You could consider pairing weaker readers with stronger readers, or read questions as a group if those who need support are on the same table.

Providing extra depth and challenge with *Power Maths*

Just as prescribed in the National Curriculum, the goal of *Power Maths* is never to accelerate through a topic but rather to gain a clear, deep and broad understanding. Here are some suggestions to help ensure all children are appropriately challenged as you work with the resources.

Overall approaches

First of all, remember that the materials are designed to help you keep the class together, allowing all children to master a concept while those who grasp it quickly have time to explore it in more depth. Use the Deepen support in the Teacher Guide (see below) to challenge children who work through the questions quickly. Here are some questions and ideas to encourage breadth and depth during specific parts of the lesson, or at any time (where no part of the lesson sequence is specified):

- **Discover**: 'Can you demonstrate your solution another way?'
- **Share**: Make sure every child is encouraged to give answers and engage with the discussion, not just the most confident.
- **Think together**: 'Can you model your answers using concrete materials? Can you explain your solution to a partner?'
- Practice: Allow all children to work through the full set of questions, so that they benefit from the logical sequence.
- **Reflect**: 'Is there another way of working out the answer? And another way?'
  'Have you found all the solutions?'
  'Is that always true?'
  'What's different between this question and that question? And what's the same?'

Note that the **Challenge** questions are designed so that all children can access and attempt them, if they have worked through the steps leading up to them. There may be some children in a given lesson who don't manage to do the **Challenge**, but it is not supposed to be a distinct task for a subset of the class. When you look through the lesson materials before teaching, think about what each question is specifically asking, and compare this with the key learning point for the lesson. This will help you decide which questions you feel it's essential for all children to answer, before moving on. You can at least aim for all children to try the **Challenge**!

Deepen activities and support

The Teacher Guide provides valuable support for each stage of the lesson. This includes Deepen tips for the guided and independent practice sections, which will help you provide extra stretch and challenge within your lesson, without having to organise additional tasks. If you have a Teaching Assistant, they can also make use of this advice. There are also suggestions for the lesson as a whole in the 'Going Deeper' section on the first page of the Teacher Guide section for that lesson. Every class is different, so you can always go a bit further in the direction indicated, if appropriate, and build on the suggestions given.

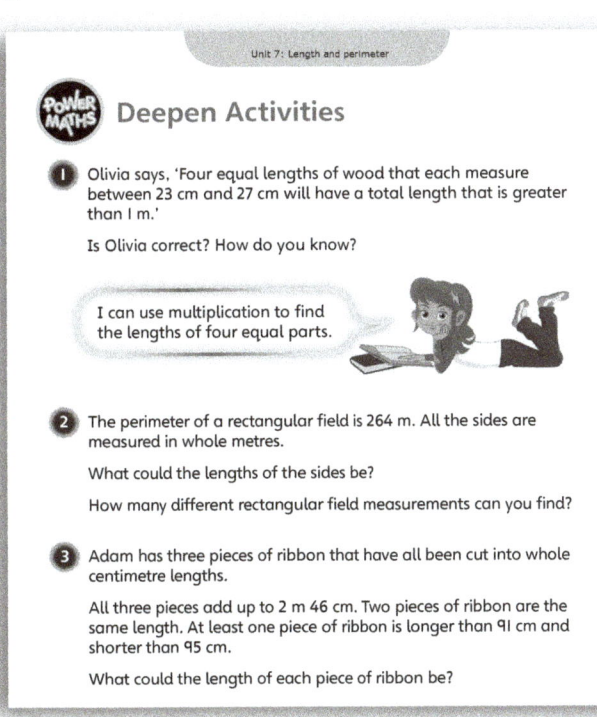

There is a Deepen activity for each unit. These are designed to follow on from the End of unit check, stretching children who have a firm understanding of the key learning from the unit. Children can work on them independently, which makes it easier for the teacher to facilitate the Strengthen activity for children who need extra support. Deepen activities could also be introduced earlier in the unit if the necessary work has been covered. The Deepen activities are on *ActiveLearn* on the Planning page for each unit, and also on the Resources page).

Using the questions flexibly to provide extra challenge

Sometimes you may want to write an extra question on the board or provide this on paper. You can usually do this by tweaking the lesson materials. The questions are designed to form a carefully structured sequence that builds understanding step by step, but, with careful thought about the purpose of each question, you can use the materials flexibly where you need to. Sometimes you might feel that children would benefit from another similar question for consolidation before moving on to the next one, or you might feel that they would benefit from a harder example in the same style. It should be quick and easy to generate 'more of the same' type questions where this is the case.

When you see a question like this one (from Unit 3, Lesson 3), it's easy to make harder examples to do afterwards if you need them. What if the 7 was also blotted out – how many possibilities would there be for the ones digits? For a trickier example, if you set up a similar addition with a tens digit blotted out (and the tens part of the answer given), children would have to factor in the exchanged 10 in order to work out what was missing.

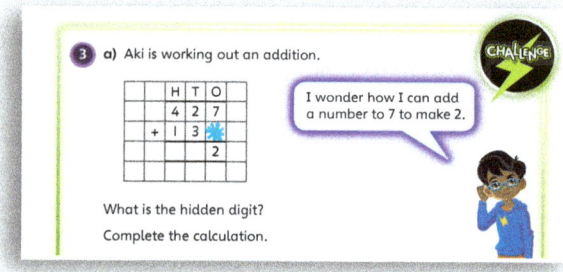

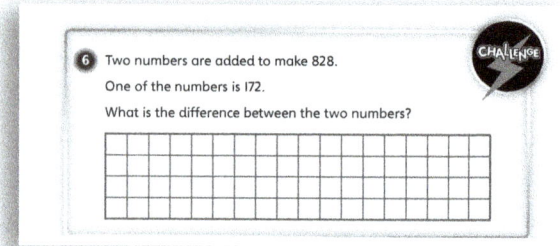

For this example (from Unit 3, Lesson 6), you could ask children to make up their own question(s) for a partner to solve. They could even make up questions to solve themselves! (In fact, for any of these examples you could ask early finishers to create their own question for a partner.)

Here's an example (from Unit 3, Lesson 12) where some of the combinations in the picture feature as questions in the lesson, but others don't. Clearly there are plenty of multi-step problems you could ask using the same information. Children could choose what to buy for their family, or you could tell them about your family and the bikes, helmets and lights you would need. You could also give them a budget to work out the change from buying a list of equipment.

Besides creating additional questions, you should be able to find a question in the lesson that you can adapt into a game or open-ended investigation, if this helps to keep everyone engaged. It could simply be that, instead of answering 5 × 5 etc. on the page, they could build a robot with 5 lots of 5 cubes.

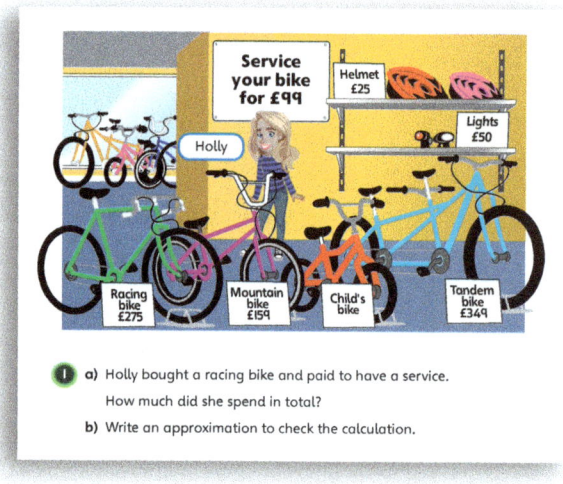

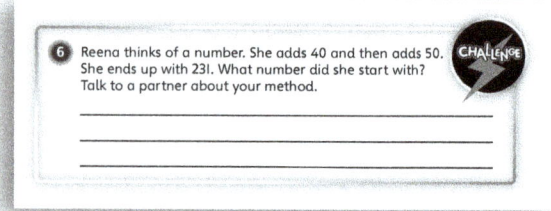

With a question like this (Unit 2, Lesson 9), children could play a game where they say the additions and the end number, and their partner has to work out the start number.

See the bullets above for some general ideas that will help with 'opening out' questions in the books, e.g. 'Can you find all the solutions?' type questions.

Other suggestions

Another way of stretching children is through mixed ability pairs, or via other opportunities for children to explain their understanding in their own way. This is a good way of encouraging children to go deeper into the learning, rather than, for instance, tackling questions that are computationally more challenging but conceptually equivalent in level.

Using *Power Maths* with mixed age classes

Overall approaches

There are many variables between schools that would make it inadvisable to recommend a one-size-fits-all approach to mixed age teaching with *Power Maths*. These include how year groups are merged, availability of Teaching Assistants, experience and preference of teaching staff, range in pupil attainment across years, classroom space and layout, level of flexibility around timetables, and overall organisational structure (whether the school is part of a trust).

Some schools will find it best to timetable separate maths lessons for the different year groups. Others will aim to teach the class together as much as possible using the mixed age planning support on *ActiveLearn* (see the lesson exemplars for ways of organising lessons with strong/medium/weak correlation between year groups). There will also be ways of adapting these general approaches. For example, offset lessons where Year A start their lesson with the teacher, while Year B work independently on the practice from the previous lesson, and then start the next lesson with the teacher while Year A work independently; or teachers may choose to base their provision around the lesson from one year group and tweak the content up/down for the other group.

Key strategies for mixed age teaching

The mixed age teaching webinar on *ActiveLearn* provides advice on all aspects of mixed age teaching, including more detail on the ideas below.

Developing independence over time
Investing time in building up children's independence will pay off in the medium term.

Clear rationale
If someone asked, 'Why did you teach both Unit 3 and 4 in the same lesson/separate lessons?', what would your answer be?

Designing a lesson
1. Identify the core learning for each group
2. Identify any number skills necessary to access the core
3. Consider the flow of concepts and how one core leads to the other

Challenging all children
The questions are designed to build understanding step by step, but with careful thought about the purpose of each question you can tweak them to increase the challenge.

Multiple years combined
With more than two years together, teachers will inevitably need to use the resources flexibly if delivering a single lesson.

Enjoy the positives!

Comparison deepens understanding and there will be lots of opportunities for children, as well as misconceptions to explore. There is also in-built pre-teaching and the chance to build up a concept from its foundations. For teachers there is double the material to draw on! Mixed age teachers require a strong understanding of the progression of ideas across year groups, which is highly valuable for all teachers. Also, it is necessary to engage deeply with the lesson to see how to use the materials flexibly – this is recommended for all teachers and will help you bring your lesson to life!

List of practical resources

Year 3B Mandatory resources

Resource	Lesson
Bar model	**Unit 8** Lessons 8, 9
Base 10 equipment	**Unit 6** Lessons 2, 4, 5, 6, 7, 8, 9, 12, 13
Counters	**Unit 6** Lessons 2, 3, 6, 7, 8, 10, 12
Counters (10s)	**Unit 6** Lesson 1
Cubes	**Unit 6** Lessons 7, 10
Dice	**Unit 9** Lesson 5
Drawing resources (ruler and pencil)	**Unit 6** Lesson 11
Fraction cards	**Unit 8** Lesson 10
Fraction strips	**Unit 8** Lessons 8, 9, 10,
Fraction tiles	**Unit 8** Lesson 7
Fraction wall	**Unit 8** Lesson 9
Metre sticks	**Unit 7** Lessons 4, 6
Mini whiteboards	**Unit 8** Lesson 6
Modelling clay	**Unit 9** Lessons 2, 3
Number lines	**Unit 7** Lesson 5 **Unit 8** Lessons 9, 10
Objects (real-life – for combining, e.g. hats and scarves)	**Unit 6** Lesson 11
Paper (to make paper aeroplanes)	**Unit 7** Lesson 1
Paper strips	**Unit 8** Lesson 10
Part-whole models (blank)	**Unit 6** Lesson 8
Pencils (colouring)	**Unit 6** Lesson 11
Place value counters	**Unit 6** Lessons 4, 5, 6, 9, 13 **Unit 7** Lesson 4
Place value grids	**Unit 6** Lessons 4, 5
Rulers	**Unit 7** Lessons 4, 6 **Unit 8** Lesson 2 **Unit 10** Lesson 1
Rulers (15 cm)	**Unit 7** Lesson 3
Rulers (30 cm)	**Unit 7** Lesson 3
Rulers (centimetre)	**Unit 7** Lessons 1, 2
Rulers (different types)	**Unit 7** Lesson 5
Rulers (metre)	**Unit 7** Lessons 1, 2, 3
Strips of paper	**Unit 8** Lesson 7
Strips of paper (12 cm long)	**Unit 8** Lesson 2
Ten frames	**Unit 6** Lesson 1
Weighing scales	**Unit 9** Lessons 2, 3, 4, 5, 6
Weights	**Unit 9** Lesson 4, 5

Year 3B Optional resources

Resource	Lesson
Art materials	**Unit 8** Lesson 1
Art straws	**Unit 7** Lesson 9
Balls (or bean bags)	**Unit 7** Lesson 6
Bar models (blank)	**Unit 7** Lesson 4
Base 10 equipment	**Unit 6** Lessons 1, 10 **Unit 7** Lessons 4, 5 **Unit 9** Lessons 5, 6 **Unit 10** Lessons 3, 5, 6
Birds-eye view of school playground (from internet)	**Unit 7** Lesson 10
Capacity measuring equipment	**Unit 10** Lessons 2, 3, 4, 5, 6
Capacity measuring equipment (100 ml, 500 ml and litre containers)	**Unit 10** Lesson 1
Card	**Unit 7** Lesson 9
Chalk	**Unit 7** Lesson 6
Coin	**Unit 7** Lesson 3
Counters	**Unit 7** Lesson 7 **Unit 9** Lesson 6
Counters (double sided)	**Unit 8** Lesson 3
Cups	**Unit 6** Lesson 7
Cartboard	**Unit 6** Lesson 2
Dice	**Unit 10** Lesson 4
Drinks bottles (empty)	**Unit 10** Lesson 1
Feathers	**Unit 7** Lesson 1
Fraction circles (printed)	**Unit 8** Lesson 4
Fraction rods	**Unit 8** Lesson 9
Fraction shapes	**Unit 8** Lessons 1, 2, 3, 4
Fraction strips (divided into various equal parts)	**Unit 8** Lesson 6
Fraction strips (printed)	**Unit 8** Lesson 5
Fraction tiles	**Unit 8** Lessons 4, 9
Fraction walls (printed)	**Unit 8** Lesson 8
Geoboards	**Unit 7** Lesson 9
Key vocabulary flashcards	**Unit 7** Lesson 5
Metre stick	**Unit 7** Lesson 11
Milk containers (capacity > 1 litre)	**Unit 10** Lesson 3
Modelling clay	**Unit 7** Lesson 11 **Unit 9** Lessons 4, 5, 6, 7
Multilink cubes	**Unit 7** Lessons 8, 11
Number cards	**Unit 8** Lesson 10
Number lines	**Unit 7** Lessons 3, 4, 8, 9, 10 **Unit 9** Lessons 2, 3, 4, 5, 6, 7 **Unit 10** Lesson 3
Number lines (blank)	**Unit 6** Lesson 1 **Unit 8** Lesson 7 **Unit 10** Lesson 1
Number lines (printed)	**Unit 9** Lesson 1
Objects (concrete – for sharing)	**Unit 6** Lesson 10
Objects (easily measurable)	**Unit 7** Lesson 2
Objects (real-life – to measure)	**Unit 7** Lesson 6
Objects (small: to be measured in millimetres, e.g. beads, pegs, paperclips)	**Unit 7** Lesson 5
Paper (to make paper aeroplanes)	**Unit 7** Lesson 6
Paper squares (or rectangles)	**Unit 8** Lesson 3

Year 3B Optional resources – *continued*

Resource	Lesson
Paper strips	**Unit 8** Lessons 8, 9
Paper strips (and similar materials for modelling calculations)	**Unit 7** Lesson 7
Parachute	**Unit 9** Lesson 2
Pencils (coloured)	**Unit 8** Lesson 3
Pieces of card (blank)	**Unit 7** Lesson 5
Pineapple	**Unit 9** Lesson 5
Place value columns	**Unit 10** Lesson 3
Place value counters	**Unit 7** Lesson 5 **Unit 9** Lessons 3, 4
Place value counters (to represent kg and g)	**Unit 9** Lesson 5
Place value grids (blank)	**Unit 6** Lesson 1
Play money	**Unit 6** Lesson 2
Poster paint	**Unit 10** Lesson 1
Pumpkin	**Unit 9** Lesson 5
Ribbon	**Unit 7** Lesson 7
Ribbon (or string or paper strips: to model calculations)	**Unit 7** Lesson 8
Rods (coloured)	**Unit 8** Lesson 8
Rulers	**Unit 7** Lessons 9, 10, 11
Rulers (metre)	**Unit 7** Lesson 7
Scissors	**Unit 7** Lessons 8, 9, 11
String	**Unit 7** Lessons 9, 11 **Unit 8** Lesson 9
Tape	**Unit 7** Lesson 7
Tape measure	**Unit 7** Lesson 7
Thread	**Unit 7** Lesson 5
Toy animal	**Unit 7** Lesson 8
Water	**Unit 10** Lesson 1
Weighing scales	**Unit 9** Lesson 1, 7
Wipeable number lines	**Unit 8** Lesson 6
Wool	**Unit 7** Lesson 9

Getting started with *Power Maths*

As you prepare to put *Power Maths* into action, you might find the tips and advice below helpful.

STEP 1: Train up!

A practical, up-front full day professional development course will give you and your team a brilliant head-start as you begin your *Power Maths* journey. You will learn more about the ethos, how it works and why.

STEP 2: Check out the progression

Take a look at the yearly and termly overviews. Next take a look at the unit overview for the unit you are about to teach in your Teacher Guide, remembering that you can match your lessons and pacing to match your class.

STEP 3: Explore the context

Take a little time to look at the context for this unit: what are the implications for the unit ahead? (Think about key language, common misunderstandings and intervention strategies, for example.) If you have the online subscription, don't forget to watch the corresponding unit video.

STEP 4: Prepare for your first lesson

Familiarise yourself with the objectives, essential questions to ask and the resources you will need. The Teacher Guide offers tips, ideas and guidance on individual lessons to help you anticipate children's misconceptions and challenge those who are ready to think more deeply.

STEP 5: Teach and reflect

Deliver your lesson — and enjoy!

Afterwards, reflect on how it went… Did you cover all five stages? Does the lesson need more time? How could you improve it?

Unit 6
Multiplication and division ③

Mastery Expert tip! 'Throughout this unit, I made sure to do lots of multiplying and dividing using base 10 equipment. This concrete learning activity really supported children's understanding by giving them an opportunity to manipulate and build the calculations!'

Don't forget to watch the Unit 6 video!

WHY THIS UNIT IS IMPORTANT

This unit is important because it develops children's multiplicative and divisive reasoning by linking their prior knowledge to 2-digit calculations which involve the expanded method and partitioning to divide. These concepts are closely linked to concrete and pictorial representations to scaffold and secure children's understanding. Children will compare multiplication and division statements using the <, > and = signs. They will develop their understanding of multiplication facts to link this knowledge to related multiplication and division calculations, for example, linking 2 × 3 = 6 and 2 × 30 = 60. Moving on, they will be introduced to the expanded method for multiplication and the partition method for dividing (leading to remainders). Finally, they will use their understanding of these new methods to solve mixed multi-step problems and puzzles involving all four operations.

WHERE THIS UNIT FITS

→ Unit 5 – Multiplication and division (2)
→ **Unit 6 – Multiplication and division (3)**
→ Unit 7 – Length and perimeter

In this unit children develop their understanding of the multiplicative properties of numbers. This unit follows their learning about multiplication and division and precedes their work on length and perimeter.

Before they start this unit, it is expected that children:
- are familiar with different concrete and visual representations for multiplying by 2, 3, 4, 5 and 10
- can share and group numbers that occur in the 2, 3, 4, 5 and 10 times-tables, making links between the 2 and 4 times-tables and the 4 and 8 times-tables
- can solve problems involving multiplication and division
- can solve division problems leading to remainders.

ASSESSING MASTERY

Children will demonstrate mastery of this unit by reliably solving 2-digit multiplication and division multi-step problems. They will be able to compare multiplication and division statements using inequality signs and confidently be able to find related calculation facts, for example, 6 × 30 = 180 and 180 ÷ 6 = 30. Children will be able to use the expanded method to multiply 2-digit numbers by 1-digit numbers and will also be able to partition numbers to divide them (including where the answers have remainders). Finally, children will be able to confidently link and use all of their knowledge to solve mixed multi-step problems involving all four operations.

COMMON MISCONCEPTIONS	STRENGTHENING UNDERSTANDING	GOING DEEPER
Children may confuse multiplying and dividing – especially when solving word problems.	Encourage children to use base 10 equipment and to use bar models to make sense of multi-step mixed problems.	Give children a completed bar model and ask them to think of a word problem to match it.
Children may not realise how to show remainders.	Encourage children to divide numbers by first partitioning (using a part-whole model), then dividing each part and writing down any remainder.	Challenge children to think of a division that leaves a remainder of 1.
Children may not understand the expanded method, for example, misrepresenting 54 × 3 as 4 × 3 added to 5 × 3 (rather than 4 × 3 and 50 × 3).	Ask children to use the expanded method, representing it with base 10 equipment.	Provide children with some multi-step word problems and ask them to explain their steps/solutions.

Unit 6: Multiplication and division ③

UNIT STARTER PAGES

Go through the unit starter pages of the **Pupil Textbook** with the whole class. Discuss the key learning points mentioned by the *Power Maths* characters, particularly focusing on the key vocabulary. Ask children for examples to ensure their understanding.

STRUCTURES AND REPRESENTATIONS

Place value grid and base 10 equipment: Place value grids with place value counters and base 10 equipment are used to demonstrate, and enable children to manipulate, the place value of numbers, to support the expanded method for multiplication and the partition method for division.

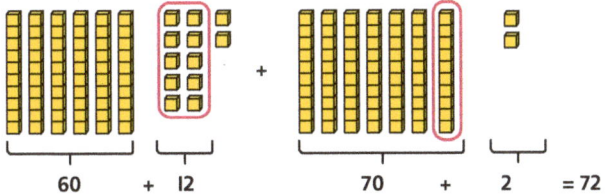

Bar model: Bar models are used in this unit to support children in solving multi-step mixed problems.

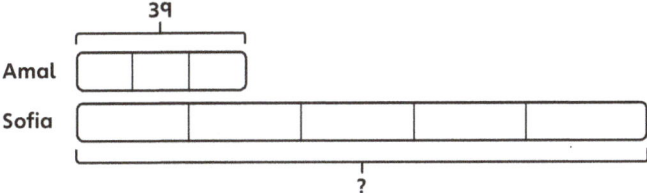

Part-whole model: Part-whole models are used in this unit to partition 2-digit numbers when dividing.

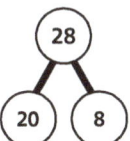

KEY LANGUAGE

There is some key language that children will need to know as part of the learning in this unit:

→ multiplication, division
→ multiplication statement, division statement, number sentence
→ greater than (>), less than (<), equal (=), compare, equally, least, most
→ remainder
→ share
→ partition, multi-step, expanded written method
→ tens (10s), ones (1s)
→ exchange

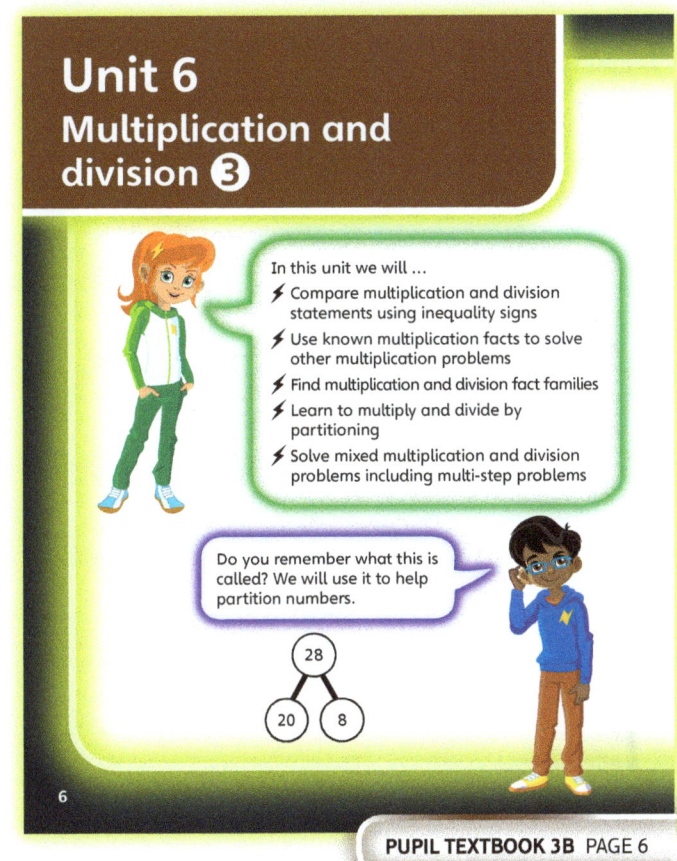

PUPIL TEXTBOOK 3B PAGE 6

PUPIL TEXTBOOK 3B PAGE 7

Unit 6: Multiplication and division (3), Lesson 1

Multiples of 10

Learning focus

In this lesson, children will understand how to find multiples of 10 by counting in 10s. They will begin to understand what happens when you multiply 1- and 2-digit numbers by 10. Children will also work out how many 10s are in 3-digit multiples of 10.

Before you teach

- Can children count in 10s from 0 to 100?
- Can children use ten frames and counters to represent numbers?
- Can children represent a number in a place value grid?

NATIONAL CURRICULUM LINKS

Year 3 Number – multiplication and division

Write and calculate mathematical statements for multiplication and division using the multiplication tables that they know, including for 2-digit numbers times 1-digit numbers, using mental and progressing to formal written methods.

ASSESSING MASTERY

Children can find multiples of 10 by counting in 10s or using other methods. Children are able to recognise 2- and 3-digit numbers that are multiples of 10.

COMMON MISCONCEPTIONS

In the lesson, children may count in 1s as opposed to counting in 10s. Explain to children that each of the bean bags or counters represents 10. Ensure that children can count in 10s from at least 0 to 100. Use a number line and base 10 equipment to support their counting.

STRENGTHENING UNDERSTANDING

When counting in 10s, children may need to see the 10 to help them. Use base 10 equipment next to each bean bag or counter to help them count. You can also use base 10 equipment to make the numbers in the place value grids and to make other multiples of 10. Children will start to see that a multiple of 10 always has a 0 in the 1s place.

GOING DEEPER

Ask children to try and generalise and come up with a method for quickly working out whether a number is a multiple of 10. Ask: *Can a multiple of 10 have any 1s? Does a multiple of 10 have to have some 10s?* These questions will help children formulate their own deeper understanding of numbers that are multiples of 10.

KEY LANGUAGE

In lesson: tens (10s), ones (1s), multiples

Other language to be used by the teacher: count, score

STRUCTURES AND REPRESENTATIONS

Ten frames, place value grids, number lines

RESOURCES

Mandatory: ten frames, 10s counters

Optional: blank place value grids, blank number lines, base 10 equipment

 In the eTextbook of this lesson, you will find interactive links to a selection of teaching tools.

Quick recap

Check that children can count on in 10s from any multiple of 10. Start with 0 and ask each child to say the next multiple of 10. See how far they can count as a class. Repeat, starting from a different multiple of 10.

Unit 6: Multiplication and division (3), Lesson 1

Discover

WAYS OF WORKING Pair work

ASK

- Question 1 a): *What game might the children have been playing? Who do you think has won the game? Why? How many beans bags has Bella collected? How many points is each bean bag worth? How many points has Bella scored?*
- Question 1 b): *How many bean bags has Andy collected? How many points has Andy scored?*

IN FOCUS To answer questions 1 a) and 1 b), children could represent the bean bags as counters on ten frames and use this to help them count in 10s. Children may want to use 10s counters to represent the bean bags or possibly use base 10 equipment if it makes it easier for children to see the 10. In question 1 b), children will need to use two ten frames to help them. Ask children if they need to count the 100.

PRACTICAL TIPS Play a game to replicate the scenario, where children win bean bags or they have to throw bean bags into a bin. Each bean bag is worth 10 points.

ANSWERS

Question 1 a): Bella has collected 7 bean bags.
Bella scored 70.

Question 1 b): Andy has collected 13 bean bags.
Andy scored 130.

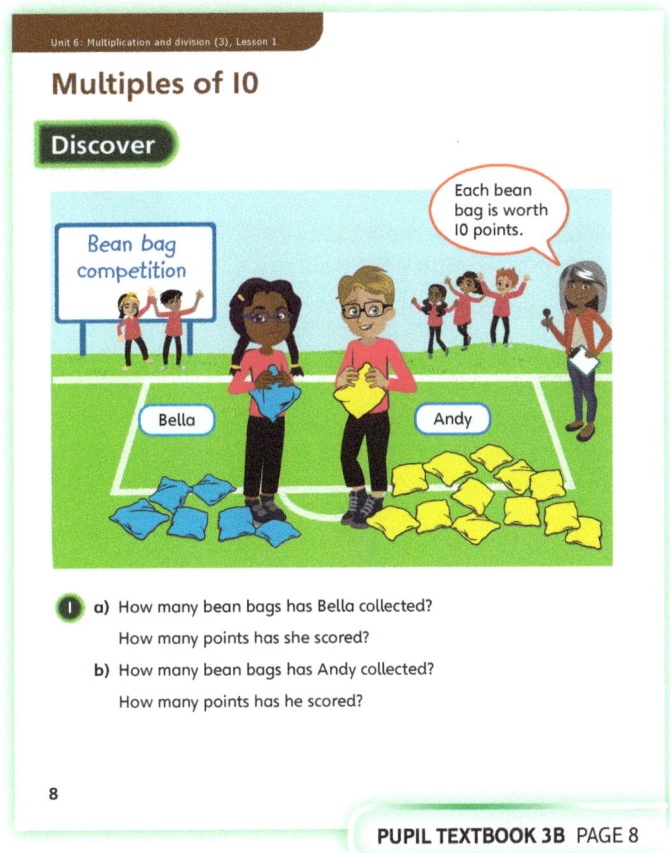

PUPIL TEXTBOOK 3B PAGE 8

Share

WAYS OF WORKING Whole class teacher led

ASK

- Question 1 a): *How many bean bags are on the ten frame? Where does the answer 70 come from?*
- Question 1 b): *How many bean bags are there across the two ten frames? What is Andy's score? How did you find the score? Did you need to count in 10s for the first ten frame or can you see it is worth 100 points?*

IN FOCUS In question 1 a), children use the ten frame to help them see that there are 7 bean bags. Some children will know this represents 70 straight away. Ensure you count from 10 to 70 with children, asking them to point and count as they go. You may want to use 10 counters instead of 10 bean bags on their ten frames, or put 1 counter over each bean bag. In question 1 b), apply a similar approach to finding out that the bean bags are worth 130 points, however, try to encourage children to see that there is 100 on the first ten frame without counting, as it is full.

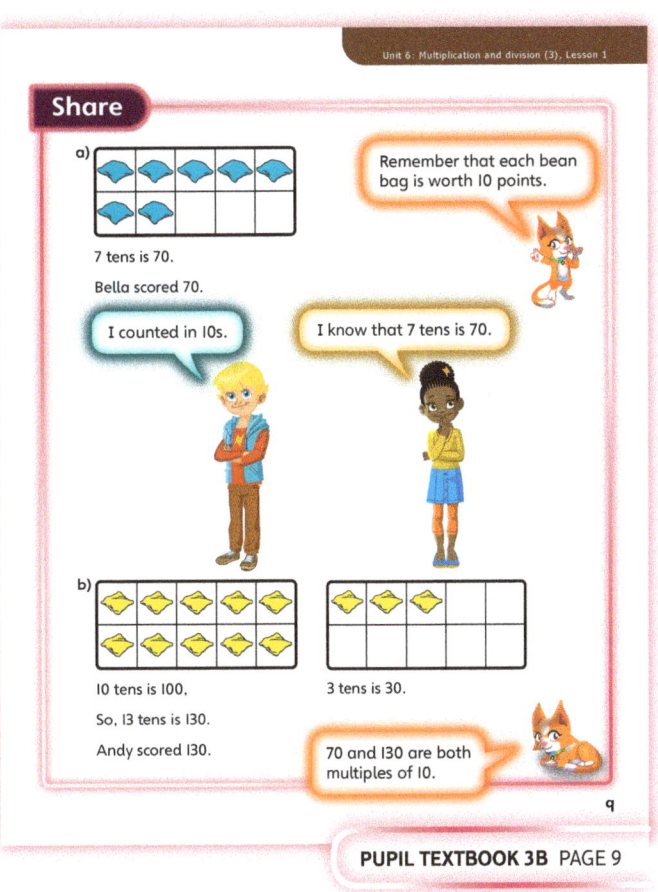

PUPIL TEXTBOOK 3B PAGE 9

47

Unit 6: Multiplication and division (3), Lesson 1

Think together

WAYS OF WORKING Whole class teacher led (I do, We do, You do)

ASK

- Question ❶: *What multiples of 10 are shown on the ten frames? How did you work it out?*
- Question ❷ a): *Which base 10 equipment can you use to represent the numbers in the place value grid? What do you notice?*
- Question ❷ b): *What do you notice about the multiples of 10?*
- Question ❸: *How do you know if a number is a multiple of 10?*

IN FOCUS In question ❶, children further practise the method they used in the **Discover** task, to work out the multiples of 10. Remind children that they can count in 100s first and then 10s. They should be able to see that 100 is ten 10s. In question ❷, children represent the numbers in the ten frame using base 10 equipment and use this to work out how many 10s there are in each number. They then go on to match multiplication sentences with their answer. They may start to notice that all multiples of 10 end in 0 and the other digits remain the same. In question ❸, children apply any patterns they have noticed earlier in the lesson to play a game with a partner. They take it in turns to cover multiples of 10. It is important that children can articulate what a multiple of 10 is before playing the game.

STRENGTHEN In question ❶, use a number line that goes up in 10s to help children count in 10s. In question ❷, use base 10 equipment to help children count in 100s and then 10s.

DEEPEN Ask children to try and generalise and come up with a method for quickly working out whether a number is a multiple of 10. Ask: *Can a multiple of 10 have any 1s? Does a multiple of 10 have to have some 10s?*

ASSESSMENT CHECKPOINT Check that children can represent a multiple of 10 using equipment and they can determine whether a 2- or 3-digit number is a multiple of 10.

ANSWERS

Question ❶ a): 170

Question ❶ b): 210

Question ❷ a): 11 tens   23 tens

Question ❷ b): Children should match:

    34 × 10   340
    13 × 10   130
    10 × 43   430

Question ❸:

●	15	105	501
●	●	●	55
93	●	●	●
●	●	301	●

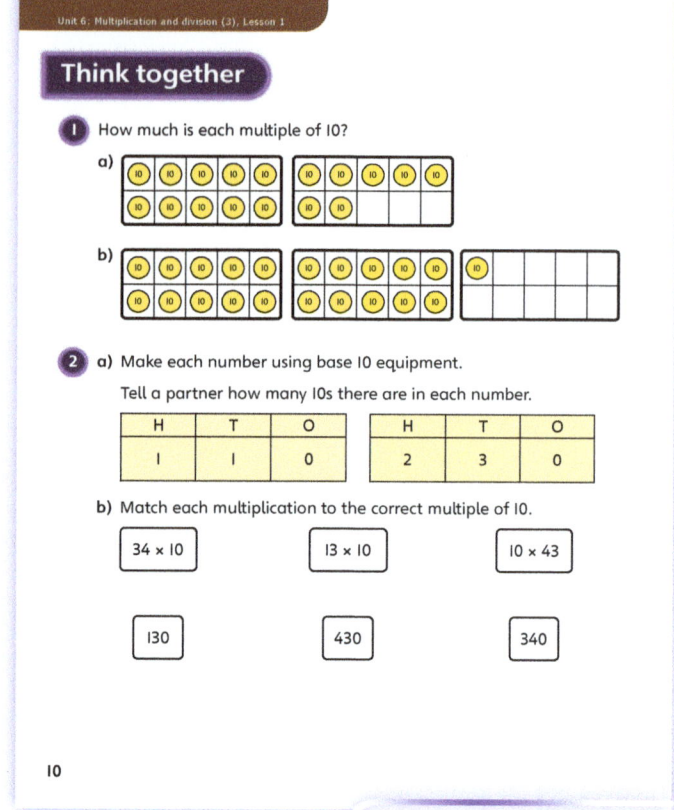

PUPIL TEXTBOOK 3B PAGE 10

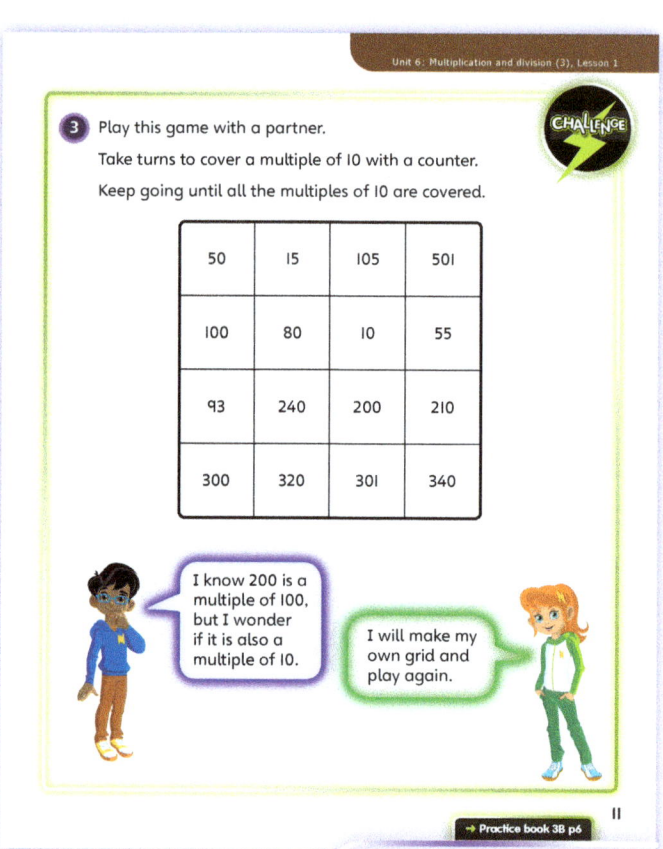

PUPIL TEXTBOOK 3B PAGE 11

48

Unit 6: Multiplication and division (3), Lesson 1

Practice

WAYS OF WORKING Independent thinking

IN FOCUS In questions ❶ and ❷, children show their understanding of 10s on a ten frame. They should recognise that the full ten frame represents 100, and then count in 10s to determine the value on the second ten frame. In question ❸, children connect multiplications where 2-digit numbers are multiplied by 10 with corresponding answers that are given in place value grids. Children should start to see that there are no 1s in any of the place value grids and that the digits are the same as in the multiplication but they have moved one place value column to the left. This will help in further questions where they have to recognise multiples of 10, such as in question ❹. In question ❺, children see a different representation for multiples of 10.

STRENGTHEN Support children in answering question ❶ by using a number line to help them count in 10s. Encourage children to be more efficient by starting from 100, rather that counting to 100 first. The ten frames should help children see that 100 is 10 tens. In question ❸, children may find it easier to make the numbers from base 10 equipment or as place value counters on ten frames to understand where the answers may come from.

DEEPEN Children should try and generalise about a multiple of 10. They should be able to convince a partner that a multiple of 10 ends in 0. They should also try to generalise about what happens to the other digits in the number.

ASSESSMENT CHECKPOINT Check that children know that multiples of 10 end in a 0 and they can work out a multiple of 10 with and without using supporting mathematical equipment.

ANSWERS Answers for the **Practice** part of the lesson can be found in the *Power Maths* online subscription.

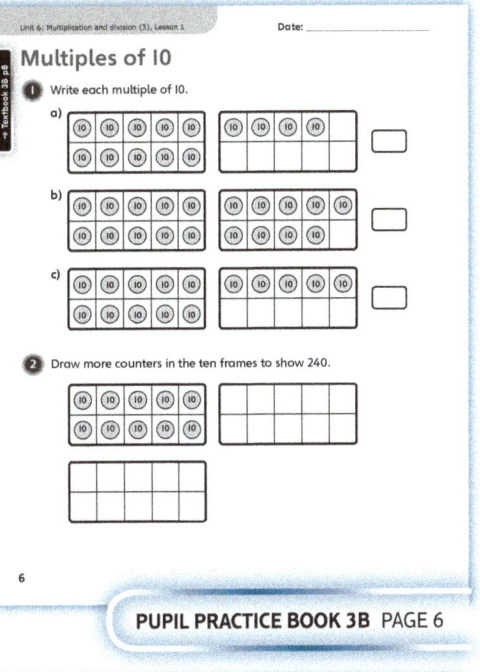

PUPIL PRACTICE BOOK 3B PAGE 6

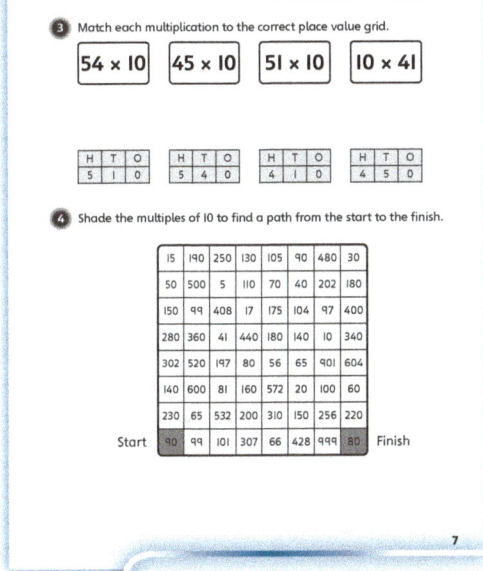

PUPIL PRACTICE BOOK 3B PAGE 7

Reflect

WAYS OF WORKING Independent thinking

IN FOCUS Children will reflect on all the answers in this lesson to help them respond to the statement. If they have not noticed the generalisation, ask them to look through their answers. Question ❹ may help them answer this question.

ASSESSMENT CHECKPOINT Check that children understand that a multiple of 10 ends with a 0. You may want them to write down some multiples of 10.

ANSWERS Answers for the **Reflect** part of the lesson can be found in the *Power Maths* online subscription.

After the lesson ⏸

- Were children able to explain that a multiple of 10 always ends in a 0?

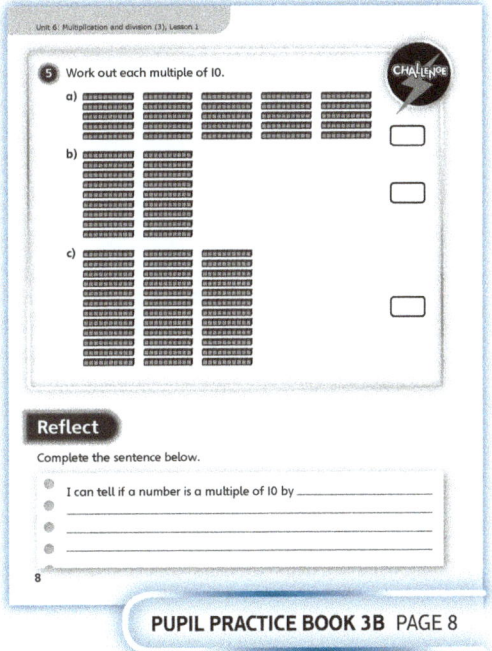

PUPIL PRACTICE BOOK 3B PAGE 8

49

Unit 6: Multiplication and division (3), Lesson 2

Related calculations

Learning focus
In this lesson, children will use known multiplication facts to solve related multiplication problems, particularly involving multiplying by 10.

Before you teach
- Can children multiply numbers by 10?
- Do children understand the commutative property of multiplication?

NATIONAL CURRICULUM LINKS

Year 3 Number – multiplication and division

Write and calculate mathematical statements for multiplication and division using the multiplication tables that they know, including for 2-digit numbers times 1-digit numbers, using mental and progressing to formal written methods.

ASSESSING MASTERY

Children can link their understanding of known multiplication facts to related calculations and can explain the link between two multiplications such as 3 × 4 and 3 × 40. They can demonstrate an understanding of commutativity in multiplication, and can fluently solve a calculation mentally without needing the related fact.

COMMON MISCONCEPTIONS

Children may not know multiplication is commutative. They might solve 5 × 4 and 5 × 40, but not 40 × 5. Ask:
- *Can you draw 5 × 4 and 4 × 5? How are they similar? How are they different?*

Children may think that to work out 40 × 5, they just do 4 × 5 and 'stick a 0 on the end'. Using place value grids, base 10 equipment and the exchange concept will help children understand multiplying by 10. Ask:
- *Can you use base 10 equipment to show why a 0 comes on the end? Does it always work?*

STRENGTHENING UNDERSTANDING

Children should know how to multiply a number by 10. Encourage children to count in 10s to secure fluency with patterns. Use exchange, place value grids and base 10 equipment to help strengthen understanding. Use commutativity and core multiplication facts that children know to connect to other multiplication facts.

GOING DEEPER

Give some numbers such as 20 and 200. How many related facts (factors) can children find? For example, 4 × 5 and 40 × 5.

KEY LANGUAGE

In lesson: multiplication, multiplication facts, total

Other language to be used by the teacher: number sentence, share, multiples

STRUCTURES AND REPRESENTATIONS

Arrays, number lines

RESOURCES

Mandatory: counters, base 10 equipment

Optional: dartboard, play money

 In the eTextbook of this lesson, you will find interactive links to a selection of teaching tools.

Quick recap

Play a game of 'multiplying by 10 bingo'. Ask children to write down six multiples of 10 between 0 and 300. Then read out questions such as 23 × 10. If they have the answer, they cross it off.

Unit 6: Multiplication and division (3), Lesson 2

Discover

WAYS OF WORKING Pair work

ASK

- Question 1 a): *Is there a connection between counting in 3s and 6s?*
- Question 1 b): *How many times more candles are there than balloons?*

IN FOCUS Question 1 is an excellent way of visually showing related multiplications. Encourage children to look for patterns. They should spot that 30 is 10 times bigger than 3, so the answer must also be 10 times bigger.

PRACTICAL TIPS Pair base 10 equipment with number lines to develop children's understanding of related multiplication calculations.

ANSWERS

Question 1 a): There are 18 balloons in total.

Question 1 b): There are 180 candles in total.
6 × 3 ones is 18 ones.
So, 6 × 3 tens is 18 tens.
6 × 3 and 6 × 30 are related facts.

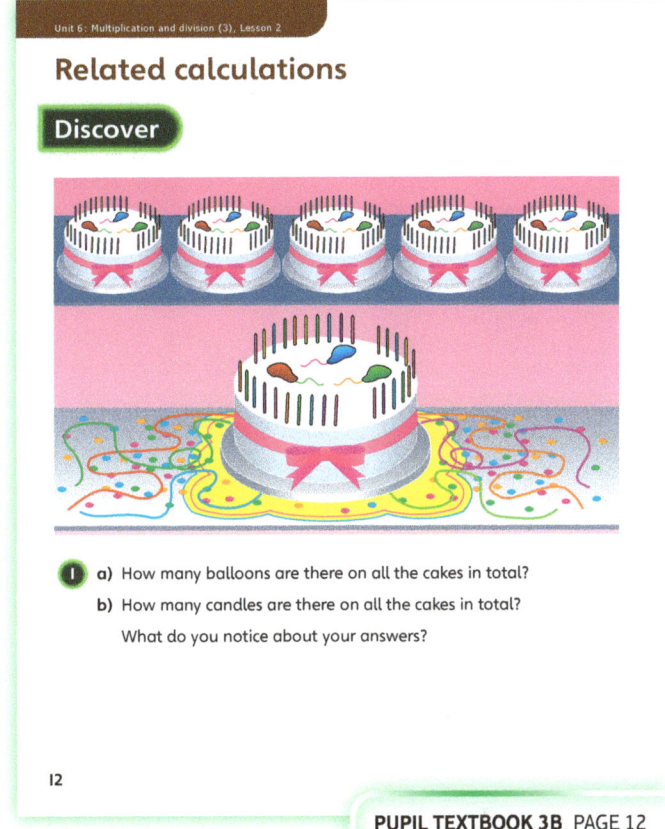

Share

WAYS OF WORKING Whole class teacher led

ASK

- Question 1 b): *Is counting each candle the easiest way to find the answer? Is there a more efficient method?*
- Question 1 b): *Is there a related multiplication fact you can use to find the answer?*
- Question 1 b): *Could multiplying the answer by 10 help?*

IN FOCUS In question 1 b), if children start counting every single candle, stop them. Ask: *Can you think of a better method? Do you remember how to multiply by 10?* This question is a great opportunity to reinforce counting skills. Ask children to count in 3s, following the sequence on the number line. Then repeat this, this time counting in 30s and following the sequence on the number line.

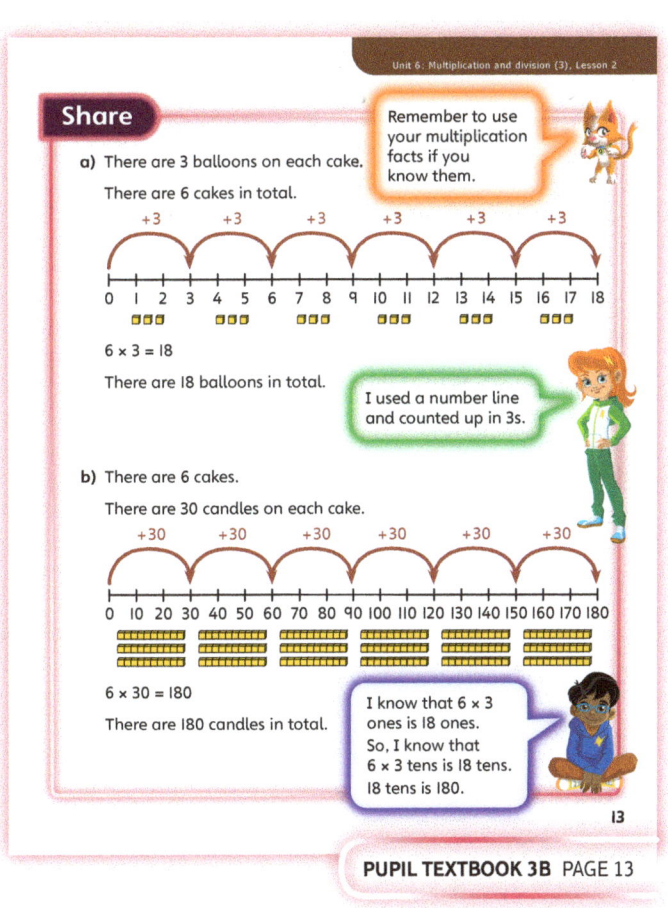

51

Think together

WAYS OF WORKING Whole class teacher led (I do, We do, You do)

ASK

- Question ① a): *How can you find the answer to the problem?*
- Question ① b): *What is the most efficient way to find the answer to this question?*
- Question ② b): *What patterns can you see? How can you use your answer to the first question to work out those below it? Can you find a fourth fact for each column of related calculations?*
- Question ③: *What is the connection between the numbers in this triangle? What multiplication facts can you find?*

IN FOCUS Question ① reinforces using efficient methods to find solutions by using related multiplication facts. Encourage children to see the connections between the multiplication facts used and to develop efficient methods when solving multiplication problems. In question ②, children solve related calculations. Challenge children to think of another related fact such as 5 × 7 = 35 (for i) and 4 × 6 = 24 (for iv) and, if scaffolding is needed, encourage use of base 10 equipment to develop the concepts via concrete representations. Ask children to spot patterns and explain what these patterns are, encouraging using efficient methods to find solutions to related multiplication problems.

STRENGTHEN Strengthen learning with counting practice as a whole class. Count in 2s then 20s, 3s then 30s, 4s then 40s, and so on. Count back, too. Encourage children to see the connection between counting in multiples of units and then the related multiples of 10s. Ask children what is the same and what is different.

DEEPEN Question ③ uses an abstract representation. Challenge children to create their own product-factor triangles for each set of calculations a) and b). Ask them if they can also find division facts related to each set of calculations.

ASSESSMENT CHECKPOINT Question ② will help you see whether children can use a known multiplication fact, such as 7 × 5, to find related multiplication facts, such as 70 × 5.

ANSWERS

Question ① a): 8 × 2 = 16

Question ① b): 8 × 20 = 160

Question ② a): 4 × 3 = 12
4 × 30 = 120

Question ② b): i) 7 × 5 = 35   iv) 6 × 4 = 24
ii) 7 × 50 = 350   v) 6 × 40 = 240
iii) 70 × 5 = 350   vi) 60 × 4 = 240

Question ③ a): 4 × 5 = 20
40 × 5 = 200
5 × 40 = 200

Question ③ b): 2 × 4 = 8
20 × 4 = 80
4 × 20 = 80

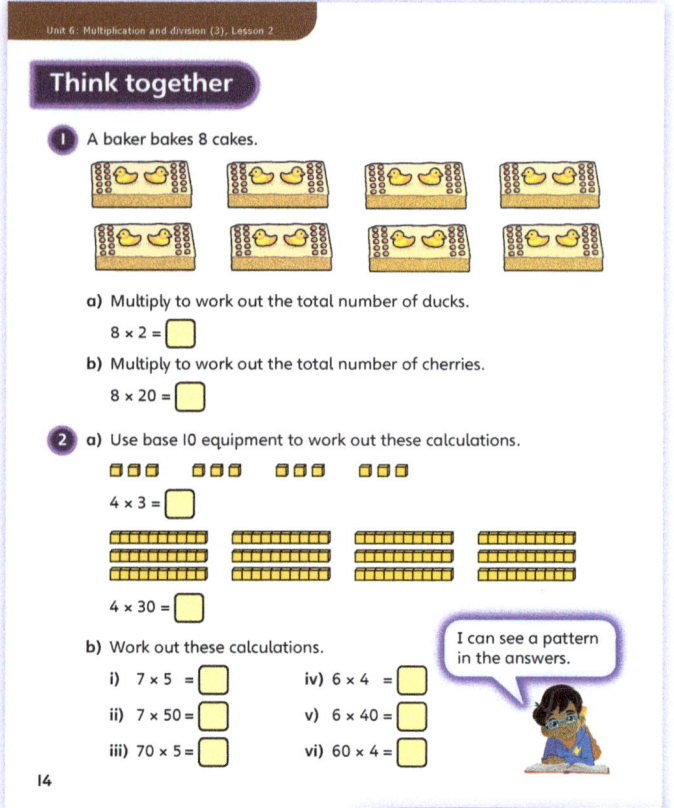

PUPIL TEXTBOOK 3B PAGE 14

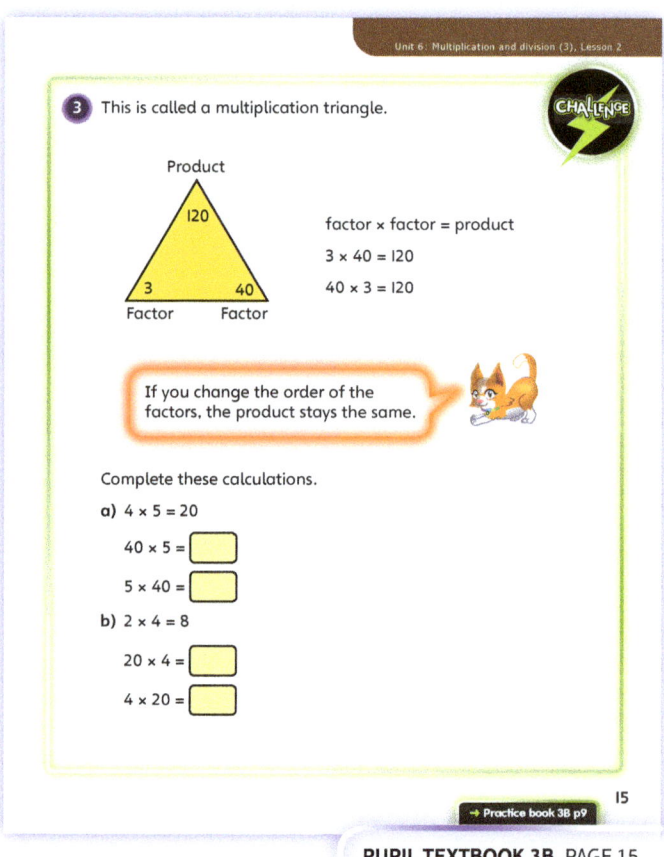

PUPIL TEXTBOOK 3B PAGE 15

Unit 6: Multiplication and division (3), Lesson 2

Practice

WAYS OF WORKING Independent thinking

IN FOCUS Questions 2 and 4 are important because they use real-life applications of related multiplication calculations. Can children think of any more real-life applications?

STRENGTHEN Following question 2, place darts on a dartboard and ask children to find and solve the related multiplications (this activity could also be completed with real-life arrays of coins, as could question 4).

For question 5 d), encourage children to think about what happens when they multiply by 0. Encourage understanding by putting it in context. Ask: *What are zero lots of 20 sweets? What are 20 lots of nothing?*

DEEPEN In question 6, challenge children to think of another statement similar to the one Holly makes, for example, *If I multiply my number by ▢, I get ▢. What do I get if I multiply my number by ▢?* Ask children to swap with a partner and solve each other's problem. This will require some deep thinking.

THINK DIFFERENTLY Question 3 asks children to write abstract multiplications that are related from pictorial representations. They may recognise the commutative nature of multiplication and that the order of the two factors does not matter.

ASSESSMENT CHECKPOINT Questions 5 c) and 5 d) will help assess whether children can quickly solve multiplications which have multiples of 10 times a single digit. Encourage children to explain their methods to ensure they have grasped the concept clearly.

ANSWERS Answers for the **Practice** part of the lesson can be found in the *Power Maths* online subscription.

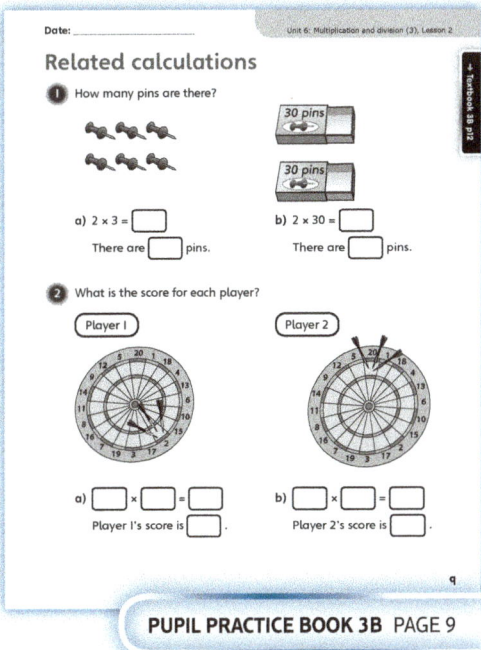

PUPIL PRACTICE BOOK 3B PAGE 9

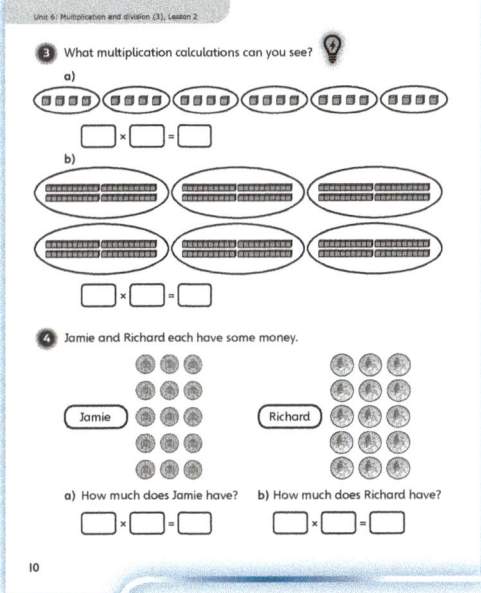

PUPIL PRACTICE BOOK 3B PAGE 10

Reflect

WAYS OF WORKING Independent thinking

IN FOCUS Children should understand that 4 × 8 can help them work out 4 × 80. Extend their learning by asking if there is another multiplication it can help them work out (40 × 8).

ASSESSMENT CHECKPOINT This reflective exercise will allow assessment of whether children can not only solve the problem but also explain their method clearly.

ANSWERS Answers for the **Reflect** part of the lesson can be found in the *Power Maths* online subscription.

After the lesson

- Can children solve a multiple of 10 × a 1-digit number independently?
- Do children still need a related multiplication fact to scaffold their learning?
- Did children show an understanding of commutativity in their learning?

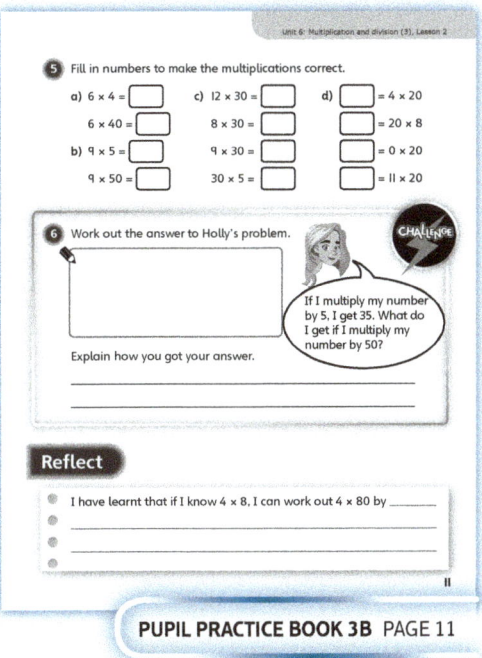

PUPIL PRACTICE BOOK 3B PAGE 11

Unit 6: Multiplication and division (3), Lesson 3

Reasoning about multiplication

Learning focus
In this lesson, children will compare multiplication statements. They will build on their understanding of commutativity in multiplication to spot patterns and will make comparisons using the < and > signs.

Before you teach
- Are children confident with times-tables?

NATIONAL CURRICULUM LINKS

Year 3 Number – multiplication and division

Solve problems, including missing number problems, involving multiplication and division, including positive integer scaling problems and correspondence problems in which *n* objects are connected to *m* objects.

ASSESSING MASTERY

Children can use and explain mental methods to solve word problems involving multiplication statements. Children understand that multiplication is commutative and can compare multiplication statements without having to work out answers, for example, they recognise that 24 × 3 is smaller than 4 × 24 because multiplication is commutative and 4 is greater than 3.

COMMON MISCONCEPTIONS

Children may not realise they can compare multiplications without working out the answer. Ask:
- *If two number sentences have a similar number, can you compare them without finding the answer?*

Children may need some support with the < and > signs and might confuse the signs. Ask:
- *What do these signs mean? Which sign means 'less than' and which sign means 'greater than'?*

STRENGTHENING UNDERSTANDING

Strengthen learning with practical activities where children develop their understanding of the pictorial and concrete representations by comparing groups of counters, while encouraging use of the terminology 'greater than' and 'less than'. Ask children to use < and > signs in number sentences and then read them out.

GOING DEEPER

Deepen learning by developing abstract concepts and notation using number sentences with missing numbers for children to fill in, for example, 4 × ☐ > 3 × ☐.

KEY LANGUAGE

In lesson: multiplication, statement, compare, more than, less than (<), greater than (>), equally, least, most

Other language to be used by the teacher: number sentence, equal

STRUCTURES AND REPRESENTATIONS

Arrays

RESOURCES

Mandatory: counters

 In the eTextbook of this lesson, you will find interactive links to a selection of teaching tools.

Quick recap
Write 6 × 4 = 24 on the board. What other multiplication facts can children write down using this calculation? For example, 4 × 6, 40 × 6, 60 × 4, etc. Ask children to write down full calculations.

Unit 6: Multiplication and division (3), Lesson 3

Discover

WAYS OF WORKING Pair work

ASK

- Question 1 a): *Do you need to work out the numbers of apples each child has to compare the amounts?*
- Question 1 a): *Which number is the same in both calculations? Does this help us compare?*
- Question 1 b): *How many apples does Richard have? How many apples does Lexi have?*

IN FOCUS In question 1 a), children look at two different multiplications and compare them without needing to find the answer to the multiplication problem. This gives children an opportunity to develop efficient methods when solving multiplication problems involving comparisons.

PRACTICAL TIPS Provide children with counters which they can use as concrete representations to support their multiplications.

ANSWERS

Question 1 a): Lexi has more apples.

Question 1 b): Richard has 12 apples. Lexi has 15 apples.
12 + 15 = 27
There are 27 apples in total.

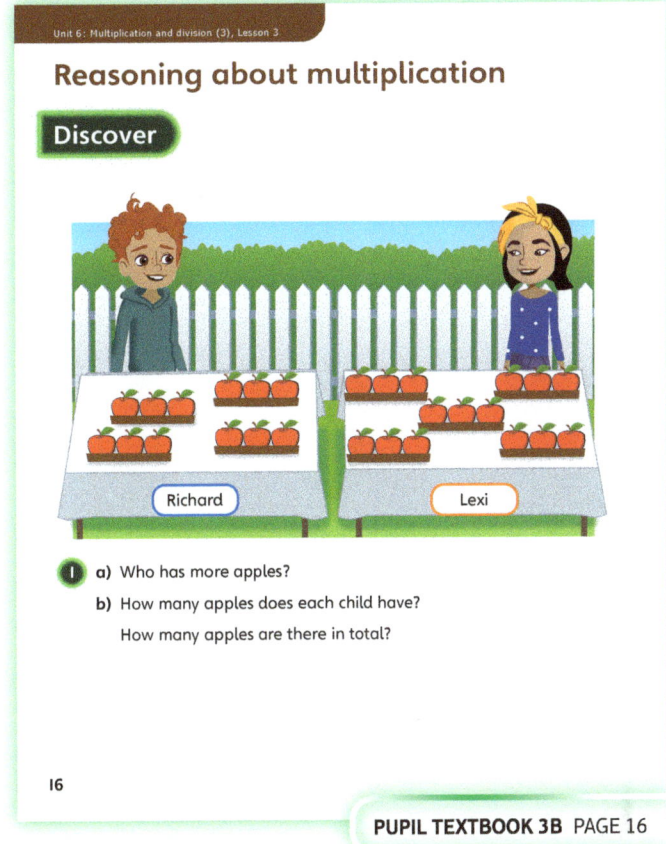

Share

WAYS OF WORKING Whole class teacher led

ASK

- Question 1 a): *What do the pictures of the arrays of apples show us? Which array is bigger?*
- Question 1 a): *Do you need to work out the answers to compare the arrays?*
- Question 1 b): *How do the bar models help to show the multiplications?*

IN FOCUS In question 1 a), encourage children to use counters to show the multiplications and to use these concrete representations to help them understand that because one array is bigger than the other it is not necessary to do the full calculation. Ensure that children know that the > sign means greater than.

In question 1 b), children need to first find out how many apples each child has before adding the two amounts together.

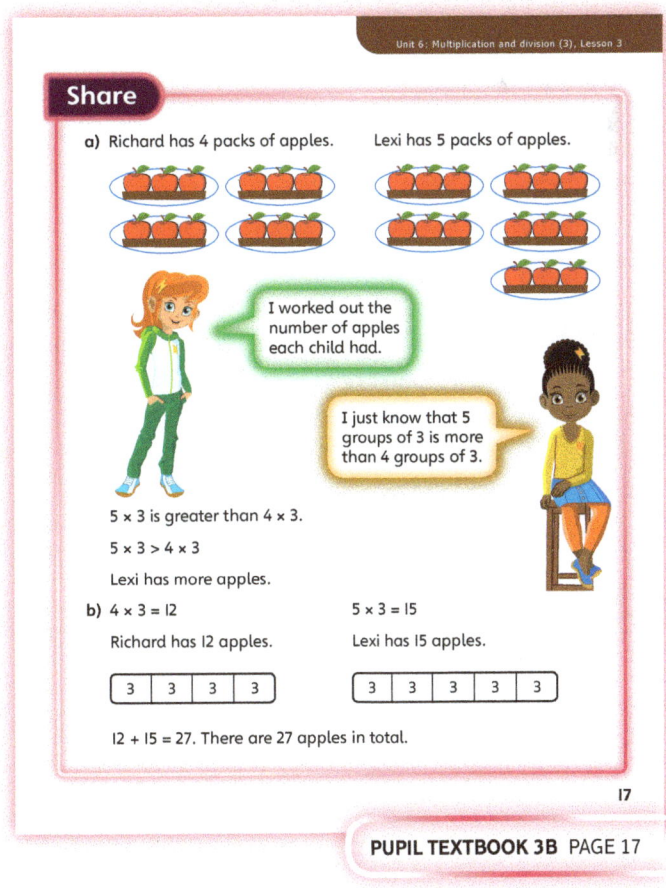

Think together

WAYS OF WORKING Whole class teacher led (I do, We do, You do)

ASK

- Question ②: *What is the same and what is different about the calculations?*
- Question ③: *What should you look for to compare the multiplications without working out the answers? Can you spot a pattern? How can you use multiplication facts to help you?*

IN FOCUS For question ①, children can visually compare the amounts. In question ②, develop the concept of solving problems using efficient methods. Encourage children to check by working out each answer and then comparing the answers.

Questions ② and ③ are a good chance to reinforce commutativity with children. Remind them that 5 × 3 = 3 × 5 and that with multiplication it does not matter which way around the numbers are.

STRENGTHEN The visual way of comparing multiplications presented in question ① is excellent for children's understanding – they will intuitively be able to compare the bags of pears. Develop this using counters to show that 5 × 4 is being compared with 2 × 4. Encourage children to use counters to represent other multiplications involving the same times-tables, and then compare them. Encourage children to express their pictorial representations verbally and then represent these using written notation to strengthen their vocabulary and abstract representation.

DEEPEN In question ③, encourage children to explain clearly which sentences they can compare without working out the answer and why. Encourage children to look for comparisons using the same multiplication facts.

ASSESSMENT CHECKPOINT In question ②, assess children on whether they can compare the number sentences without having to work out the answer, and also whether they can use the correct sign (< and >).

ANSWERS

Question ① a): Richard has more pears.

Question ① b): There are 28 pears in total.

Question ② a): 3 × 5 < 18   c) 5 × 3 = 3 × 5

Question ② b): 4 × 5 > 5 × 3   d) 8 × 2 > 3 × 5

Question ③ a): 3 × 40 > 2 × 20

Question ③ b): i) 7 × 30 = 30 × 7
 ii) 6 × 20 = 3 × 40
 iii) 4 × 30 > 5 × 20

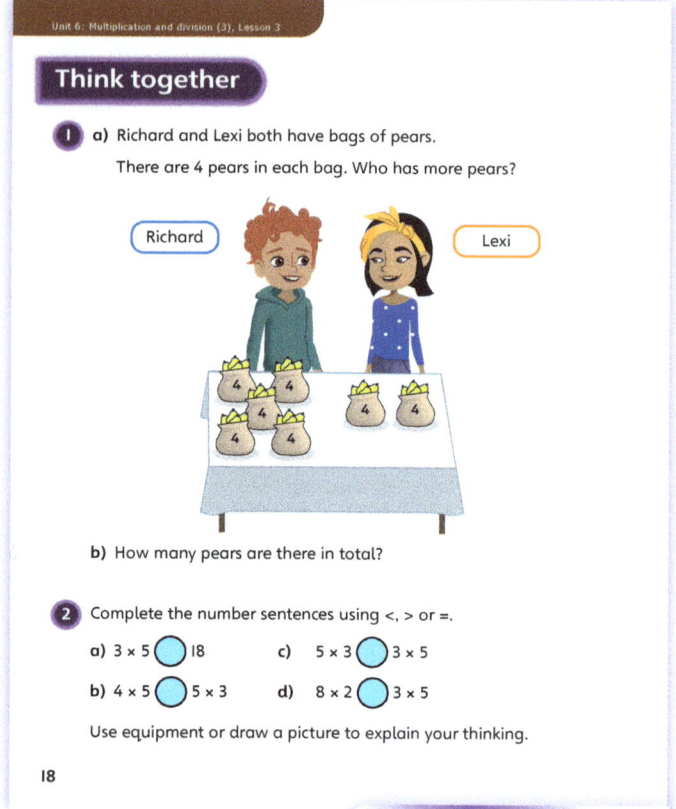

PUPIL TEXTBOOK 3B PAGE 18

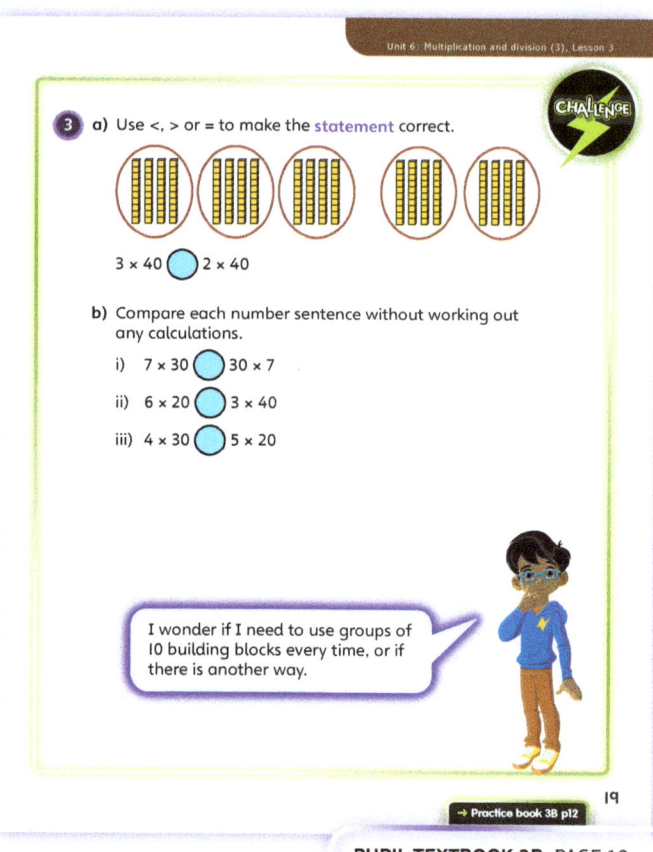

PUPIL TEXTBOOK 3B PAGE 19

Unit 6: Multiplication and division (3), Lesson 3

Practice

WAYS OF WORKING Independent thinking

IN FOCUS Question ① asks children to compare two calculations with pictorial support. They should be able to see that 6 × 10 is greater than 5 × 10 using the images. Further questions build on this by providing reduced support. The majority of questions ask children to use <, > and = signs to complete number sentences. Encourage children to use reasoning as opposed to working out the answers and then comparing. In some cases, some children may need to work out the answers to check they are correct.

STRENGTHEN For the more abstract questions, such as in questions ②, ③ and ⑥, encourage children to use base 10 equipment to help them reason why one calculation may be greater or less than another one.

DEEPEN Throughout the questions, ask children to explain their answers with reasons. The key part of this lesson is for children to correctly articulate why 40 × 3 is, for example, the same as 3 × 40.

ASSESSMENT CHECKPOINT Check that children know how to use < and > signs and can make comparisons without working out calculations.

ANSWERS Answers for the **Practice** part of the lesson can be found in the *Power Maths* online subscription.

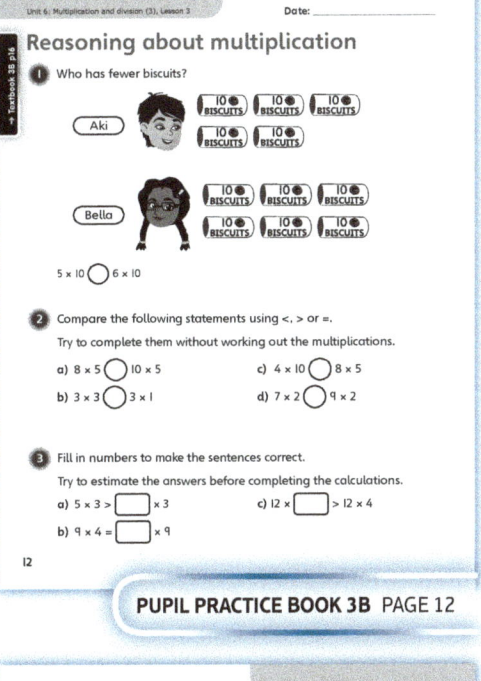

PUPIL PRACTICE BOOK 3B PAGE 12

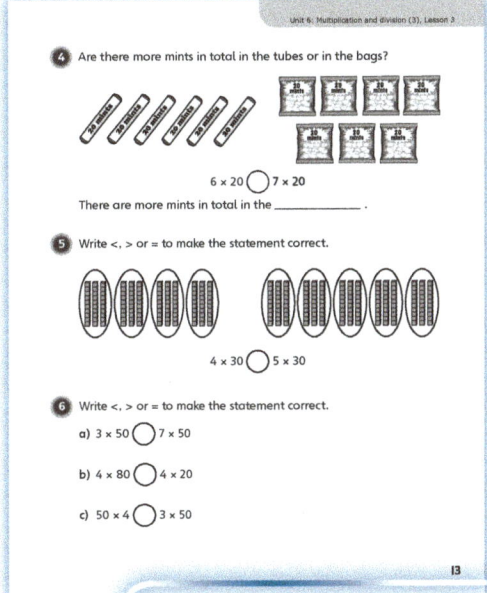

PUPIL PRACTICE BOOK 3B PAGE 13

Reflect

WAYS OF WORKING Group work

IN FOCUS Children may initially approach this question using trial and error. They should soon see that if they put the same number in each box then the statement can never be true.

ASSESSMENT CHECKPOINT Check that children can give two correct numbers to complete the puzzle and that they can explain their method.

ANSWERS Answers for the **Reflect** part of the lesson can be found in the *Power Maths* online subscription.

After the lesson

- Are children still relying on working out the answers rather than using their reasoning skills?
- Can children explain how to compare multiplications?

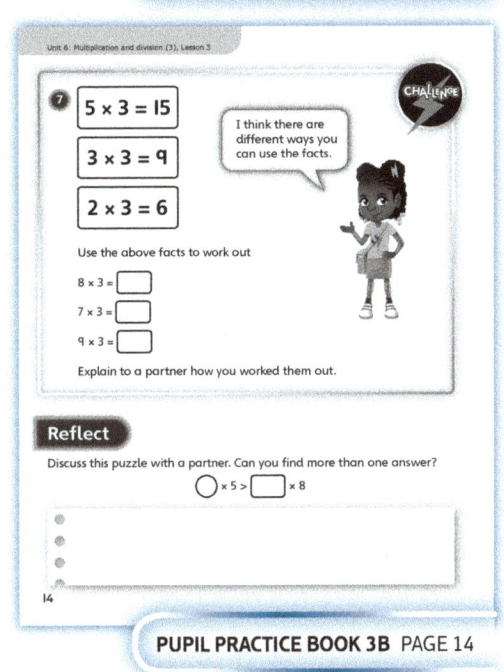

PUPIL PRACTICE BOOK 3B PAGE 14

Unit 6: Multiplication and division (3), Lesson 4

Multiply 2-digits by 1-digit – no exchange

Learning focus

In this lesson, children will use the expanded method to solve 2-digit numbers multiplied by 1-digit numbers. They will demonstrate a secure understanding of partitioning and place value in their calculations.

Before you teach

- How confident are children with using place value grids and base 10 equipment?
- How confident are children with partitioning 2-digit numbers?

NATIONAL CURRICULUM LINKS

Year 3 Number – multiplication and division

Write and calculate mathematical statements for multiplication and division using the multiplication tables that they know, including for 2-digit numbers times 1-digit numbers, using mental and progressing to formal written methods.

ASSESSING MASTERY

Children can reliably use the expanded method to calculate multiplications. They can explain their method clearly, and confidently represent their thinking using concrete and pictorial representations.

COMMON MISCONCEPTIONS

Children may not partition correctly, for example, breaking down 34 × 3 into 3 × 3 and 4 × 3, instead of 30 × 3 and 4 × 3. Ask:
- *Can you show the value of each digit using a place value grid?*

STRENGTHENING UNDERSTANDING

Encourage children to use base 10 equipment alongside a place value grid. This will help them to partition the numbers, making the expanded method easier to understand and implement.

GOING DEEPER

Provide children with multiplication sentences which contain mistakes. Challenge them to spot the errors and explain their reasoning.

KEY LANGUAGE

In lesson: multiplication, place value

Other language to be used by the teacher: partition

STRUCTURES AND REPRESENTATIONS

Number lines, place value grid

RESOURCES

Mandatory: place value counters, base 10 equipment, place value grids

 In the eTextbook of this lesson, you will find interactive links to a selection of teaching tools.

Quick recap

Give children some base 10 equipment and place value counters. Say a 2-digit number, for example 73, and ask children to make this number using their concrete resources. Repeat several times for other 2-digit numbers and also for 3-digit numbers.

Unit 6: Multiplication and division (3), Lesson 4

Discover

WAYS OF WORKING Pair work

ASK
- Question 1 a): *How can you show how many flowers each person bought using base 10 equipment?*
- Question 1 b): *Could you instead multiply the 10s first and then the 1s?*

IN FOCUS In question 1 b), children are required to multiply a 1-digit number by a 1-digit number and a 2-digit number by a 1-digit number. They need to complete these calculations before they are able to find the total number of flowers.

PRACTICAL TIPS Encourage use of base 10 equipment and place value grids to give children a fuller understanding of how the multiplication can be represented and to develop the concept of partitioning in concrete form.

ANSWERS

Question 1 a): Each person bought 23 flowers. Children should show 23 on base 10 equipment: 2 tens and 3 ones.

Question 1 b): The 3 people bought 69 flowers altogether.

PUPIL TEXTBOOK 3B PAGE 20

Share

WAYS OF WORKING Whole class teacher led

ASK
- Question 1 b): *How does the place value grid help you understand the multiplication?*
- Question 1 b): *How does the base 10 equipment make the multiplication clear?*
- Question 1 b): *Why multiply the 1s first, then the 10s?*
- Question 1 b): *Why is it more efficient to count in 3s? Why is it more efficient to count in 20s?*

IN FOCUS Using the place value grid in question 1 b) breaks down the multiplication and encourages children to see how partitioning helps with multiplication. As a whole class, count the 1s first, encouraging children counting in 1s to count in 3s since this is a more efficient method. Move on to counting the 10s (and if appropriate counting in 20s since this is even quicker) and then combining the 1s and 10s to give the answer. Encourage children to think about counting efficiently, using base 10 equipment to scaffold this activity.

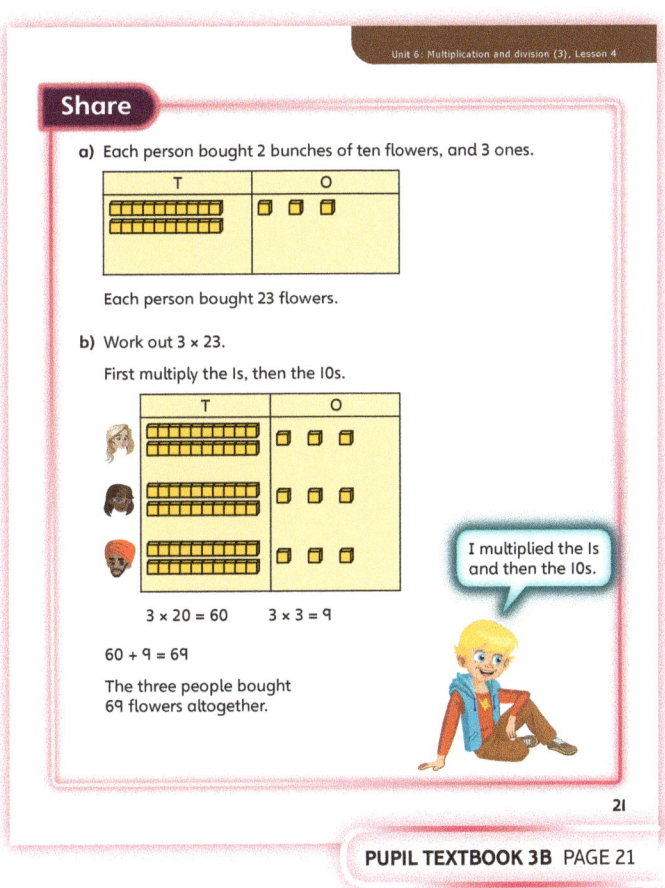

PUPIL TEXTBOOK 3B PAGE 21

Unit 6: Multiplication and division (3), Lesson 4

Think together

WAYS OF WORKING Whole class teacher led (I do, We do, You do)

ASK

- Question ❶ a): *Can you explain why the place value grid shows 4 × 12? Where is the 12? Where is the 4? How can you work out the answer?*
- Question ❷: *What calculation does each place value grid show? How do you know? What will you work out first?*
- Question ❸: *Do you notice anything about the answers? Can you work any of the answers out in your head?*

IN FOCUS In these questions, children practise multiplying a 2-digit number by a 1-digit number without exchange, using base 10 equipment and place value counters to support them. In question ❶, children can use base 10 equipment on a place value grid to help them. Encourage children to start by finding the total number of 1s and then the total number of 10s. This will help them when it comes to multiplications where they need to exchange and will set them up for efficient column methods. In addition, ensure that children understand why each calculation is important. In question ❷, children replace base 10 equipment with place value counters. Question ❸ does not provide pictorial support that aids the multiplication, although children should be encouraged to use place value equipment, where necessary, to support them.

STRENGTHEN Use place value equipment to support children. Point to the place value equipment as you work out any calculations. Encourage children to write down the multiplications to work out the number of 10s and 1s. This may help them with the whole calculation. Ensure that children have a structured method.

DEEPEN Deepen children's understanding by encouraging them to do calculations involving more than 10 ones, requiring exchanging 1 ten for 10 ones. Challenge them to work both pictorially and mentally (without written working or a place value grid), for multiplications such as 4 × 12, 13 × 5, and so on.

ASSESSMENT CHECKPOINT Questions ❶ and ❷ will check whether children can use place value equipment to work out simple 2-digit by 1-digit multiplications

ANSWERS

Question ❶ a): 48

Question ❶ b): 84

Question ❷ a): 3 × 13 = 39

Question ❷ b): 3 × 22 = 66

Question ❸: 11 × 3 = 33 m
21 × 3 = 63 m
31 × 3 = 93 m

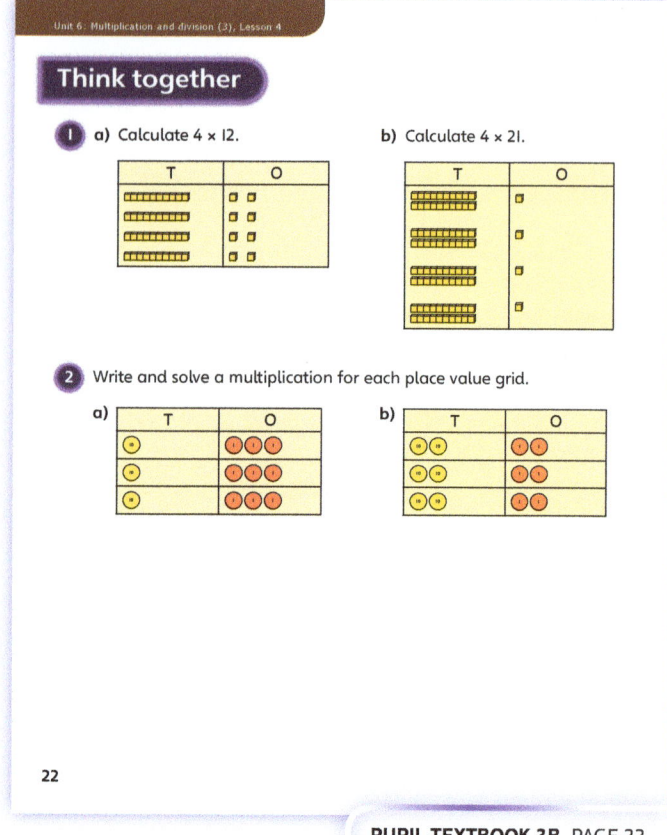

PUPIL TEXTBOOK 3B PAGE 22

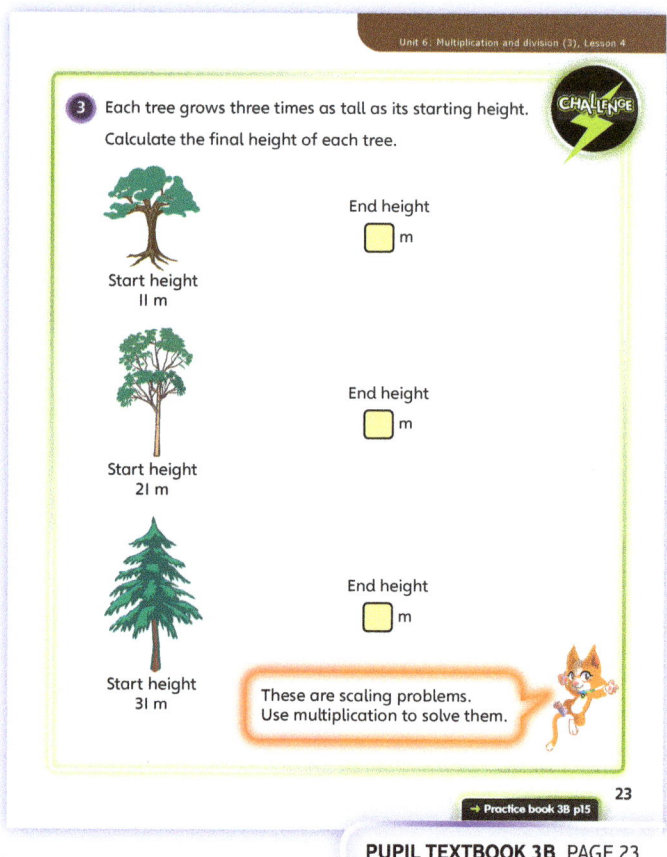

PUPIL TEXTBOOK 3B PAGE 23

Unit 6: Multiplication and division (3), Lesson 4

Practice

WAYS OF WORKING Independent thinking

IN FOCUS In question ②, children use base 10 equipment to help them multiply. By partitioning a 2-digit number into 10s and 1s, they can multiply each part separately and add the answers. Question ③ develops partitioning using place value counters.

It strengthens strengthening understanding of the connection between place value counters and base 10 equipment by demonstrating to children the equivalence between base 10 equipment and place value counters (for example, one 10s rod equals one 10s counter).

STRENGTHEN Use of base 10 equipment will support children with partitioning. After question ③, encourage children to manipulate the place value counters and challenge them to find 'reverse' problems. For example, give children some place value counters such as 4 tens and 4 ones. Can children think of a multiplication (2 digits × 1 digit) that makes 44?

DEEPEN Question ⑤ will deepen children's learning by developing their ability to use the expanded method for multiplication mentally. Support children by encouraging them to practise mentally 'holding' a number (the total of the 10s) whilst they are working out the 1s. If needed, provide more multiplication problems for children to practise with.

ASSESSMENT CHECKPOINT Question ④ will assess children's ability to multiply 2-digit numbers by a 1-digit number. Check that children can make the calculations using the expanded method. It is worth observing how much reliance they place on using concrete or pictorial representations.

ANSWERS Answers for the **Practice** part of the lesson can be found in the *Power Maths* online subscription.

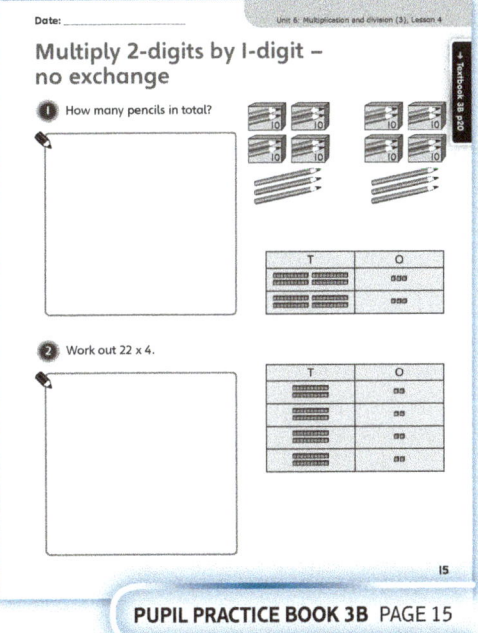

PUPIL PRACTICE BOOK 3B PAGE 15

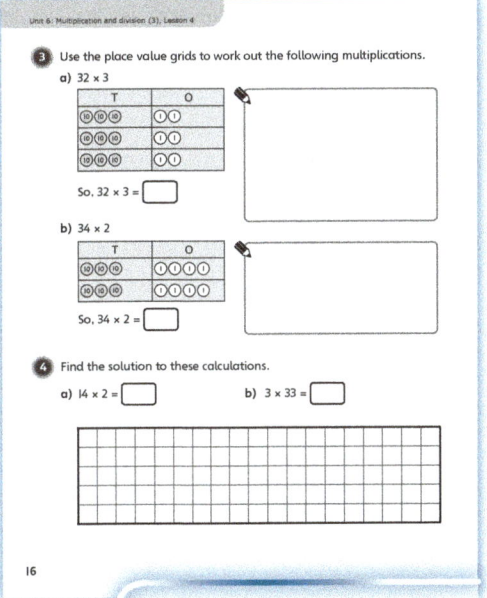

PUPIL PRACTICE BOOK 3B PAGE 16

Reflect

WAYS OF WORKING Independent thinking

IN FOCUS This question gives children an opportunity to explain the expanded method to multiply 2-digit numbers by 1-digit numbers. Children should be able to explain and use their knowledge of place value to confidently describe the steps needed to complete the calculation.

ASSESSMENT CHECKPOINT Look for children confidently and fluently explaining the steps needed to carry out the expanded method of multiplication.

ANSWERS Answers for the **Reflect** part of the lesson can be found in the *Power Maths* online subscription.

After the lesson
- Can children carry out the expanded method of multiplication?
- Can any children complete the method mentally?
- Do children still rely on place value counters or place value grids?

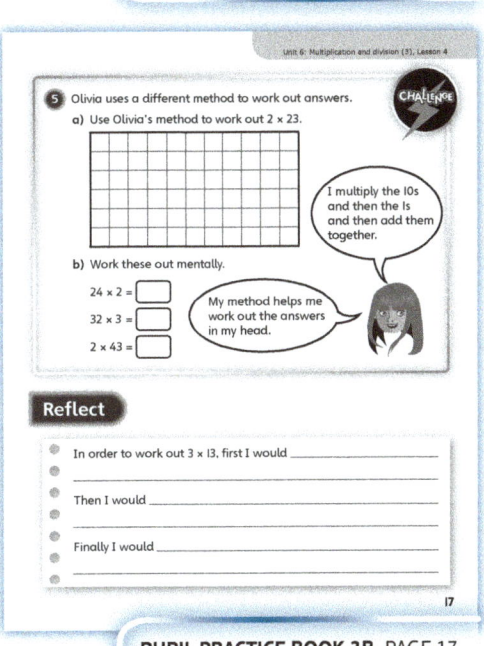

PUPIL PRACTICE BOOK 3B PAGE 17

61

Unit 6: Multiplication and division (3), Lesson 5

Multiply 2-digits by 1-digit – exchange

Learning focus
In this lesson, children will continue to use the expanded method to solve more complicated 2-digit number by 1-digit number multiplications requiring the exchange of ones into the tens column.

Before you teach
- Are children secure with using the expanded method for multiplication when grouping is not required?
- Are children confident in multiplying by 10?

NATIONAL CURRICULUM LINKS

Year 3 Number – multiplication and division

Write and calculate mathematical statements for multiplication and division using the multiplication tables that they know, including for 2-digit numbers times 1-digit numbers, using mental and progressing to formal written methods.

ASSESSING MASTERY

Children can reliably use the expanded method to multiply a 2-digit number by a 1-digit number involving grouping and exchange. They can link their understanding of place value to their calculations and represent these using concrete, pictorial and abstract representations. They will become increasingly efficient at using this method and will be able to do some calculations mentally.

COMMON MISCONCEPTIONS

Children may not yet be confident multiplying by a multiple of 10. Ask:
- How can 3 × 5 = 15 help us solve 3 × 50?

STRENGTHENING UNDERSTANDING

Encourage children to use concrete resources such as place value counters to represent the multiplications. Support children in partitioning the 2-digit numbers and grouping the tens and the ones columns. This will help improve their understanding of the expanded method.

GOING DEEPER

Children could be encouraged to find solutions to 'reverse' problems by giving them answers to multiplications, such as 120, and then asking them to think of the questions (6 × 20, 20 × 6, 3 × 40, and so on).

KEY LANGUAGE

In lesson: multiplication

STRUCTURES AND REPRESENTATIONS

Place value grid

RESOURCES

Mandatory: place value counters, base 10 equipment, blank place value grids

 In the eTextbook of this lesson, you will find interactive links to a selection of teaching tools.

Quick recap
Ask children to set out the calculation 13 × 3 in a place value grid using place value counters or base 10 equipment. Ask them to solve the multiplication. Remind them to work out the total number of 1s first and then the total number of 10s.

Discover

WAYS OF WORKING Pair work

ASK
- Question 1 a): *How can you tell what Max is working out?*
- Question 1 a): *Where did Max go wrong?*
- Question 1 b): *How can you tell what Jamilla is working out?*

IN FOCUS In question 1 a), children are required to match a pictorial representation (place value grid with counters) to a multiplication. They must also spot the error that Max has made – children should be able to tell because 12 × 3 is the only calculation involving 1 ten (the others are 20s). In question 1 b), children must match base 10 equipment to a multiplication. It is important to note that question 1 b) is different to question 1 a) since multiplication of the ones column results in an answer greater than 10 and the need to exchange the 10 ones for 1 ten.

PRACTICAL TIPS This introduction to multiplying 2-digit numbers by 1-digit numbers with exchange lends itself to the use of base 10 equipment and place value counters to represent the multiplications. Encourage children to recreate the multiplications to secure understanding of the use of place value and the expanded method when multiplying.

ANSWERS

Question 1 a): Max is working out the calculation 3 × 12.
3 × 12 = 36, so Max is not correct.

Question 1 b): Jamilla is working out 3 × 24.
3 × 24 = 72

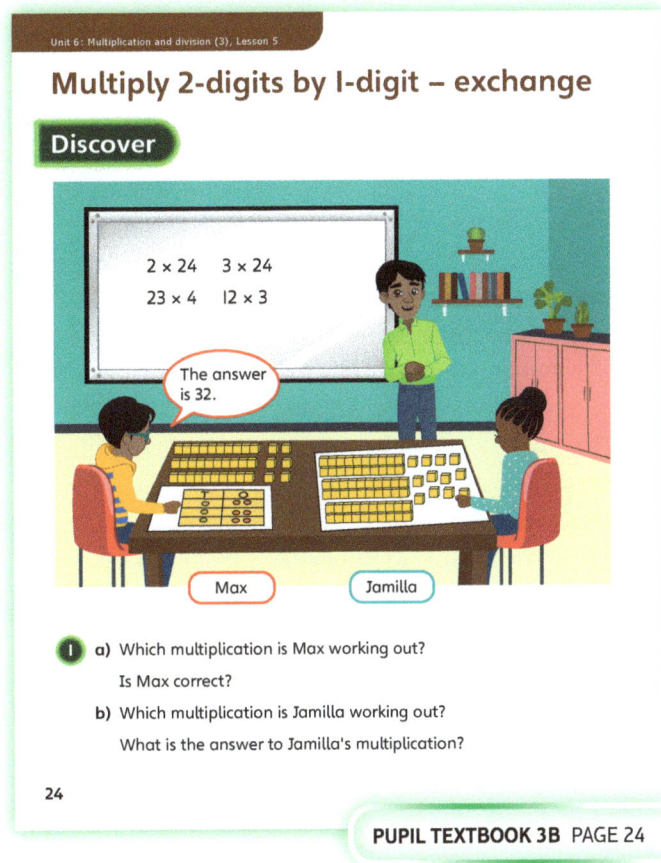

Share

WAYS OF WORKING Whole class teacher led

ASK
- Question 1 b): *How do you know Jamilla is working out 3 × 24?*
- Question 1 b): *What is similar about these methods? What is different?*
- Question 1 b): *Why is it useful to group the 10 ones together when adding 60 + 12?*

IN FOCUS It is important when working on these two questions to ensure that children recognise that in question 1 b) the 1s have been grouped together, making it easier to mentally calculate (60 + 10 = 70; 70 + 2 = 72) and that grouping was not needed in question 1 a). Support children by using concrete resources so they can manipulate and experience the need to exchange the 10 ones for 1 ten when multiplying 2-digit numbers by 1-digit numbers.

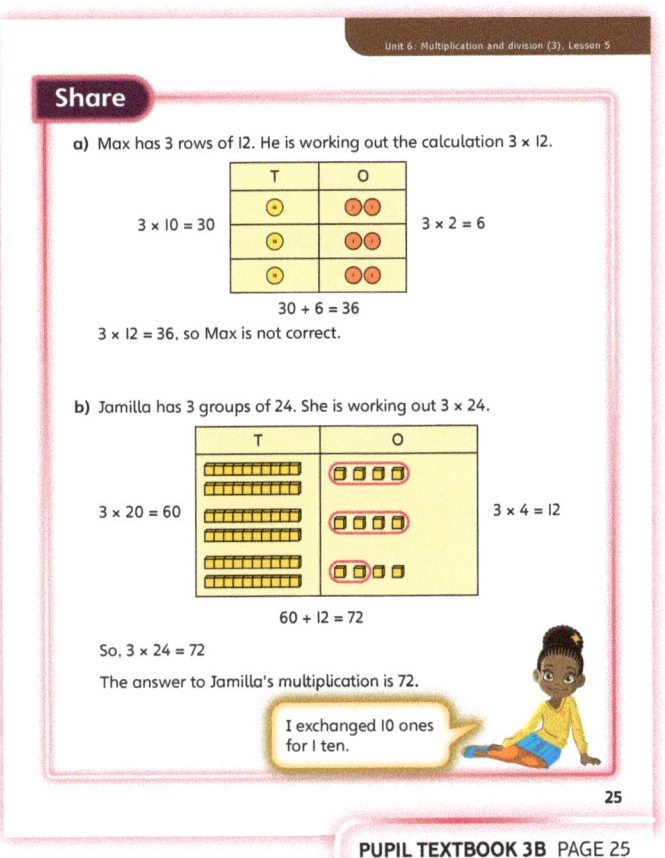

63

Think together

WAYS OF WORKING Whole class teacher led (I do, We do, You do)

ASK

- Question ❶: *Can you explain how the place value grid shows the multiplication? Why should you work out the 1s first? Then what should you work out?*
- Question ❷: *Can you explain how the place value grid shows 17 × 2? What should you work out first? Can you talk through the steps to work out the answer?*
- Question ❸: *What is your prediction? Why did you choose this?*

IN FOCUS In all the questions, children should be prompted to work out the total number of 1s first and then the total number of 10s, following the method used in the **Discover** and **Share** sections. Children can then add the answers. Encourage children to exchange 10 ones for 1 ten as this shows why the answer is what it is. In question ❸, children have to first predict which answers they think will have the greatest and smallest product. The word 'product' may need to be explained to them. Children may choose 45 × 2 as this is the largest number, but when they see this does not necessarily provide the largest product, they may realise that they need to look at the full calculation.

STRENGTHEN Provide children with base 10 equipment and place value counters to support them in their working. Make the method that children are using explicit by explaining the parts of the calculation they are doing, for example 4 × 3 ones = 12 ones.

DEEPEN In question ❸, children are asked to think about which calculations give the greatest and smallest products. Listen for reasons that children give. Some children may be able to approximate answers. Children will then need to work out the answers to check their predictions. They should reflect on their predictions and actual answers to determine how their prediction may be different if asked a similar question in the future. To further deepen understanding, ask children which ones they think need more than one exchange. How do they know?

ASSESSMENT CHECKPOINT Question ❶ and ❷ should be used to check whether children can work out 2-digit number by 1-digit number multiplications where they have to exchange either the 1s or the 10s or both. Look for the methods that children use and try to encourage as efficient working as possible.

ANSWERS

Question ❶: 4 × 3 = 12   4 × 20 = 80   4 × 23 = 92

Question ❷: 10 × 2 = 20   7 × 2 = 14   17 × 2 = 34

Question ❸: Children should predict the greatest and smallest products.
14 × 5 = 70
15 × 4 = 60   smallest product
24 × 5 = 120   greatest product
25 × 4 = 100
45 × 2 = 90

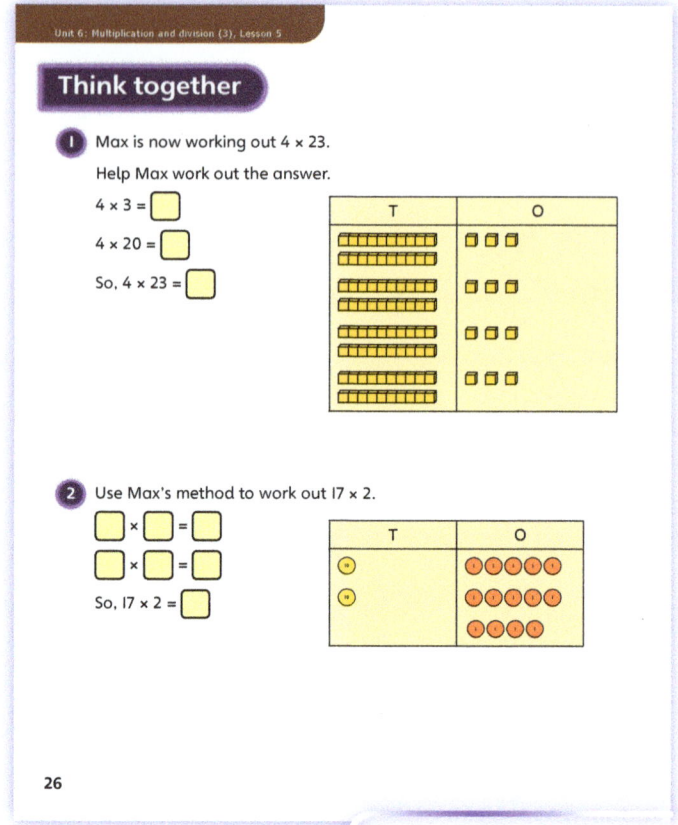

PUPIL TEXTBOOK 3B PAGE 26

PUPIL TEXTBOOK 3B PAGE 27

Practice

WAYS OF WORKING Independent thinking

IN FOCUS To secure understanding, question ⑥ is a matching question involving children matching written multiplication calculations to answers (without finding the answers). This is important, since some children may unnecessarily work out the answers to each multiplication to match them. Encourage children to think about how they could match them without calculating the answers. Some children may use rounding and estimate, for example, 56 × 3 becomes 60 × 3 = 180 (close to 185), while others may just look at the units: 37 × 5 … 7 × 5 = 35, and conclude that the answer must be 185.

STRENGTHEN If children are struggling with question ①, encourage them to build the calculations first using their own base 10 equipment. Repeat this activity for a range of other calculations.

DEEPEN Question ⑤ b) is an excellent opportunity to deepen learning since it requires children to work backwards and use parts of expanded method calculations to recreate the original multiplication. Encourage children to explain their reasoning as they work through the problem. Ask: *What does the first calculation tell you? What do you need to divide the parts by? How do you know?*

ASSESSMENT CHECKPOINT Question ④ will assess children's ability to complete multiplication calculations in the format of word problems. Observe if children can understand the word problem and if they can use the expanded method to work out 33 × 5 with no representation.

ANSWERS Answers for the **Practice** part of the lesson can be found in the *Power Maths* online subscription.

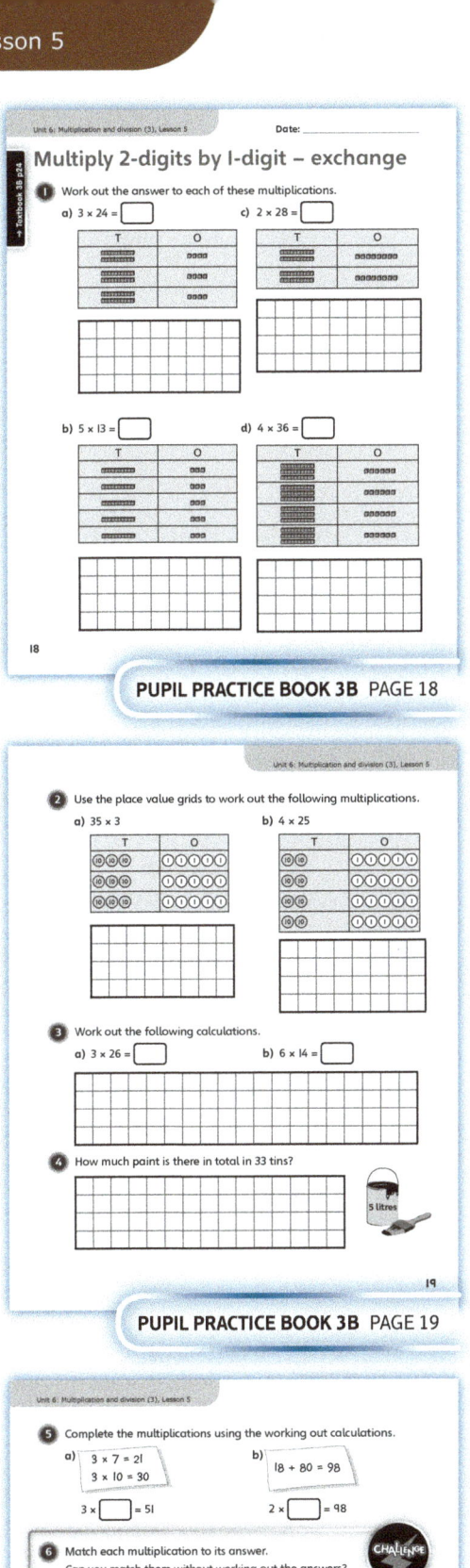

Reflect

WAYS OF WORKING Independent thinking

IN FOCUS This question allows children to explore the relationship between two multiplications. Children should notice that 72 is double 36 but 2 is half of 4 – so the answers must be the same. Encourage children to represent the two multiplications with base 10 equipment if they are not quite sure why they are equal.

ASSESSMENT CHECKPOINT Look for children confidently explaining their answer, since this will indicate that they have a very good understanding of multiplication, and what a multiplication represents.

ANSWERS Answers for the **Reflect** part of the lesson can be found in the *Power Maths* online subscription.

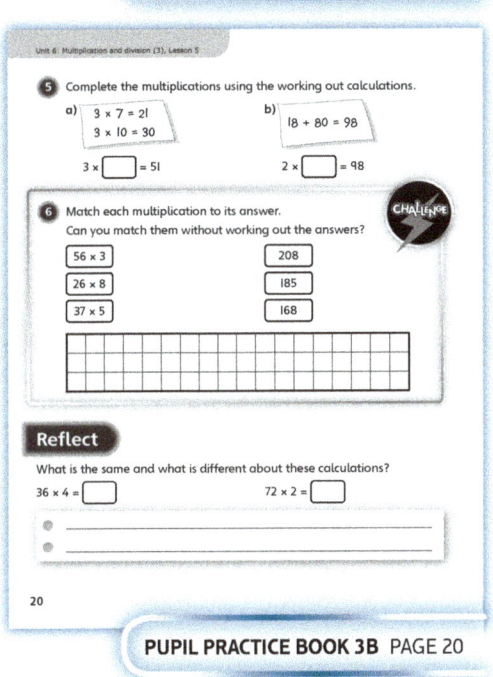

After the lesson

- Are children confident in using the expanded method of multiplication?
- Can children explain the expanded method of multiplication in clear steps?
- Do some children still rely on base 10 equipment, place value counters or place value grids for support?

Unit 6: Multiplication and division (3), Lesson 6

Expanded written method

Learning focus
In this lesson, children will apply their understanding of place value to written methods for solving multiplications using the expanded method in column format. They will solve mixed problems, including those in context.

Before you teach
- Are children confident using place value to write 2-digit numbers?
- Are children confident using the expanded method for multiplication using pictorial and concrete representations?

NATIONAL CURRICULUM LINKS

Year 3 Number – multiplication and division

Write and calculate mathematical statements for multiplication and division using the multiplication tables that they know, including for 2-digit numbers times 1-digit numbers, using mental and progressing to formal written methods.

ASSESSING MASTERY

Children can use the expanded method of multiplication in written format to solve a range of problems showing fluency by placing digits in columns and multiplying in steps before adding columns of digits. They will be able to confidently explain their methods and reasoning.

COMMON MISCONCEPTIONS

Children may have some place value misconceptions and may not be able to place digits in their correct columns, for example, they may think that 25 × 3 is 5 × 3 and 2 × 3. Ask:
- *What does the 2 stand for? Can you show it using base 10 equipment?*

STRENGTHENING UNDERSTANDING

Encourage children to use base 10 equipment alongside the calculations to support their understanding of which digit to place in which column.

GOING DEEPER

Encourage children to solve expanded multiplications with missing digits requiring them to do 'reverse' calculations using inverse operations. Ask: *How can you check your answers?*

KEY LANGUAGE

In lesson: expanded written method, multiplication, multiply, column method, column, place value, ones (1s), tens (10s), hundreds (100s)

STRUCTURES AND REPRESENTATIONS

Place value grid

RESOURCES

Mandatory: counters, base 10 equipment, place value counters

 In the eTextbook of this lesson, you will find interactive links to a selection of teaching tools.

Quick recap

Write the calculation 23 × 4 on the board. Ask children to use place value equipment to work out the answer. Ask them to talk through the steps with a partner.

Unit 6: Multiplication and division (3), Lesson 6

Discover

WAYS OF WORKING Pair work

ASK

- Question 1 a): *How many trips does Mrs Dean make in total?*
- Question 1 b): *How far does she travel to the gym? How far is it from the gym to the school?*

IN FOCUS Questions 1 a) and 1 b) give children an opportunity to multiply 2-digit numbers by a 1-digit number in a real life context. Encourage children to make multiplication calculations rather than just add the distances.

PRACTICAL TIPS Make a classroom display in which a multiplication is solved using the expanded method (in columns). Showing an example is a good reference point for children. The calculations could involve distances children or teachers travel to school or other local places of interest.

ANSWERS

Question 1 a): Mrs Dean travels 92 km in total.

Question 1 b): The total distance is 32 km.

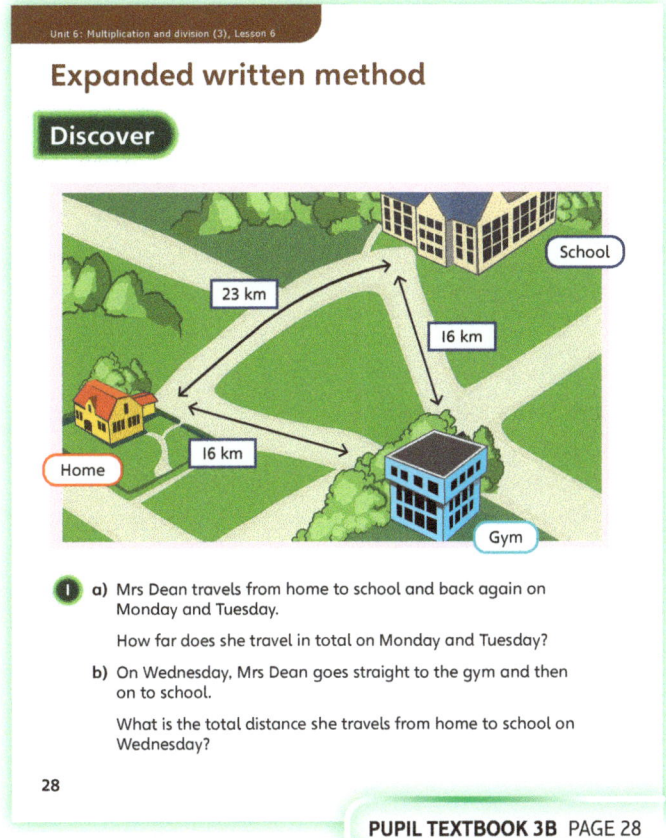

PUPIL TEXTBOOK 3B PAGE 28

Share

WAYS OF WORKING Whole class teacher led

ASK

- Question 1 a): *How do the columns help organise the multiplication?*
- Question 1 a): *Do you calculate the 1s or the 10s first?*
- Question 1 a): *In which column do you write the 1? 2? 8? 0?*
- Question 1 b): *What does each line of the expanded written method show?*

IN FOCUS Question 1 a) introduces the expanded method in written form. Explore how using columns for written multiplication is an efficient way to lay out the calculation by explaining how the columns help organise the workings. Remind children that they calculate the 1s first and then the 10s by starting on the right-hand column. Highlight the fact that the 10 ones make 10, for example, with the number 12, explain the importance of putting the 1 in the tens column and the 2 in the ones column. If needed, support this by using concrete resources such as base 10 equipment. In question 1 b), children may recognise the link between multiplication and repeated addition. Point out that 2 lots of 16 (2 × 16) is the same as 16 + 16.

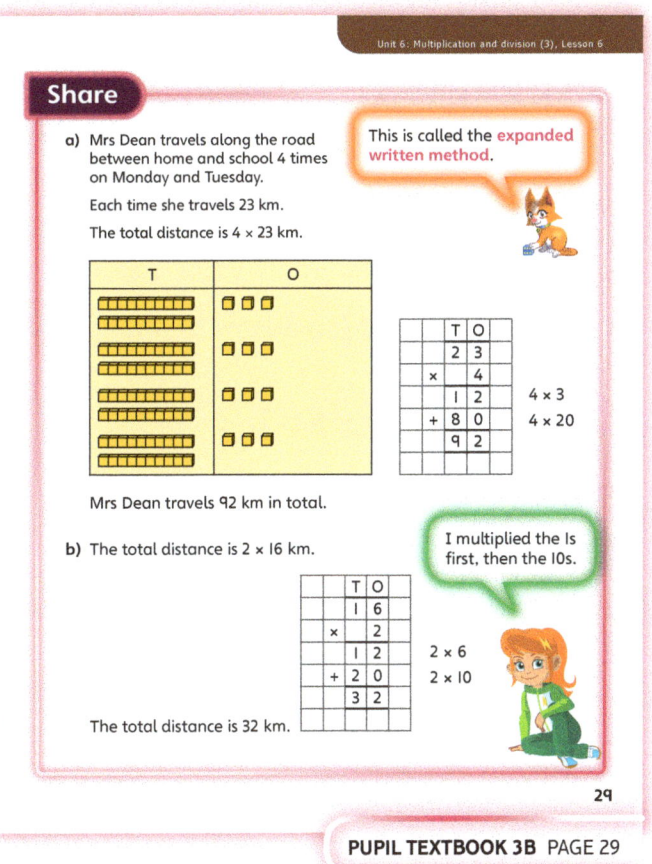

PUPIL TEXTBOOK 3B PAGE 29

67

Unit 6: Multiplication and division (3), Lesson 6

Think together

WAYS OF WORKING Whole class teacher led (I do, We do, You do)

ASK

- Question ❶: *Can you explain why the place value grid shows 6 × 15? Where is the 15? Where is the 6? How do you work this out? What calculation do you need to do first?*
- Question ❷: *What calculation do you need to do first? What do you then need to do? Then what do you do with your answers? Can you remember what this method is called?*
- Question ❸: *What calculation did you do? What is the answer? How did you work it out? What does the word 'product' mean? How many different products can you find?*

IN FOCUS Questions ❶ and ❷ provide two examples of working out an answer to a multiplication using expanded form. Base 10 equipment is provided to help explain where the abstract calculations come from. Make sure children understand the connection between the base 10 equipment and the calculations. Question ❸ requires children to translate a word problem into an abstract multiplication. For each different multiplication, children should lay out the column multiplication without prompting, moving to a more abstract representation. Encourage children to recognise what numbers need to go into the relevant columns and where to begin the calculation. This question is important because all products are above 100. Explore with children what they think should happen once the 10s have been calculated. Show them how to add the hundreds column and finish the calculation. If children need support, use concrete resources such as base 10 equipment to demonstrate that 10 tens make 100.

STRENGTHEN Support children in writing the calculations in columns by using base 10 equipment. Encourage them to translate these concrete representations into written form.

DEEPEN In question ❸, challenge children to explore how they might predict which calculation gives the greatest or the least product before they do the multiplications. Can they justify their reasoning, using statements such as *54 × 3 will give a greater product than 45 × 3*?

ASSESSMENT CHECKPOINT Question ❸ checks that children can set up their own expanded method multiplication and work out the correct answer.

ANSWERS

Question ❶: 15 × 6 = 90 km

Question ❷: 15 × 5 = 75

Question ❸: 54 × 3 = 162
53 × 4 = 212
45 × 3 = 135
43 × 5 = 215   greatest product
34 × 5 = 170
35 × 4 = 140

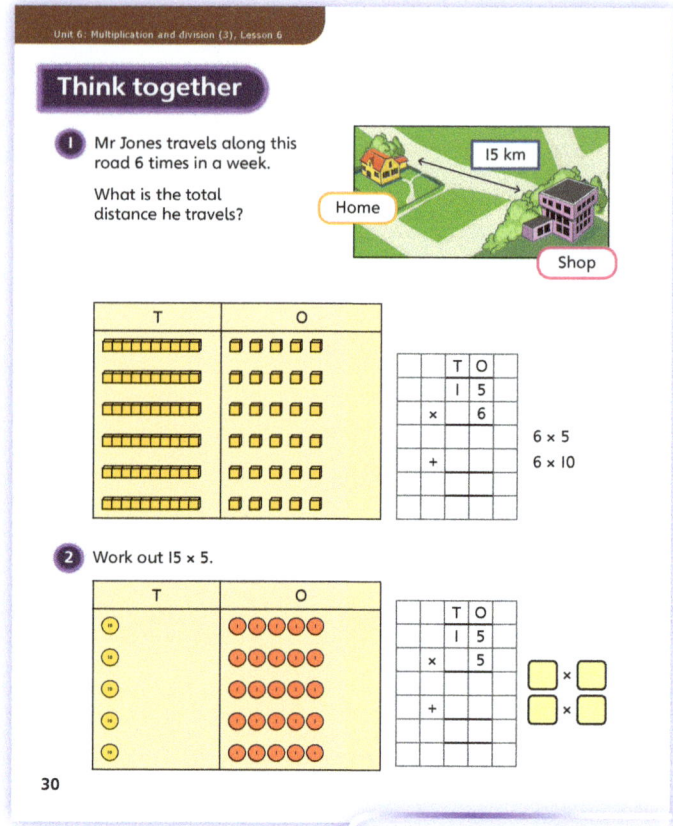

PUPIL TEXTBOOK 3B PAGE 30

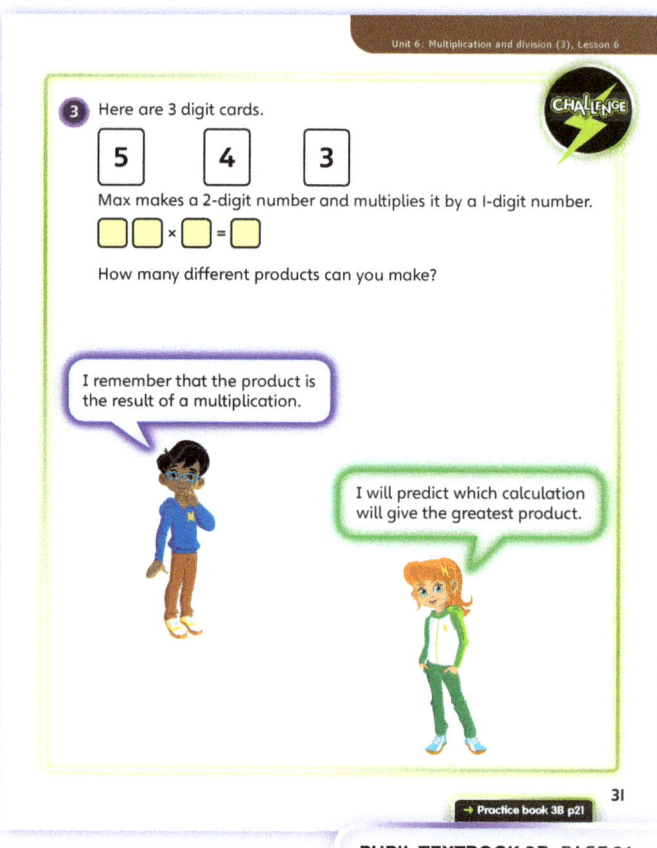

PUPIL TEXTBOOK 3B PAGE 31

Unit 6: Multiplication and division (3), Lesson 6

Practice

WAYS OF WORKING Independent thinking

IN FOCUS Question ④ encourages children to really explore the column method of multiplication via a digit placing exercise. As children progress through this question they are required to think carefully about multiples and use the inverse operation of division to place the digits accordingly.

STRENGTHEN Question ⑥ will strengthen learning since children are required to remember the expanded method and make the calculation (without concrete or pictorial representations for support). To support understanding of the column method in abstract form, provide children with base 10 equipment to reinforce the concepts but ensure that children can complete the calculations in column format.

DEEPEN Question ⑦ is an excellent opportunity for children to develop and demonstrate abstract thinking. Encourage children to think carefully about the circle in the ones column. Ask: *What do the circles represent? What number multiplied by 3 gives the starting number? What does the heart represent?*

THINK DIFFERENTLY Question ⑤ requires children to multiply by 1. It will be valuable here to explore with children that multiplication does not always make a number bigger and that a written method is not always the best approach.

ASSESSMENT CHECKPOINT Question ③ gives the opportunity to assess children's ability to solve written multiplication problems without the prompt of a column layout. Look for children being able to translate the problems into column format and accurately apply the expanded method.

ANSWERS Answers for the **Practice** part of the lesson can be found in the *Power Maths* online subscription.

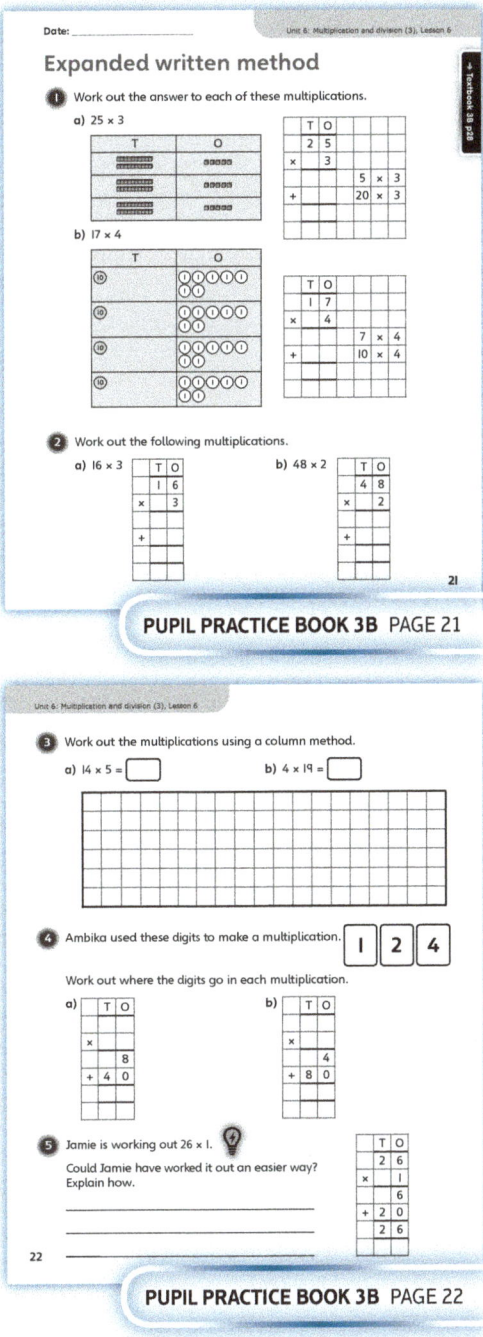

PUPIL PRACTICE BOOK 3B PAGE 21

PUPIL PRACTICE BOOK 3B PAGE 22

Reflect

WAYS OF WORKING Independent thinking

IN FOCUS This question requires children to clearly explain the method of multiplication used in this lesson. Encourage children to write instructions in numbered steps and explain their reasoning. Children should confidently be able to write the multiplication in column format showing fluency with place value and describe and complete the steps to solve the problem.

ASSESSMENT CHECKPOINT Look for children clearly explaining their method and reasoning, confidently being able to multiply a 2-digit number by a 1-digit number using the expanded method in column format without using pictorial or concrete representations.

ANSWERS Answers for the **Reflect** part of the lesson can be found in the *Power Maths* online subscription.

After the lesson

- Can children write down the expanded method of multiplication in columns (including 100s in the answer)?
- Do children still rely on base 10 equipment, place value counters or place value grids for support?

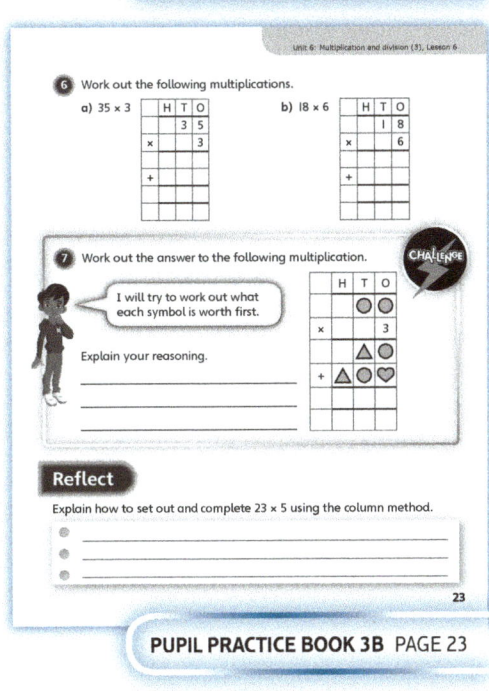

PUPIL PRACTICE BOOK 3B PAGE 23

69

Unit 6: Multiplication and division (3), Lesson 7

Link multiplication and division

Learning focus
In this lesson, children will explore the link between multiplication and division. Children should be able to write down related division facts for a given multiplication fact and vice versa.

Before you teach
- Are children confident with the multiplication sign?
- Do children know the connection between an array and a multiplication fact?
- Do children know their multiples of 10?

NATIONAL CURRICULUM LINKS

Year 3 Number – multiplication and division

Solve problems, including missing number problems, involving multiplication and division, including positive integer scaling problems and correspondence problems in which *n* objects are connected to *m* objects.

ASSESSING MASTERY

Children can write down a division fact from a given multiplication fact. For example, if they are given 6 × 4 = 24, they know that 24 ÷ 6 = 4 and that 24 ÷ 4 = 6. Children should be able to understand the relationship between multiplication and division.

COMMON MISCONCEPTIONS

Children often think that division is commutative, because multiplication is. For example, they often know that 3 × 4 = 12 and so 4 × 3 = 12, children then go on think that 4 ÷ 12 = 3. To support children, ask them to make a 4 × 3 array with counters. Ask:
- *What does the 12 represent? What do the 4 and 3 represent? Why do we need to start with the 12?*

STRENGTHENING UNDERSTANDING

Use counters, cubes and base 10 equipment to support understanding of the connection between multiplication and division. For example, children may make arrays of the multiplication facts to help them with the division facts. Encourage children to talk through what each of the numbers mean in the calculation in relation to the array. Explain that they do not need to calculate answers to certain divisions if they know the related multiplication fact. This makes calculating more efficient rather than having to share and group equipment each time.

GOING DEEPER

Children could make their own factor-product triangles for known multiplication facts. Ask them to write down as many related facts as possible, including ones where they may have multiples of 10 too, for example, 4 × 3 = 12 and so 40 × 3 = 120. This brings in knowledge from a previous lesson in this unit.

KEY LANGUAGE

In lesson: multiple, fact, division, multiply, divide, related, product, factor, **equally**

Other language to be used by the teacher: share, group, fact family

STRUCTURES AND REPRESENTATIONS

Factor-product triangles

RESOURCES

Mandatory: counters, cubes and base 10 equipment
Optional: cups

 In the eTextbook of this lesson, you will find interactive links to a selection of teaching tools.

Quick recap
Play 'What's my multiplication?'. On the board, write ☐ × ☐ = 18. Ask children to write down as many answers as they can. How many can they find? Repeat for other 2-digit numbers, such as 24, 27, 36, 45 and 60.

70

Discover

WAYS OF WORKING Pair work

ASK

- Question 1 a): *How many bones can you see? How many dogs are the bones being shared between? What multiplication fact might help you work this out?*
- Question 1 b): *How many dog treats are there? How can you work out how many treats each dog will get?*

IN FOCUS Questions 1 a) and 1 b) should be approached initially as a division. Children are likely to use place value counters or base 10 equipment to represent the bones and treats. They can then share them into three equal groups. Encourage children to see if they could have worked them out using known multiplication facts. This will make the division more efficient.

PRACTICAL TIPS To replicate this scenario in the classroom, the bones and treats could be replaced with everyday classroom objects, shared out between three children.

ANSWERS

Question 1 a): Each dog will get 2 bones.

Question 1 b): Each dog will get 20 treats.

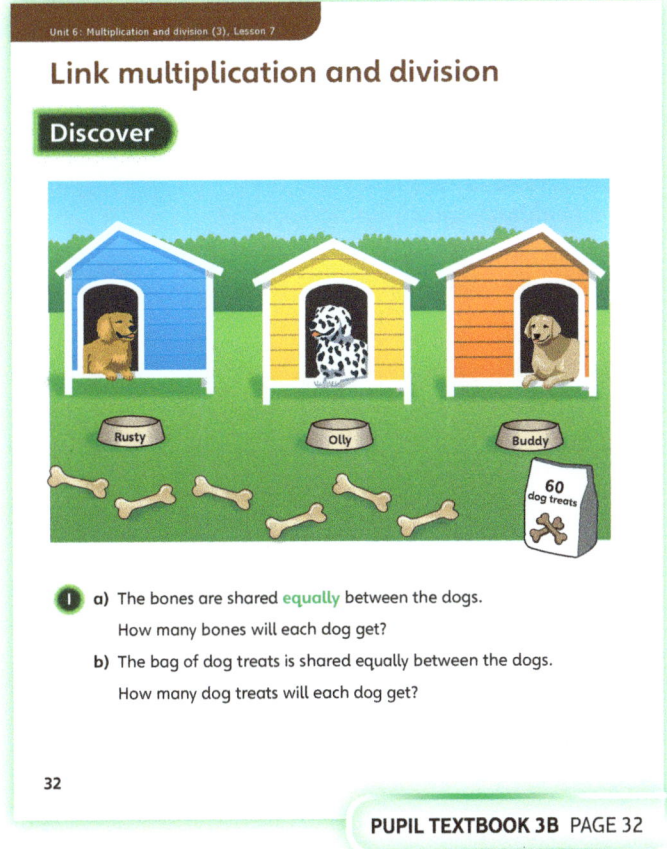

PUPIL TEXTBOOK 3B PAGE 32

Share

WAYS OF WORKING Whole class teacher led

ASK

- Question 1 a): *How did you work out the answer? How can you use the multiplication 3 × 2 = 6 to help you?*
- Question 1 b): *How did you work out the answer? What equipment did you use? How can you use 3 × 2 = 6 (or a different multiplication) to help you?*

IN FOCUS In question 1 a), the most likely method that children will use to answer this is to share the 6 bones between the 3 dogs. Use counters or cubes and share them equally into 3 cups or groups. Model the method and ask children to follow along. They should be able to see that there are 2 bones in each group. Discuss how they can use the fact that 3 × 2 = 6 to know that 6 ÷ 3 = 2. In question 1 b), children should use base 10 equipment to help them with sharing, but then remind children of the work they did on related calculations; if 2 × 3 = 6, then 20 × 3 = 60 and so 60 ÷ 3 = 20.

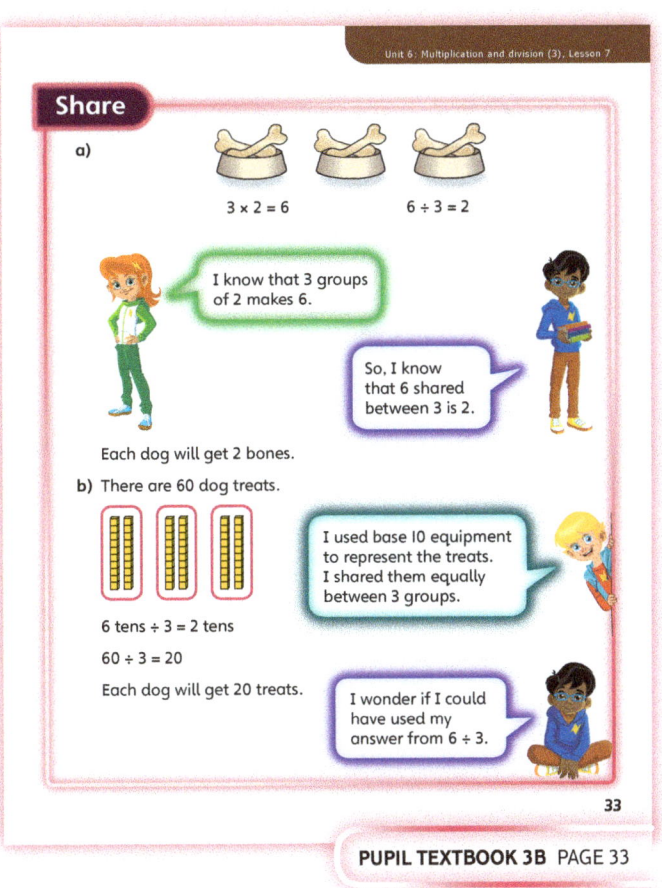

PUPIL TEXTBOOK 3B PAGE 33

Unit 6: Multiplication and division (3), Lesson 7

Think together

WAYS OF WORKING Whole class teacher led (I do, We do, You do)

ASK
- Question ❶: *What multiplication is shown in each of the diagrams? What division could be shown?*
- Question ❷: *How can you use the multiplications to work out the divisions?*
- Question ❸: *What is the connection between the numbers in the corners of the triangle? Can you explain why there are four facts? Can you make your own factor-product triangle?*

IN FOCUS Throughout these questions, children will be making the connection between multiplication calculations and related division calculations. Question ❶ uses pictorial representations to help children see the connection between the calculations. Question ❷ takes a more abstract approach, using known facts that children should be confident with. Finally, question ❸ uses a more abstract factor-product triangle to determine facts. Children will see that there are four facts that can come from a factor-product triangle.

STRENGTHEN Throughout this exercise children should use concrete objects to support them, like in question ❶. Children may find it useful to make arrays to help them in questions ❷ and ❸.

DEEPEN Children should make up their own factor-product triangles to write down a multiplication and division fact family. Children may then create related fact families too, such as using 6 × 3 = 18 to work out the fact family for 60 × 3 = 180.

ASSESSMENT CHECKPOINT Question ❷ checks that children can use multiplications to work out divisions. Watch for children working out the divisions using equipment or other methods. Instead of these methods, children should be using the multiplications to help them. This shows that children understand the connection between multiplication and division.

ANSWERS

Question ❶: 5 × 3 = 15   15 ÷ 3 = 5
2 × 4 = 8   8 ÷ 4 = 2

Question ❷: 4 × 5 = 20   3 × 10 = 30
50 × 2 = 100   20 ÷ 5 = 4
30 ÷ 10 = 3   100 ÷ 2 = 50

Question ❸ a): 5 × 6 = 30
6 × 5 = 30
30 ÷ 6 = 5
30 ÷ 5 = 6

Question ❸ b): 12 × 3 = 36
3 × 12 = 36
36 ÷ 12 = 3
36 ÷ 3 = 12

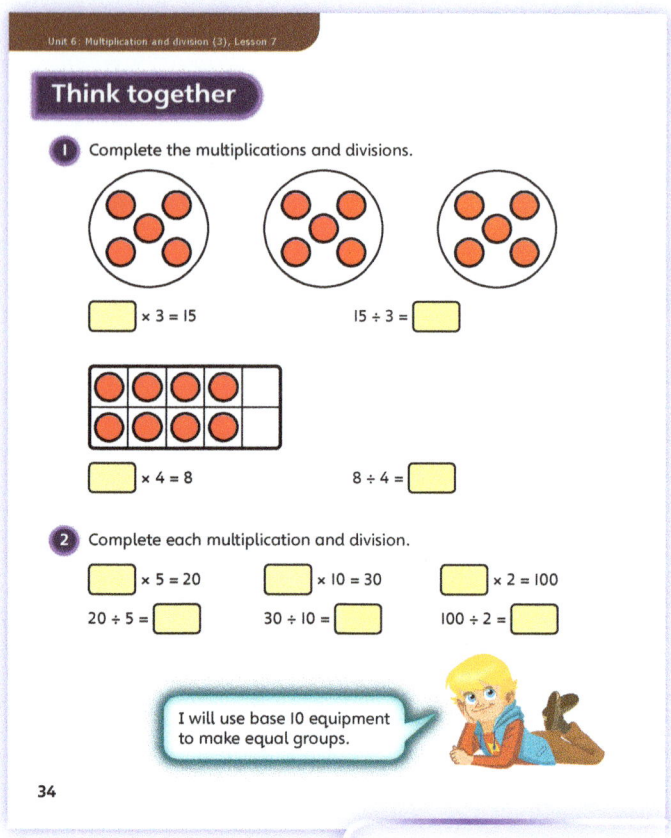

PUPIL TEXTBOOK 3B PAGE 34

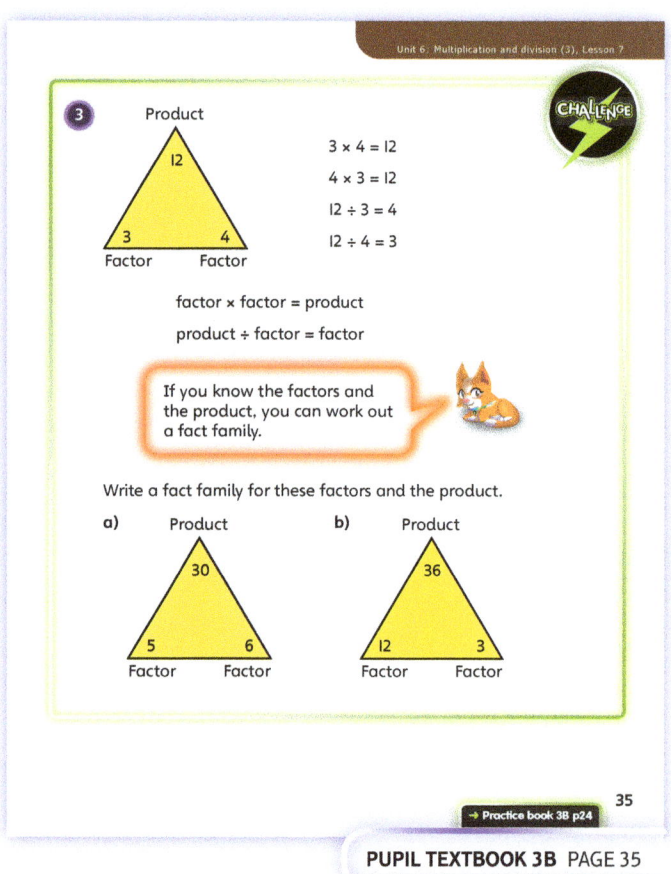

PUPIL TEXTBOOK 3B PAGE 35

72

Unit 6: Multiplication and division (3), Lesson 7

Practice

WAYS OF WORKING Independent thinking

IN FOCUS Question ❶ provides a simple division to ease children into the concept. Encourage them to use a sharing method or use their known multiplication facts to work out the missing value. Question ❷ asks children to complete the related multiplication and division facts, with images to support them. Question ❸ then moves on to factor-product triangles, with children first finding missing numbers and then using the triangle to complete the fact family for these related numbers. Question ❹ shows place value counters in arrays, where children have to work out firstly what is shown and then write down related calculations. Throughout the later questions, children should be using known facts to work out other facts rather than calculating. For example, to work out 24 ÷ 3, they use knowledge of 8 × 3 = 24, rather than sharing 24 objects equally between 3 groups.

STRENGTHEN For all questions, children should use concrete objects for support if needed, particularly for the factor-product triangles in question ❸.

DEEPEN Children should make up their own factor-product triangles to write down a multiplication and division fact family. Children may then create related fact families too, such as using 6 × 3 = 18 to work out the fact family for 60 × 3 = 180.

ASSESSMENT CHECKPOINT Question ❸ should be used to assess children's understanding of the connection between multiplication and division facts. Watch for children working out the divisions using equipment or other methods. Instead of these methods, children should be using the multiplications to help them. This shows that children understand the connection between multiplication and division.

ANSWERS Answers for the **Practice** part of the lesson can be found in the *Power Maths* online subscription..

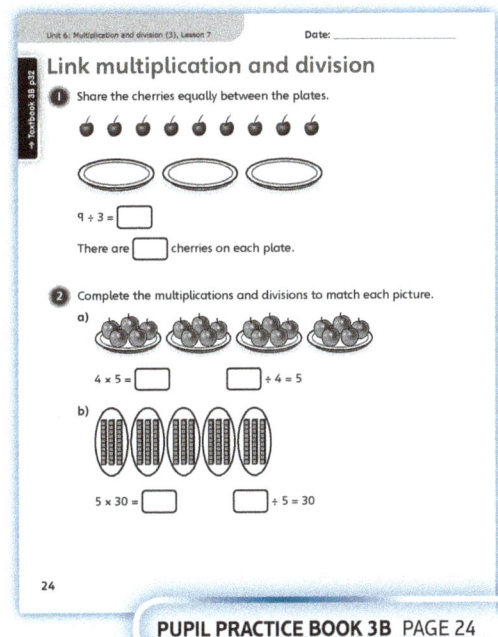

PUPIL PRACTICE BOOK 3B PAGE 24

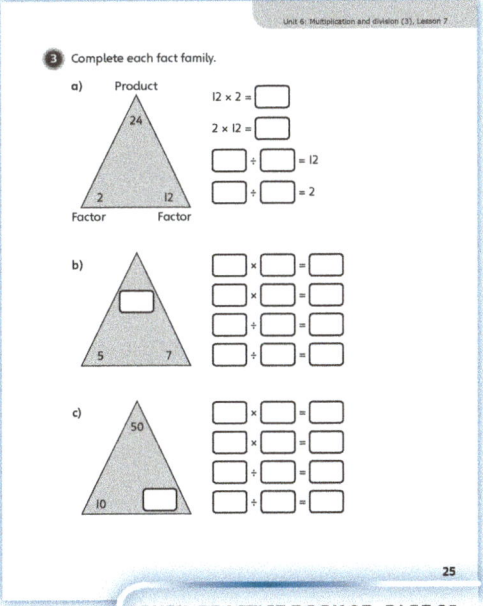

PUPIL PRACTICE BOOK 3B PAGE 25

Reflect

WAYS OF WORKING Independent thinking

IN FOCUS This question gets to the heart of what children need to take away from this lesson. Children need to be able to use a given calculation to write down other facts. Although children may use concrete resources to support them, encourage them to write down the other related division and multiplication facts without calculating.

ASSESSMENT CHECKPOINT Check that children can write down related multiplication and division calculations without the need for calculating. As a minimum, children should be able to write down the three other facts that form the fact family. Some children may then write down other related facts involving multiples of 10.

ANSWERS Answers for the **Reflect** part of the lesson can be found in the *Power Maths* online subscription.

After the lesson

- Are children able to find the answers to divisions using related multiplication facts?
- Can children write down a multiplication/division fact family given a factor-product triangle?

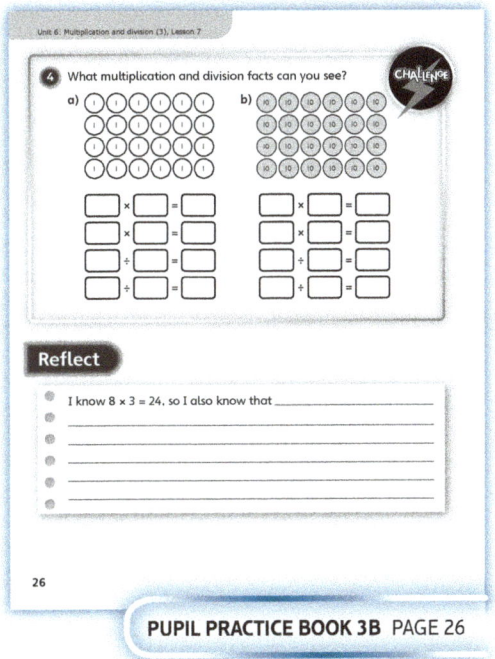

PUPIL PRACTICE BOOK 3B PAGE 26

Unit 6: Multiplication and division (3), Lesson 8

Divide 2-digits by 1-digit – no exchange

Learning focus

In this lesson, children will use their understanding of place value, partitioning and division to divide a 2-digit number by a 1-digit number.

Before you teach

- Can children partition and divide 2-digit numbers?
- Can children use multiplication and division as inverse operations?

NATIONAL CURRICULUM LINKS

Year 3 Number – multiplication and division

Write and calculate mathematical statements for multiplication and division using the multiplication tables that they know, including for 2-digit numbers times 1-digit numbers, using mental and progressing to formal written methods.

ASSESSING MASTERY

Children can accurately partition a number into 10s and 1s, and then divide both parts before adding the two amounts together. They can link their understanding of place value to their method and can confidently explain the steps in their method and begin to divide mentally.

COMMON MISCONCEPTIONS

Children may not be confident with partitioning. Ask:
- *Could a place value grid help here?*

Some children may carry out the method correctly, but forget the final step of adding the divided amounts. Ask:
- *Does 48 divided by 2 equal 204? Why not?*

STRENGTHENING UNDERSTANDING

Encourage children to use part-whole models to support partitioning and base 10 equipment to then divide 10s and 1s, therefore using pictorial and concrete representations.

GOING DEEPER

Encourage children to investigate division problems with missing numbers, for example, ☐ ÷ 5 = ☐, and use multiplication to check their answer. Ask: *Can you find out what the divisions are? What are the possibilities?*

KEY LANGUAGE

In lesson: multiplication, division, partition, equally, share, place value, tens (10s), ones (1s)

Other language to be used by the teacher: inverse operation, efficient method

STRUCTURES AND REPRESENTATIONS

Place value grid, part-whole model

RESOURCES

Mandatory: counters, base 10 equipment, blank part-whole models

 In the eTextbook of this lesson, you will find interactive links to a selection of teaching tools.

Quick recap

Ask children to divide 16 objects between 2 people. Ask: *How many does each person get?* Now ask them to divide 15 by 3. Explain the different methods they could use, including using counters or drawing dots on a page. For each calculation, ask them what each number represents.

Discover

WAYS OF WORKING Pair work

ASK

- Question 1 a): *How many sheep are there in total? How many farmers?*
- Question 1 a): *What is another name for sharing?*
- Question 1 b): *How can you check your answer?*

IN FOCUS Questions 1 a) and 1 b) give children an opportunity to divide numbers in a real-life context. This will give an opportunity to pre-assess children's confidence with division before starting the main part of the lesson.

PRACTICAL TIPS Print some large part-whole models and give children base 10 equipment to partition into 10s and 1s. After this, show them how to divide the 10s then the 1s. This practical work will reinforce understanding.

ANSWERS

Question 1 a): There are 48 sheep.
48 ÷ 2 = 24
Each farmer has 24 sheep.

Question 1 b): 48 ÷ 4 = 12
12 sheep go in each pen.

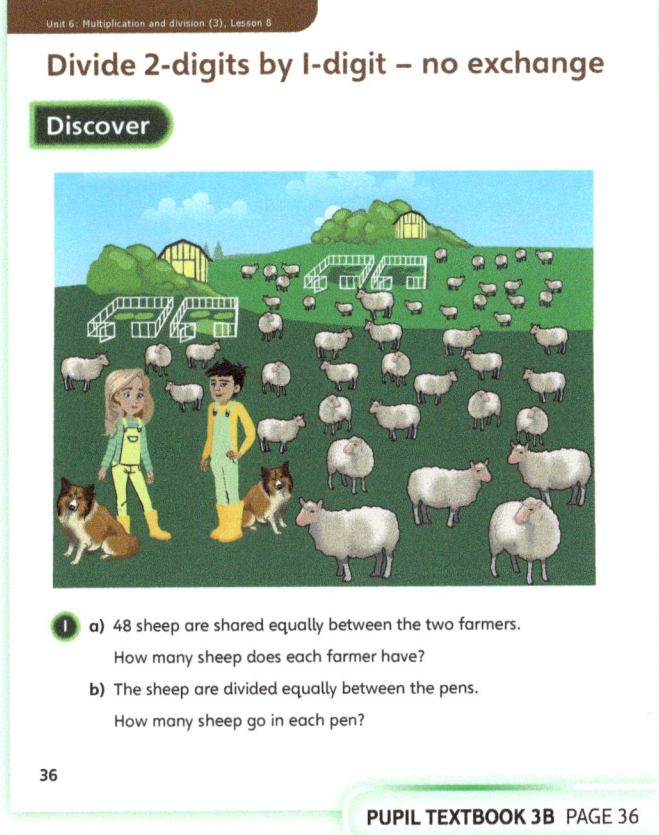

PUPIL TEXTBOOK 3B PAGE 36

Share

WAYS OF WORKING Whole class teacher led

ASK

- Question 1 a): *Did you remember to add the 10s and 1s?*
- Question 1 a): *Why is it better to count the total number of sheep and divide by 2 than to share the sheep one by one?*
- Question 1 b): *How can you check the answer?*
- Question 1 b): *What is the inverse of division?*

IN FOCUS In question 1 a), explore with children why sharing the sheep one at a time is not an efficient method, encouraging children to recognise why mental strategies are important. This could be emphasised with a 'challenge the teacher' contest: children can repeat the method with another question, while the teacher shares one by one. Children will enjoy 'beating' the teacher, while the message sinks in. The use of base 10 equipment will support children, if needed, by allowing them to experience physically sharing the blocks to reinforce the concept of dividing.

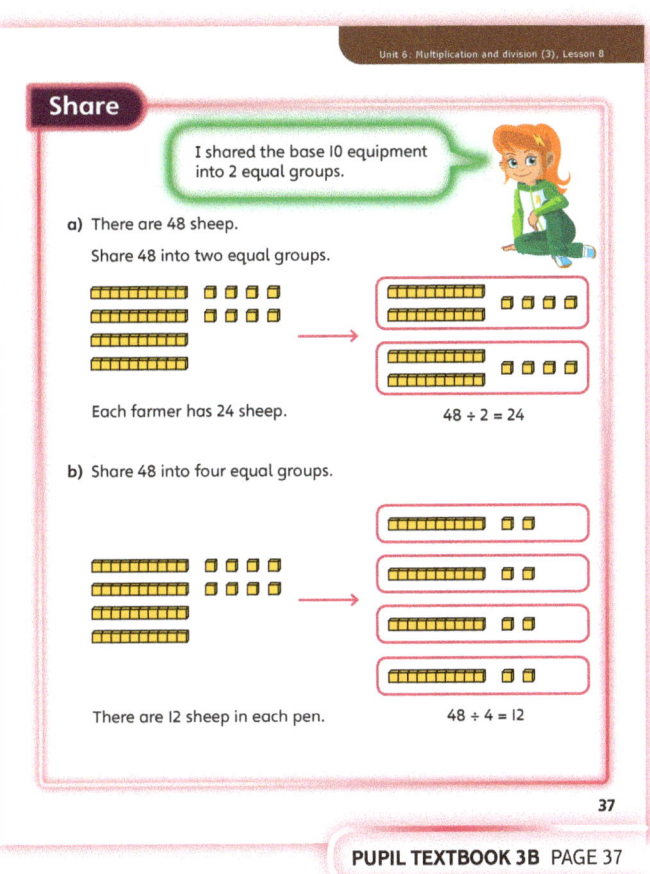

PUPIL TEXTBOOK 3B PAGE 37

Think together

WAYS OF WORKING Whole class teacher led (I do, We do, You do)

ASK
- Question ①: *How does the diagram help you to partition the number?*
- Question ②: *How can you check your answer?*
- Question ③: *Whose method is more efficient?*
- Question ③: *What advice would you give to Mo?*

IN FOCUS Question ① reinforces the method of using partitioning to divide 2-digit numbers by 1-digit numbers, while using the pictorial representation of base 10 equipment. Question ③ explores with children why the method of division modelled in this lesson is efficient.

STRENGTHEN Encourage children to complete the calculations using base 10 equipment to manipulate and experience the divisions. Ask: *What step do you need to do first? How can you divide up the 10s and the 1s? What is the final step you need to do?*

DEEPEN When solving question ②, encourage children to complete some calculations in their heads – retaining the partitioned numbers is a good skill for children to practise and develop. Encourage children to check their answer using multiplication.

ASSESSMENT CHECKPOINT Question ① will assess children's ability to use the partitioning method of division (with scaffolding) while question ② engages the same method but with less scaffolding. Look for children showing an understanding of using partitioning to divide numbers into 10s and 1s and then adding the 10s and 1s together. In particular, look for children who are able to check their work using multiplication as the inverse operation.

ANSWERS

Question ①: 39 ÷ 3 = 13

Question ②: 68 ÷ 2 = 34

Question ③: 84 ÷ 4 = 21
Alex's method is more efficient for such a large number.

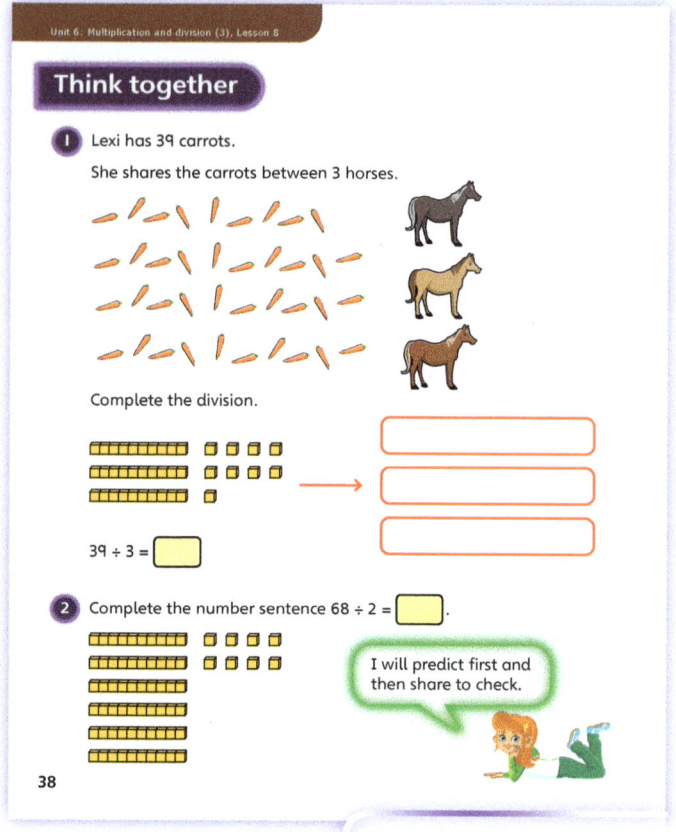

PUPIL TEXTBOOK 3B PAGE 38

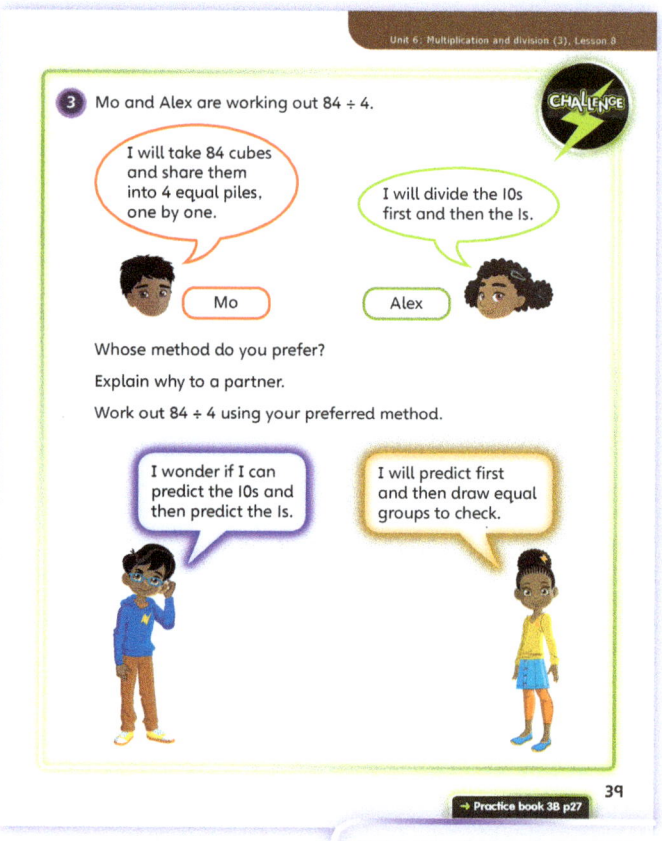

PUPIL TEXTBOOK 3B PAGE 39

Unit 6: Multiplication and division (3), Lesson 8

Practice

WAYS OF WORKING Independent thinking

IN FOCUS In questions ❶ to ❸, children use the method of division they used in the main lesson. They divide numbers (represented by base 10 equipment) into differing numbers of groups. Children should draw lines and dots to represent the base 10 equipment. In question ❹, the base 10 equipment is removed. Children are encouraged to use the method where they divide the 10s and divide the 1s. None of the examples require an exchange. In question ❹, children may still continue to draw or use base 10 equipment if they still need this for support. Question ❺ provides a word problem that children need to translate into a division.

STRENGTHEN Throughout these questions, encourage children to use base 10 equipment or counters on their desk to support their understanding.

DEEPEN In question ❻, children should realise that each division statement tells them the number of groups (the number after the division symbol) and the number in each group (the answer). They have to work out the total before division. They should realise that they need to do a multiplication to help them. They can use arrays to support their understanding if necessary.

ASSESSMENT CHECKPOINT Children should have a method to help them answer question ❹. This method may involve using base 10 equipment or other equipment, although children should start to be gaining confidence in using written (but non-formal) written methods.

ANSWERS Answers for the **Practice** part of the lesson can be found in the *Power Maths* online subscription.

Reflect

WAYS OF WORKING Independent thinking

IN FOCUS Children may not initially see how multiplication relates to the divisions. Ask them to think about some of the divisions they did and how they knew the answers. For example, how did they know that 8 divided by 2 is equal to 4?

ASSESSMENT CHECKPOINT Children should be able to explain that they can use their multiplication facts to help work out divisions.

ANSWERS Answers for the **Reflect** part of the lesson can be found in the *Power Maths* online subscription.

After the lesson

- Are children able to partition the 2-digit numbers without pictorial scaffolding?
- Are children able to complete all the steps of the division method?
- Are children able to use the method to work out answers mentally?

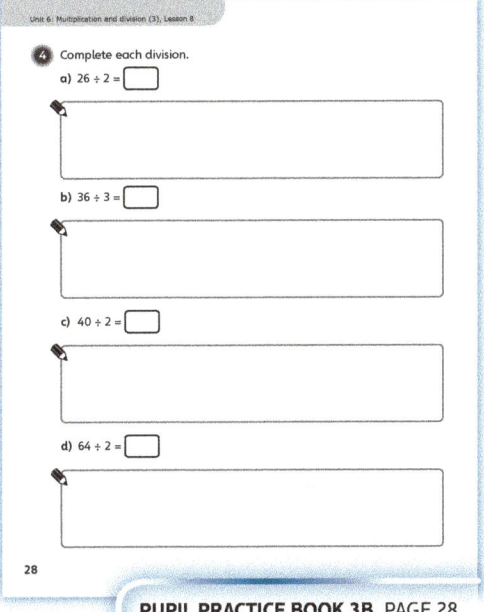

PUPIL PRACTICE BOOK 3B PAGE 27

PUPIL PRACTICE BOOK 3B PAGE 28

PUPIL PRACTICE BOOK 3B PAGE 29

77

Unit 6: Multiplication and division (3), Lesson 9

Divide 2-digits by 1-digit – flexible partitioning

Learning focus
In this lesson, children will continue to use their understanding of place value and division to divide 2-digit numbers by 1-digit numbers using partitioning, while using exchange to simplify calculations.

Before you teach
- Are children confident with finding multiples and recognising numbers divisible by 2, 3 and 5?
- Are children confident with using exchange in place value?

NATIONAL CURRICULUM LINKS

Year 3 Number – multiplication and division

Write and calculate mathematical statements for multiplication and division using the multiplication tables that they know, including for 2-digit numbers times 1-digit numbers, using mental and progressing to formal written methods.

ASSESSING MASTERY

Children can accurately partition a number using exchange if necessary to assist the division, and then divide both parts before adding the two amounts together. They can link their understanding of place value to their method and can confidently explain the steps in their method and explain why exchange is needed.

COMMON MISCONCEPTIONS

Children may not realise a number can be partitioned in different ways. When dividing using partitioning, it is useful to use multiples so that the division can be completed using known multiplication facts. Ask:
- *What would be a good way to partition so both parts can be divided?*

STRENGTHENING UNDERSTANDING

Strengthen understanding of partitioning into multiples of a number by partitioning numbers in different ways, for example, 28 = 20 + 8 or 10 + 18. If needed, use base 10 equipment to support with concrete representations, encouraging children to partition numbers in as many ways as possible.

GOING DEEPER

Explore with children why partitioning 42 into 40 and 2 does not help when dividing by 3. Repeat with other numbers and challenge children to find the most useful way to partition.

KEY LANGUAGE

In lesson: multiplication, division, share, partition, tens (10s), ones (1s), exchange

STRUCTURES AND REPRESENTATIONS

Part-whole model

RESOURCES

Mandatory: place value counters, base 10 equipment

 In the eTextbook of this lesson, you will find interactive links to a selection of teaching tools.

Quick recap

Ask children to work out divisions such as 24 ÷ 2 and 69 ÷ 3, where each of the digits is a multiple of the number you are dividing by. Children may use place value equipment to help them. Ask them to explain how they divide the 10s and divide the 1s.

Unit 6: Multiplication and division (3), Lesson 9

Discover

WAYS OF WORKING Pair work

ASK

- Question 1 a): *How many lanterns were released by the boats?*
- Question 1 b): *How many boats are there?*
- Question 1 b): *Is 40 divisible by 3?*

IN FOCUS Question 1 b) gives children an opportunity to discuss and experiment with dividing 2-digit by 1-digit numbers where exchange is needed. When partitioned, 42 gives 4 tens and 2 ones, but 40 when divided by 3 does not give a whole number. Instead, children will need to partition 42 into 3 tens and 12 ones so that they can divide by 3.

PRACTICAL TIPS Ask children to share items in multiples of 10 between 3 people, for example, 10 stickers between 3 children. It is important for children to notice that the numbers cannot be divided equally without splitting them into smaller pieces.

ANSWERS

Question 1 a): 42 lanterns were released.
Children should use base 10 equipment to show 42: 4 tens and 2 ones.

Question 1 b): 42 ÷ 3 = 14
Each boat released 14 lanterns.

PUPIL TEXTBOOK 3B PAGE 40

Share

WAYS OF WORKING Whole class teacher led

ASK

- Question 1 a): *What different ways can you partition 42?*
- Question 1 a): *Does 40 + 2 help?*
- Question 1 b): *What number are you dividing by?*
- Question 1 b): *What numbers do you know are divisible by 3?*
- Question 1 b): *How can you use exchange to split 42 into numbers that are divisible by 3?*

IN FOCUS Question 1 b) is important because children should recognise that partitioning 42 into 40 + 2 does not help them solve the problem. Encourage children to look for multiples of 3 that make up 42. It will be important for children to have a good understanding of multiples of 3. If needed, revisit looking for multiples of 2, 3, 4, 5 and 10.

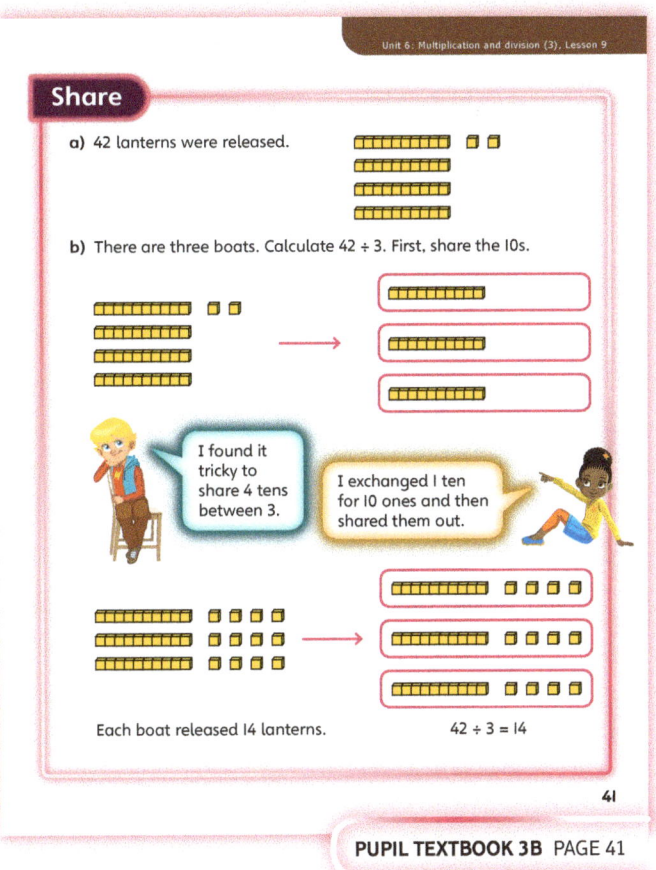

PUPIL TEXTBOOK 3B PAGE 41

79

Think together

WAYS OF WORKING Whole class teacher led (I do, We do, You do)

ASK
- Question 1: *Why is it easier to divide 20 and 16 by 2?*
- Question 2: *Which times-tables help with this question?*
- Question 3: *Why do you use the number 50 in the part-whole models? How does this help solve the division problem?*

IN FOCUS Question 1 demonstrates a range of ways to partition before dividing. Encourage children to recognise that partitioning 36 into 20 and 16 is a good strategy because they are both known multiplication facts from the 2 times-table. In question 3, children partition multiples of 5 using part-whole models rather than base 10 equipment. Unlike in questions 1 and 2, the part-whole models in question 3 do not partition the number into 10s and 1s but each have 50 as one part. Children will need to think about what number should go in the other part, and then how to divide each part by 5.

STRENGTHEN In question 1, encourage children to use base 10 equipment to represent a range of numbers in different partitions. For example, 45 = 40 + 5 or 30 + 15. This will help develop flexibility in partitioning and be good practice for later in the lesson. Question 3 can be supported with base 10 equipment. It may be beneficial to set similar questions in which children are required to partition numbers with a section of the part-whole model completed.

DEEPEN When solving question 3, encourage children to solve more division calculations in which the number to be divided can be partitioned into 3 parts. Use a part-whole model with 3 parts to support them with this.

ASSESSMENT CHECKPOINT Question 2 provides an opportunity to assess whether children can divide 2-digit numbers by a 1-digit number using partitioning with an exchange.

ANSWERS

Question 1 a): 36 ÷ 2 = 18

Question 1 b): 48 ÷ 3 = 16

Question 2: 56 ÷ 2 = 28   75 ÷ 3 = 25

Question 3 a): 65 ÷ 5 = 13

Question 3 b): 75 ÷ 5 = 15

Question 3 c): 85 ÷ 5 = 17

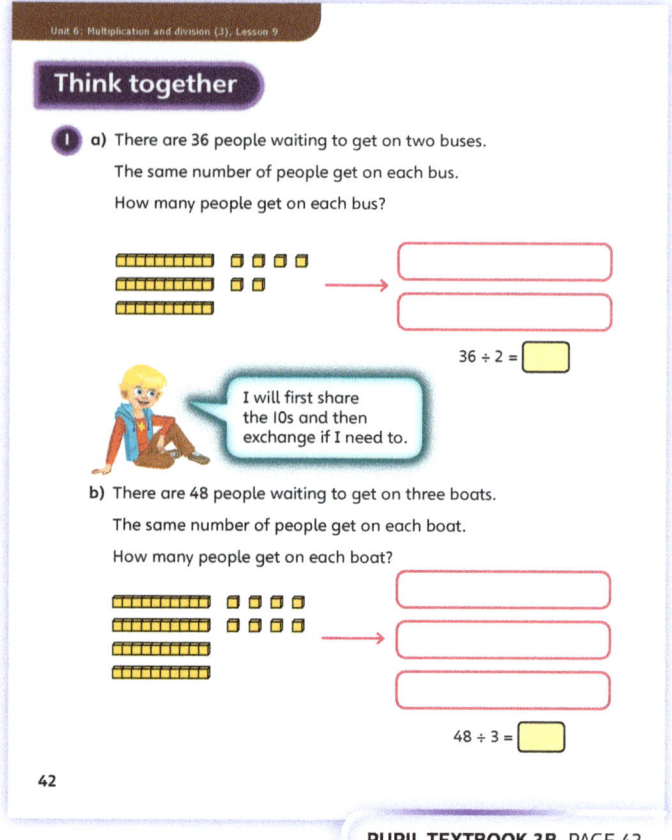

PUPIL TEXTBOOK 3B PAGE 42

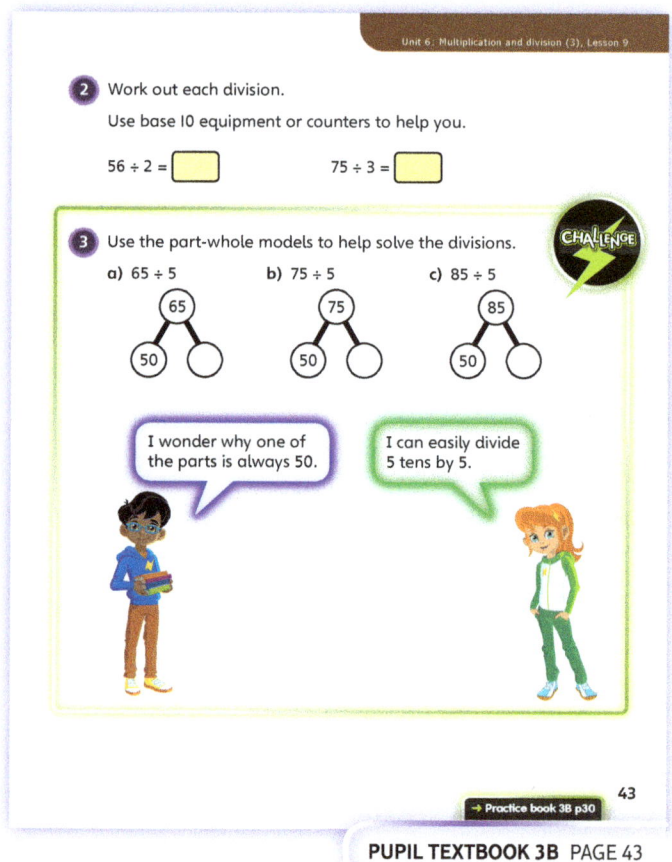

PUPIL TEXTBOOK 3B PAGE 43

Practice

WAYS OF WORKING Independent thinking

IN FOCUS Question ① develops children's ability to divide 2-digit numbers by 1-digit numbers while scaffolding their understanding using pictorial representations of base 10 equipment. In question ④, children should realise that each division statement tells them the number of groups (the number after the division symbol) and the number in each group (the answer). They have to work out the total before division. They should start to realise that they need to do a multiplication to help them. They might use base 10 equipment to help them see what each number in the division sentence represents.

STRENGTHEN To strengthen understanding of partitioning in any of these questions, ask: *Can you use multiples of numbers you find it easier to work with?*

DEEPEN When considering question ④, deepen children's learning by asking: *What does each number in the division sentence represent? How can you check your answers?*

THINK DIFFERENTLY Question ③ is an excellent opportunity for children to use their learnt method to solve a real-life problem. Some children may use multiplication (27 × 3) to find the answer. If this is the case, encourage them to check their answer by doing the inverse (a division).

ASSESSMENT CHECKPOINT Question ③ will assess children's ability to apply the division method in a real-life context. If children can solve this and confidently explain why their solution is correct, it is likely that they have mastered the lesson.

ANSWERS Answers for the **Practice** part of the lesson can be found in the *Power Maths* online subscription.

Reflect

WAYS OF WORKING Pair work

IN FOCUS Encourage children to have a go at each question using base 10 equipment with a partner. Ask: *What do you notice?* Encourage them to see that in the first example they did not need to make an exchange, but in the second problem they needed to exchange 1 ten for 10 ones as they needed to partition the 10.

ASSESSMENT CHECKPOINT Check that children can do the two divisions, using base 10 equipment to support them, and are able to record any workings alongside.

ANSWERS Answers for the **Reflect** part of the lesson can be found in the *Power Maths* online subscription.

After the lesson

- Can children partition numbers to make them easier to divide?
- Did children show an understanding of using inverse operations to check their work?

Unit 6: Multiplication and division (3), Lesson 10

Divide 2-digits by 1-digit with remainders

Learning focus

In this lesson, children will learn that some division calculations have a remainder. They will use concrete and pictorial methods to determine the remainder.

Before you teach

- Do children understand grouping and sharing for division?
- Are children confident dividing a 2-digit number by 1-digit number where there is no remainder?

NATIONAL CURRICULUM LINKS

Year 3 Number – multiplication and division

Write and calculate mathematical statements for multiplication and division using the multiplication tables that they know, including for 2-digit numbers times 1-digit numbers, using mental and progressing to formal written methods.

ASSESSING MASTERY

Children can divide a 2-digit number by a 1-digit number involving a remainder, using known multiplication facts to predict remainders without having to calculate. They can explain the division rules involved.

COMMON MISCONCEPTIONS

Children may not use their understanding of multiplication facts to predict when there will be a remainder. Ask:
- *How can you tell when a number can be divided by 2? 5? 10?*

Children may leave a remainder that is larger than the divisor. Ask:
- *Can you make another group from the remainder or can you share the one still left?*

STRENGTHENING UNDERSTANDING

Strengthen understanding of the division rules for the 2, 3, 4, 5 and 10 times-tables by encouraging children to use their known multiplication facts to find patterns and create divisibility rules. Encourage the use of base 10 equipment so children can manipulate and experience how dividing can create remainders. Children should use sharing and/or grouping with concrete materials to show that there is a remainder.

GOING DEEPER

Encourage children to use known multiplication facts to explain how, from looking at a division, they can tell there is a remainder. Can they work out what the remainder is without needing to share or group? For example, can they work out the remainder when 17 is divided by 5? Challenge children with reverse 'What is the number?' puzzles. Ask: *The remainder of a division is 2. What number could be the starting number? Can you find a range of divisions that match this?*

KEY LANGUAGE

In lesson: multiplication, division, equal, equally, even, odd, remainder, share, partition, tens (10s), ones (1s), exchange

RESOURCES

Mandatory: counters, cubes

Optional: other concrete objects that can be shared, base 10 equipment

 In the eTextbook of this lesson, you will find interactive links to a selection of teaching tools.

Quick recap

Ask children to solve some divisions of 2-digit numbers by 1-digit numbers that do not have a remainder and do not require an exchange. For example, 75 ÷ 5, 28 ÷ 2, 39 ÷ 3. Repeat with divisions that do require an exchange. For example, 56 ÷ 4, 18 ÷ 3, 72 ÷ 8. You might want to talk about different ways you could partition each number to help with the division.

Unit 6: Multiplication and division (3), Lesson 10

Discover

WAYS OF WORKING Pair work

ASK
- Question 1 a): *How many children are there? How can you divide the children into two groups?*
- Question 1 b): *How many children are there now? Can you divide them into two equal groups?*

IN FOCUS In questions 1 a) and 1 b), children use objects such as counters to represent the children in the scenario. Look for children using a sharing method to divide the 6 counters into 2 groups. In question 1 b), they should realise that they cannot share the children equally and have 1 child left over.

PRACTICAL TIPS This concept could be introduced in PE while setting up team games. Alternatively, use 6 or 7 children in the classroom to act out how they might divide into 2 groups.

ANSWERS

Question 1 a): 6 ÷ 2 = 3
There are 2 equal teams with 3 players on each team.

Question 1 b): 7 cannot be shared equally by 2. There is one left out.
We say 7 ÷ 2 = 3 remainder 1.

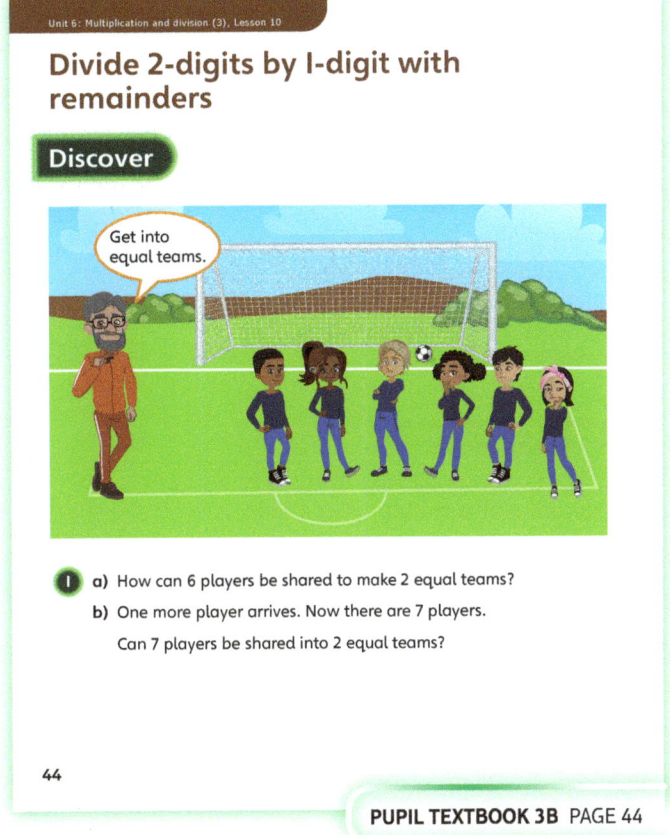

PUPIL TEXTBOOK 3B PAGE 44

Share

WAYS OF WORKING Whole class teacher led

ASK
- Question 1 a): *How many children are there? How can you divide the children into 2 groups?*
- Question 1 b): *How do you know you cannot divide the children into 2 groups? What is the number of children left over called?*

IN FOCUS In question 1 a), use counters or images of children (or act out as a class) how to share the 6 children into 2 groups. Use the method of sharing 1 at a time into each group until there are no children left. In question 1 b), repeat from the start with 7 children and show that there is 1 child left over that cannot be shared, otherwise we would have unequal groups. Use Ash's comment to help you explain that the 1 child left over is called the remainder and that 7 divided by 2 is equal to 3 remainder 1.

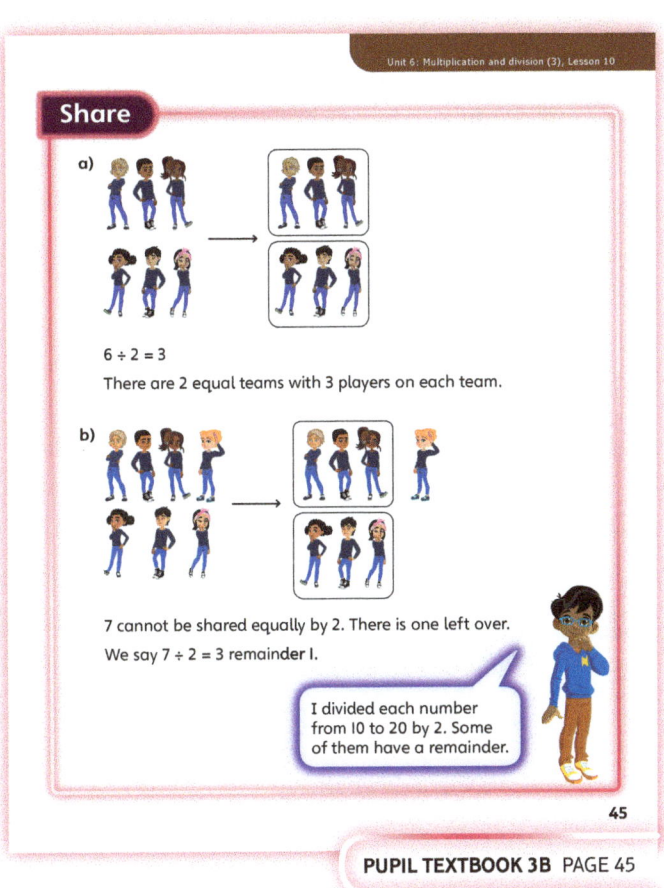

PUPIL TEXTBOOK 3B PAGE 45

Unit 6: Multiplication and division (3), Lesson 10

Think together

WAYS OF WORKING Whole class teacher led (I do, We do, You do)

ASK

- Question ❶: *What can you use to help you with the divisions? Which divisions have a remainder? Can you tell if they have a remainder before working out the answers?*
- Question ❷: *What is the same and what is different about the calculations? Do you know which ones will not have a remainder? How do you know that the others will have a remainder? Can you work out the remainder using multiplication facts?*

IN FOCUS Question ❶ asks children to do some simple divisions to see which ones have remainders. Children should use counters, cubes or drawings to help them with the divisions. They may realise that numbers in the 3 times-table that they met earlier in the year will not have a remainder. In question ❷, children take 21 counters and try and share them equally into 3, 4 and 5 groups. They should notice that they can divide 21 into 3 equal groups as 21 is a multiple of 3. 21 cannot be divided by 4 or 5 equally as 21 is not in the relevant times-tables. These discussions are important to help children with understanding division with remainders.

STRENGTHEN Throughout the questions, use concrete materials to support the divisions and show when there is and is not a remainder.

DEEPEN Question ❸ further develops the concept of division leading to a remainder, but with a larger number. Instead of taking 38 counters, children may look to use methods they used in other lessons, such as using base 10 equipment. Throughout the questions, encourage children to use their knowledge of multiples to make predictions about whether there will be a remainder.

ASSESSMENT CHECKPOINT Use question ❷ to assess whether children have an understanding of when a division will have a remainder. They will use concrete methods as a minimum to work out whether there is a remainder from a division.

ANSWERS

Question ❶: 9 ÷ 3 = 3   10 ÷ 3 = 3 remainder 1
11 ÷ 3 = 3 remainder 2

Question ❶: 21 ÷ 3 = 7   21 ÷ 4 = 5 remainder 1
21 ÷ 5 = 4 remainder 1

Question ❸: 38 ÷ 3 = 12 remainder 2

Question ❸ a): 12 children stand in each line.

Question ❸ b): 2 children are left over.

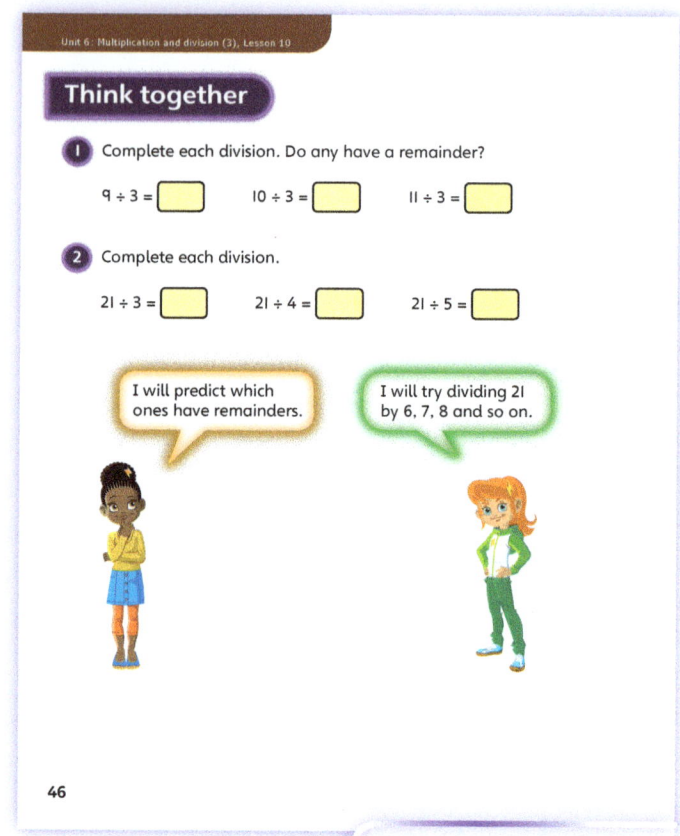

PUPIL TEXTBOOK 3B PAGE 46

PUPIL TEXTBOOK 3B PAGE 47

Unit 6: Multiplication and division (3), Lesson 10

Practice

WAYS OF WORKING Independent thinking

IN FOCUS Questions ❶ and ❷ encourage children to use counters or drawings to divide small numbers and determine whether a division has a remainder or not. Children write their answers in full, for example 4 remainder 1. In question ❹, children keep the 25 constant and divide by different numbers. Ask children to predict whether they think an answer will have a remainder and, if so, how big they think the remainder will be. Children may be able to find the answer (including any remainder) to a division by considering their multiplication facts. This should be encouraged if children are able to do this at this stage. In question ❺, children should be able to say whether there is a remainder by considering what they know about multiples of 5.

STRENGTHEN In questions ❸ and ❹, children should use counters or cubes to support their divisions. Make sure children understand what each number in the division means before using a sharing or grouping method. Allow children to do the practical part of the division and, when they get to the point where they can no longer share, explain that this is the remainder (if there are any).

DEEPEN Deepen children's understanding by giving larger numbers that have a remainder. They can then use the methods used in the previous lessons where they divide the 10s first and then the 1s, using base 10 equipment to support if necessary and using any exchanges. Question ❺ provides an example of this.

Children should also be becoming increasingly confident in working out whether a number will have a remainder or not. They use their knowledge of multiplication facts to help them. For example, for 25 ÷ 4, children should be encouraged to try and work out the remainder by considering how far it is from the closest multiple of 4 less than 25 (24).

ASSESSMENT CHECKPOINT Questions ❸ and ❹ will assess children's ability to divide 2-digit numbers by 1-digit numbers involving remainders with limited scaffolding.

ANSWERS Answers for the **Practice** part of the lesson can be found in the *Power Maths* online subscription.

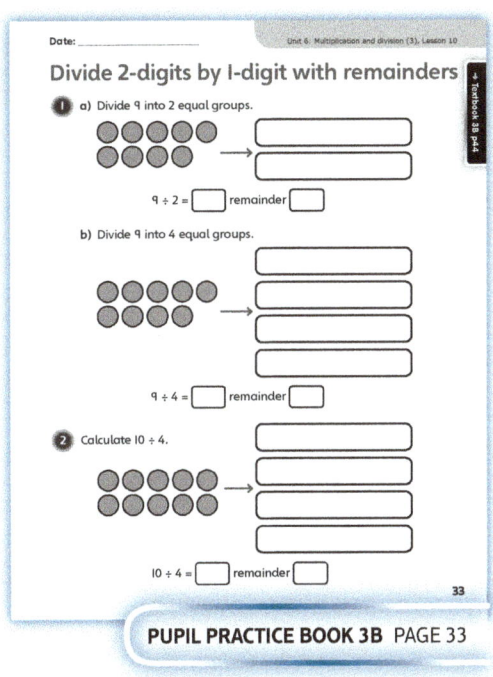

PUPIL PRACTICE BOOK 3B PAGE 33

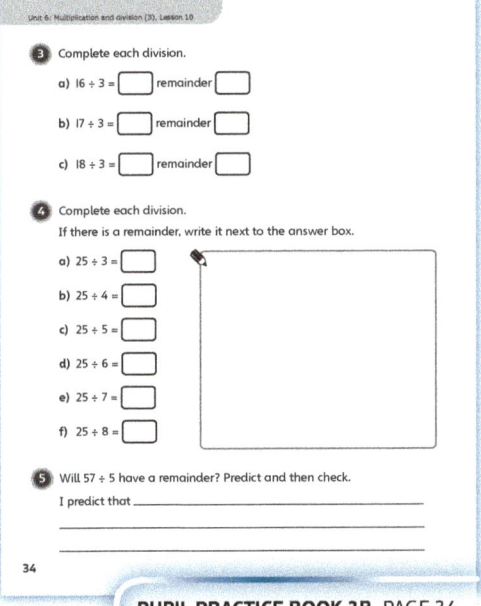

PUPIL PRACTICE BOOK 3B PAGE 34

Reflect

WAYS OF WORKING Independent thinking

IN FOCUS Some children may consider multiplication facts and knowledge of multiples to determine whether there is a remainder, others may use equipment or pictorial methods.

ASSESSMENT CHECKPOINT Look for children confidently predicting what the remainders will be and giving clear explanations of why, for example, 27 ÷ 2 has a remainder because 27 is an odd number. Observe if children are using mental methods, or using some jottings or equipment to find the answers. It is important at this stage that children have some method to find a division that includes a remainder.

ANSWERS Answers for the **Reflect** part of the lesson can be found in the *Power Maths* online subscription.

After the lesson
- Are children confident about what a remainder is?
- Can children confidently work out a division that includes a remainder?
- Do children know the division rules for the multiplication tables?

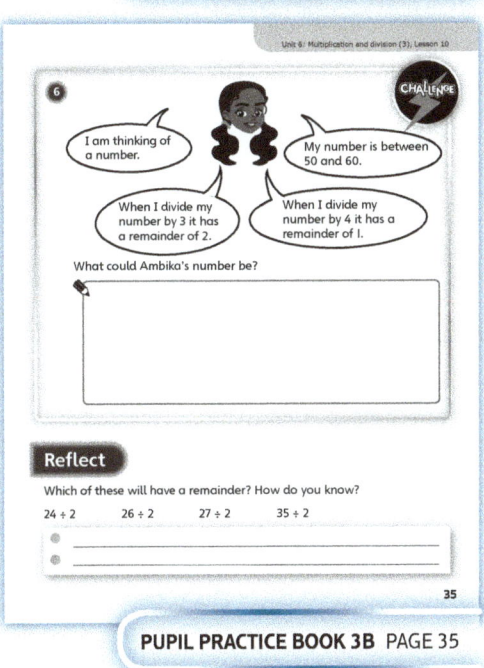

PUPIL PRACTICE BOOK 3B PAGE 35

85

Unit 6: Multiplication and division (3), Lesson 11

How many ways?

Learning focus
In this lesson, children will calculate the number of ways that n objects can be connected to m objects and will use the multiplication rule for correspondence problems.

Before you teach
- Are children able to work methodically?
- Are children able to show solutions pictorially by drawing them?

NATIONAL CURRICULUM LINKS

Year 3 Number – multiplication and division

Solve problems, including missing number problems, involving multiplication and division, including positive integer scaling problems and correspondence problems in which n objects are connected to m objects.

ASSESSING MASTERY

Children have a secure understanding of how many different ways one group of objects can be connected to another group of objects. They can work systematically, and show their explanations using a diagram or table and use the multiplication rule for correspondence problems.

COMMON MISCONCEPTIONS

Children may not work in a methodical way to find solutions. Ask:
- *Would a table help?*

Children may miss some combinations. Ask:
- *Have you found all of the solutions?*

STRENGTHENING UNDERSTANDING

Children can be encouraged to use real-life objects, such as hats and scarves, to find the total number of combinations and draw pictorial representations to show these. Strengthen understanding by encouraging children to draw tables or diagrams of the combinations (using lines to connect the objects) to explore the total number of combinations possible.

GOING DEEPER

Ask children to work with 3 groups of objects exploring why, if there are 3 groups of 3 objects, there will be $3 \times 3 = 9$ combinations. Challenge children to investigate if this rule always works.

KEY LANGUAGE

In lesson: multiply, combinations, pairs, table, different, same

Other language to be used by the teacher: groups, order

STRUCTURES AND REPRESENTATIONS

Tables, line diagrams

RESOURCES

Mandatory: drawing resources (ruler and pencil), real-life objects for combining such as hats and scarves, colouring pencils

 In the eTextbook of this lesson, you will find interactive links to a selection of teaching tools.

Quick recap
Ask children multiplication facts from the 2, 3 and 5 times-tables. Ask them to write their answers down as quickly as possible on a white board and then show you. You may want to also ask them to draw arrays to show the multiplications.

Unit 6: Multiplication and division (3), Lesson 11

Discover

WAYS OF WORKING Pair work

ASK

- Question 1 a): *How many scarves are can you choose from to put on the snowman? How many hats?*
- Question 1 b): *Can you find a link between the number of hats and scarves and the number of ways of dressing the snowman?*

IN FOCUS Question 1 introduces the concept of correspondence problems where a number of objects are linked to a number of different objects. The context here is of a snowman with hats and scarves and gives children an opportunity to explore the number of ways of combining different objects while making a conjecture about possible patterns.

PRACTICAL TIPS This introduction lends itself to similar practical activities involving dressing up. Children could be encouraged to dress the teacher or each other with a number of hats and scarves, investigating how many different combinations are possible. They could draw their solutions to create a visual reference.

ANSWERS

Question 1 a):

Hat	A	A	B	B	C	C	D	D
Scarf	1	2	1	2	1	2	1	2

There are 8 different ways.

Question 1 b): The total number of ways to dress the snowman is 4 × 2 = 8 ways.

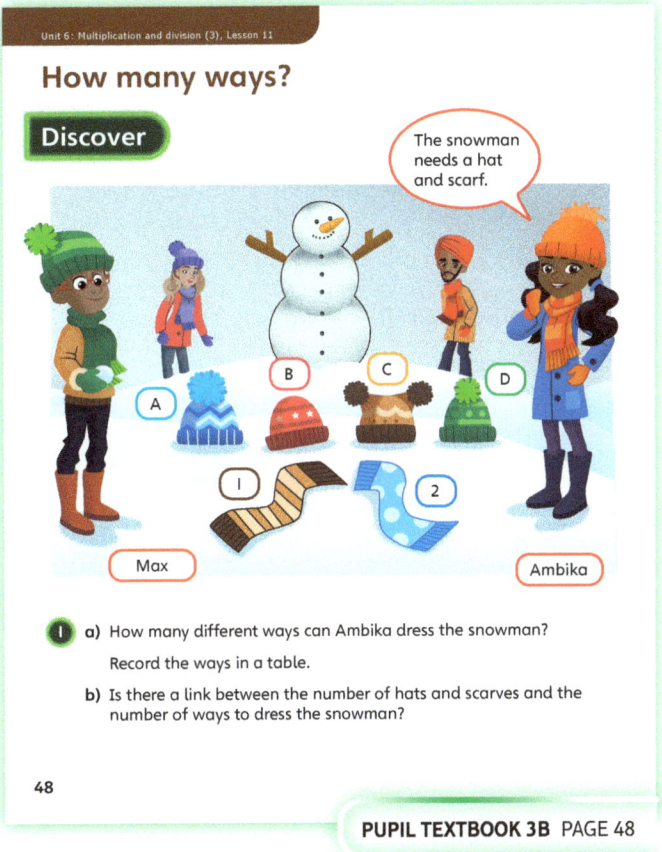

PUPIL TEXTBOOK 3B PAGE 48

Share

WAYS OF WORKING Whole class teacher led

ASK

- Question 1 a): *How do you know you have found all of the answers?*
- Question 1 a): *How does the table help find the answer?*
- Question 1 b): *How is multiplication linked to the number of different ways?*
- Question 1 b): *What is the most efficient way of finding the answer?*
- Question 1 b): *Does this multiplication rule always work? How can you find out?*

IN FOCUS In question 1 a), it is important to encourage children to find all the different combinations for dressing the snowman. When children are finished, ask them to convince you that they have found all of the correct answers. Question 1 b) links the number of ways to a multiplication rule. Challenge children to investigate this in other situations, such as 3 scarves and 2 hats. Encourage children to understand that using the multiplication rule is an efficient way of solving correspondence problems.

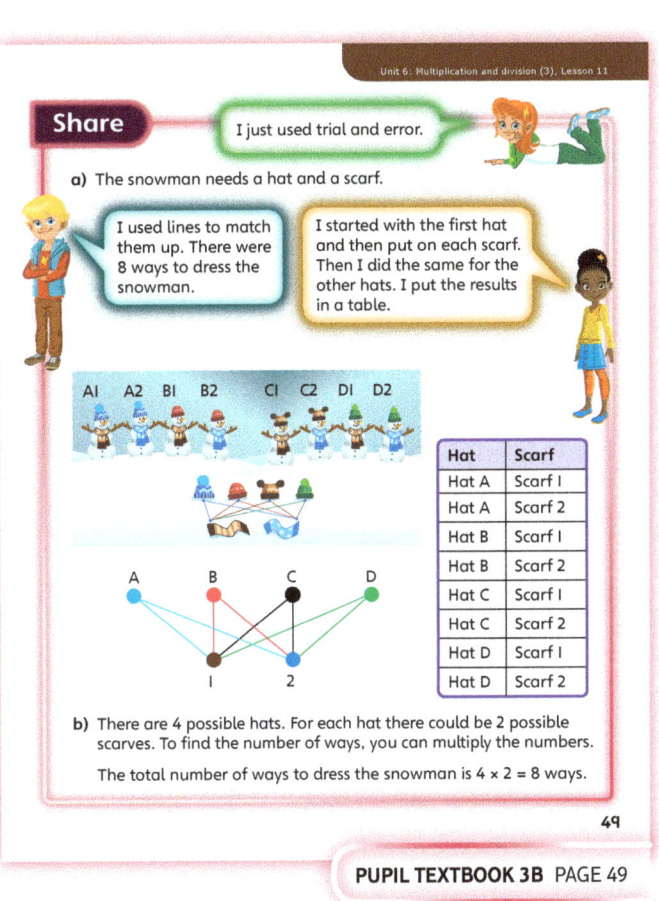

PUPIL TEXTBOOK 3B PAGE 49

Think together

WAYS OF WORKING Whole class teacher led (I do, We do, You do)

ASK

- Question ❶: *What rule can you apply to find the solution?*
- Question ❸ a): *Can you use a table to put in all of the solutions in a logical order?*
- Question ❸ b): *Since there are 2 children and 4 options, is the answer 2 × 4?*

IN FOCUS In question ❸, look for children understanding why there are 6 ways for the 2 children to buy different items. You might explain it like this:

- If one chose the sandwich, the other could choose one of the strawberry, the water or the crisps. That's 3 different pairs.
- If one chose the strawberry, the other could choose the water or the crisps to give us a new pair of items – since the sandwich and the strawberry have already been chosen. That's 2 more pairs.
- The last new pair is the water and the crisps.

Question ❸ explores children's ability to adapt the number of items being matched and their fluency with finding combinations in different situations. Look for children understanding why, for example, if there are 2 groups of 4 objects, there will be 16 combinations because 4 × 4 = 16.

STRENGTHEN If children are still struggling to find all of the connections following on from question ❷, encourage them to do some practical sorting with real objects making real combinations.

DEEPEN Question ❸ deepens children's understanding by changing the number of items to be combined. Encourage children to explore why 2 × 4 = 8 does not give the solution.

ASSESSMENT CHECKPOINT Questions ❶ and ❷ will assess children's ability to find the total number of combinations of 2 groups of objects. Look for children confidently applying the multiplication rule to find and check their answers.

ANSWERS

Question ❶ a): 5 × 2 = 10 ways

Question ❶ b):

Coats	A	A	B	B	C	C	D	D	E	E
Shoes	1	2	1	2	1	2	1	2	1	2

Question ❷ a): There are 2 × 3 = 6 possible pairs.

Question ❷ b): The possible pairs are:
- book / crayon
- book / ruler
- book / sharpener
- rucksack / crayon
- rucksack / ruler
- rucksack / sharpener

Question ❸ a): There are 6 different ways to buy different items:

Question ❸ b): There are 4 more ways if they can choose the same item.
There are 10 ways altogether.

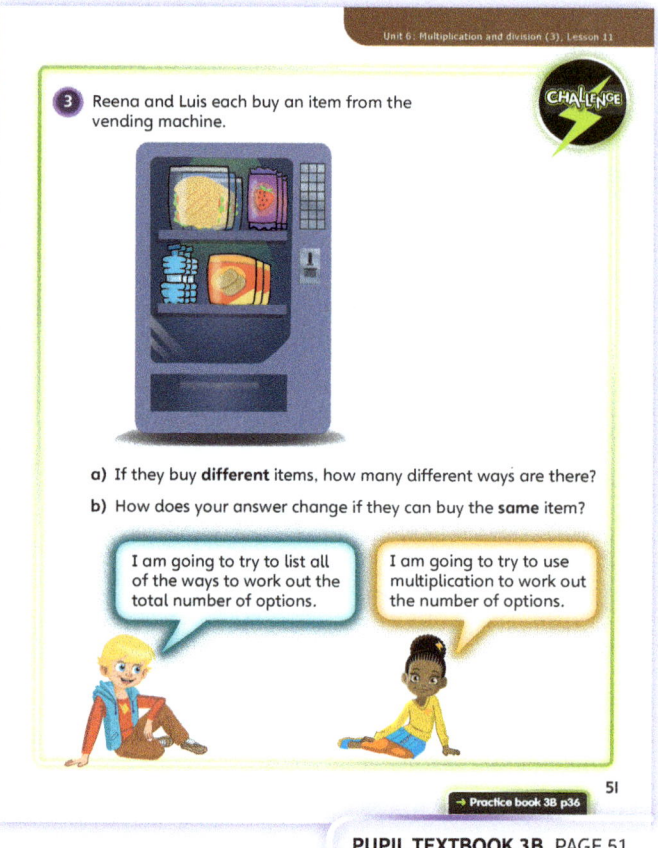

PUPIL TEXTBOOK 3B PAGE 51

Unit 6: Multiplication and division (3), Lesson 11

Practice

WAYS OF WORKING Independent thinking

IN FOCUS Questions ① and ② develop children's ability to find the number of combinations of two different groups of objects linking this to the multiplication rule while scaffolding via the use of tables.

Question ③ reduces scaffolding and requires children to find the number of combinations using pictures and the multiplication rule.

STRENGTHEN For children struggling with finding combinations, question ② is a good opportunity to recap 'drawing' the solution. Encourage children to use a ruler and pencil to connect each card to each letter card so that they can count the connections, then link this to the multiplication rule to strengthen understanding.

DEEPEN Question ④ is a good opportunity to deepen understanding. Encourage children to find the solution using a table or making a drawing. Ask: *How many solutions would there be if the colours were the same?*

ASSESSMENT CHECKPOINT Questions ① and ② give the opportunity to assess children's ability to find the total number of combinations of two groups of objects; in particular, if they can explain how they know they have found all of the correct solutions to question ① b), it is likely that they have mastered the lesson.

Question ③ will assess children's ability to work abstractly on correspondence problems. Look for children relying on drawings and/or tables and those who can use the multiplication rule to find the answer using only mental methods.

ANSWERS Answers for the **Practice** part of the lesson can be found in the *Power Maths* online subscription.

Reflect

WAYS OF WORKING Independent thinking

IN FOCUS This question gives an important opportunity to recap the methods used in this lesson. Children should confidently be able to describe the steps to finding the total number of different ways and link this to using the multiplication rule. Encourage children to include the following words: order, table, drawing, multiplication.

ASSESSMENT CHECKPOINT Look for children fluently explaining how to solve a correspondence problem, including using a table or linking diagram, and referring to the multiplication rule.

ANSWERS Answers for the **Reflect** part of the lesson can be found in the *Power Maths* online subscription.

After the lesson

- Did children recognise the usefulness of using tables and/or diagrams to organise their work?
- Did children rely on pictures or tables rather than use the multiplication rule?
- Were children confident in using the multiplication rule?

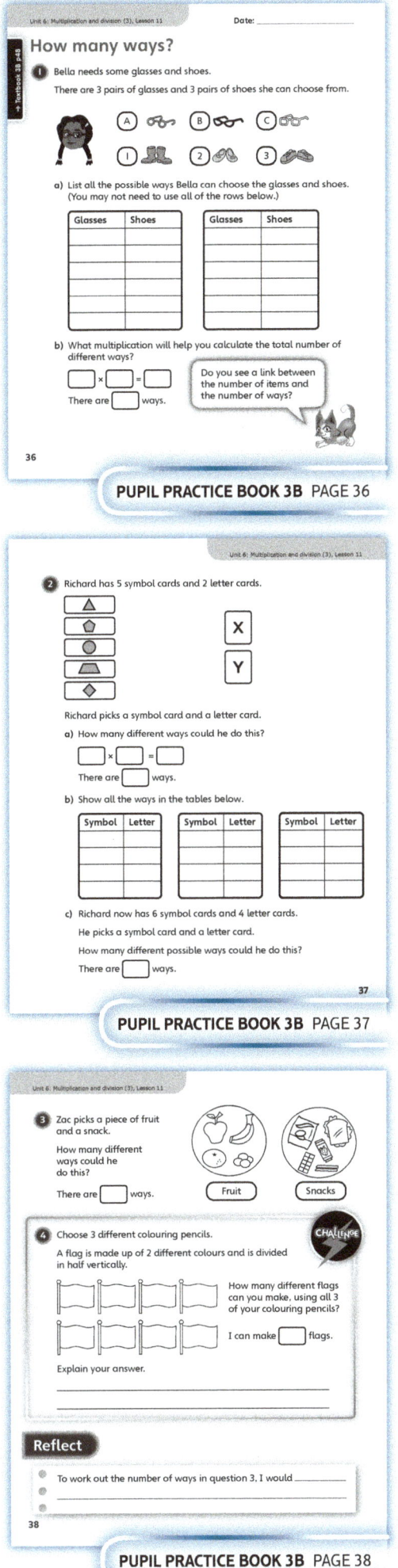

89

Unit 6: Multiplication and division (3), Lesson 12

Problem solving – mixed problems ①

Learning focus
In this lesson, children will solve mixed problems involving multiplication and division of 2-digit numbers.

Before you teach
- How will you use representations to explain problems?
- Are children confident multiplying and dividing 2-digit numbers?
- Are children confident with using inverse operations to check their answers?

NATIONAL CURRICULUM LINKS

Year 3 Number – multiplication and division

Solve problems, including missing number problems, involving multiplication and division, including positive integer scaling problems and correspondence problems in which *n* objects are connected to *m* objects.

Write and calculate mathematical statements for multiplication and division using the multiplication tables that they know, including for 2-digit numbers times 1-digit numbers, using mental and progressing to formal written methods.

ASSESSING MASTERY

Children can reliably interpret a range of problems and puzzles, decoding whether a problem involves multiplication or division, and calculate the answer. Finally, they can fluently use the inverse operation to check their answers.

COMMON MISCONCEPTIONS

Children may not interpret a question correctly, for example, dividing instead of multiplying. Ask:
- *Can you use a place value grid or bar model? What does the language used in the question tell you to do?*

STRENGTHENING UNDERSTANDING

Encourage children to use a range of representations (bar model, place value grid, arrays, part-whole models) to visually show the word problem. This will increase and strengthen their understanding.

GOING DEEPER

Challenge children to think of their own word problem based on a given multiplication or division, perhaps working in pairs. They can solve each other's problem and then check their answers using inverse operations.

KEY LANGUAGE

In lesson: multiplication, division, tens (10s), ones (1s)

Other language to be used by the teacher: number sentence, partition

STRUCTURES AND REPRESENTATIONS

Place value grid, part-whole model, bar model

RESOURCES

Mandatory: base 10 equipment, counters

 In the eTextbook of this lesson, you will find interactive links to a selection of teaching tools.

Quick recap

Ask children how to work out 2-digit by 1-digit multiplications, such as 23 × 3, 37 × 3 and 42 × 6. Children need to be able to use a written method, supported if necessary, by place value equipment. A reminder of these methods will help them with the calculations they need in order to solve the word problems in this lesson.

Unit 6: Multiplication and division (3), Lesson 12

Discover

WAYS OF WORKING Pair work

ASK
- Questions 1 a) and b): *What operation do you need to do?*

IN FOCUS Questions 1 a) and b) introduce children to using their understanding of multiplying and dividing 2-digit numbers in real-life contexts. Encourage children to use visual representations and to explain what mathematical operations they need to complete to solve the problems, giving reasons for their methods.

PRACTICAL TIPS This concept lends itself to a variety of real-life contexts involving 2-digit numbers, for instance, commuting to school: given the distance travelled to school by a teacher in one day, what is the distance travelled in one week and likewise if another teacher travels a certain distance in one week how far do they travel in one day?

ANSWERS
Question 1 a): Amal makes 56 wooden horses in 4 days.
Question 1 b): Sofia makes 23 wooden giraffes each day.

PUPIL TEXTBOOK 3B PAGE 52

Share

WAYS OF WORKING Whole class teacher led

ASK
- Question 1 a): *How can you tell that the calculation needed is is a multiplication?*
- Question 1 a): *Would using base 10 equipment help?*
- Question 1 a): *How can you check your answer?*
- Question 1 b): *How can you tell that the calculation needed is a division?*
- Question 1 b): *Is there a remainder?*

IN FOCUS Question 1 a) sets a real-life multiplication problem giving children an opportunity to use their knowledge of multiplying 2-digit numbers contextually. In question 1 b), children solve a similar problem but involving division. In both questions, encourage children to use place value counters or base 10 equipment to visually represent the numbers in the problems. Encourage children to sort equipment into groups: this will give them a clear, concrete representation. For question 1 a), encourage children to use counters in groups of 14 to help them realise that it is 14 lots of 4 = 14 × 4.

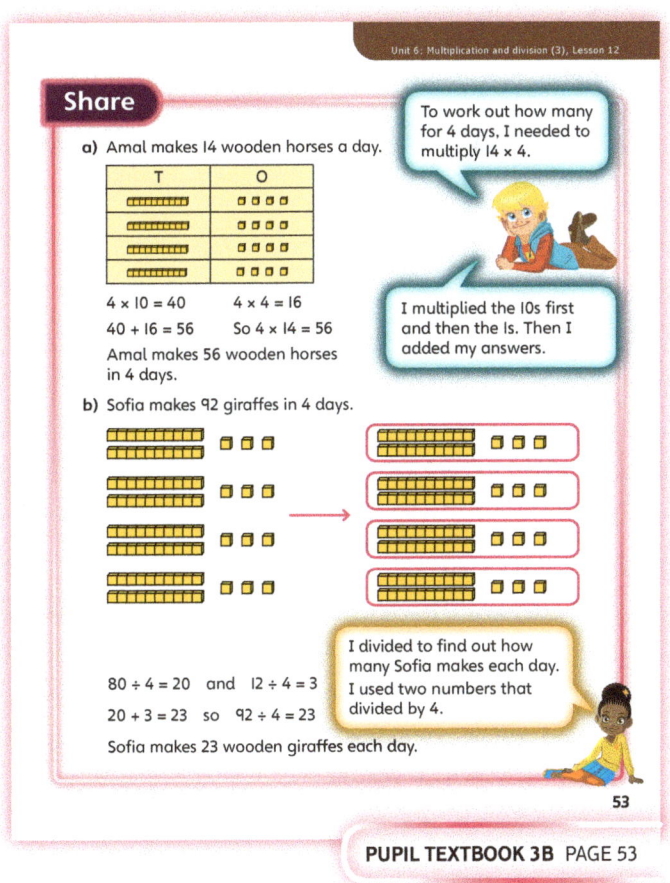

PUPIL TEXTBOOK 3B PAGE 53

Think together

WAYS OF WORKING Whole class teacher led (I do, We do, You do)

ASK

- Question ❶: *What operation do you need to use to solve this?*
- Question ❶: *How does the place value grid help you? How many 1s equal 1 ten? How many 1s are there in total? How many 10s are there in total?*
- Questions ❷ and ❸: *How can you check your answer?*
- Question ❸: *How does the bar model help you understand the question? Do you need to divide and multiply in this question? How many steps do you need to do in this question?*

IN FOCUS Question ❸ involves proportional reasoning and requires children to think carefully about the operations needed in this multi-step problem, which involves multiplication and division. Encourage children first to find out how many toy people Amal makes in 1 hour (39 ÷ 3 = 13), and then use this to find how many toy people Sofia makes in an hour – she makes twice as many, so 26 toy people in 1 hour. The final step is to multiply this by 5 (26 × 5 = 130). Encourage children to check their answer using the inverse operations.

STRENGTHEN When solving question ❶, encourage children to use the place value grid, using concrete resources if needed. Encourage children to explore the vocabulary used in the questions by asking: *What words used in the question give you a clue about whether you need to do a division or multiplication?*

DEEPEN When considering question ❸, children may realise that they can do 13 × 10 to work out the answer (because 13 × 10 = 13 × 2 × 5 = 26 × 5). This shows deeper thinking. Encourage them to explain their reasoning.

ASSESSMENT CHECKPOINT Question ❸ gives a good opportunity to assess children's fluency with solving mixed 2-digit number problems; if children can confidently solve this, explain their answer and check their calculations using inverse operations, they have likely mastered the lesson.

ANSWERS

Question ❶: There are 72 wheels in total on 6 trains.

Question ❷: 16 helicopters can be made in 48 hours.

Question ❸: Sofia makes 130 toy people in 5 hours.

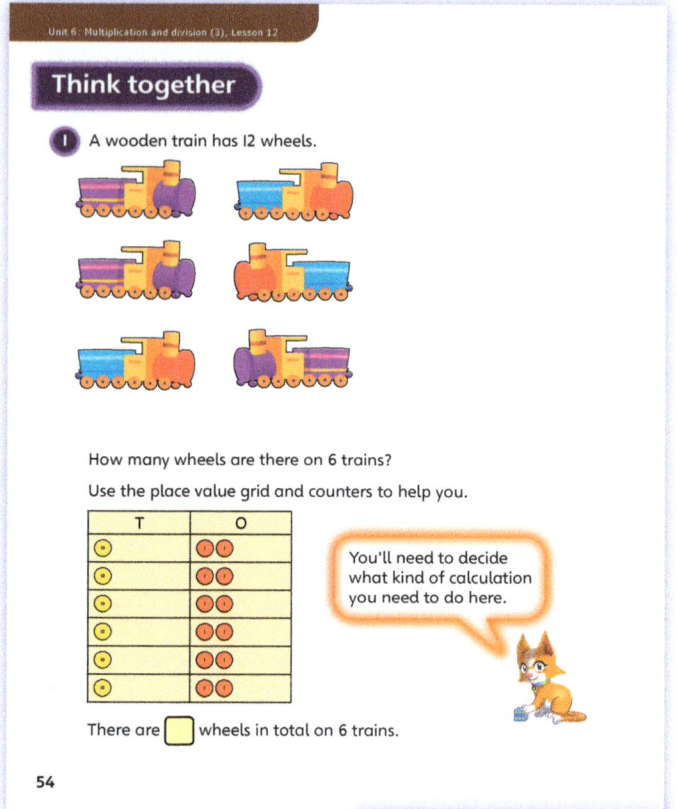

PUPIL TEXTBOOK 3B PAGE 54

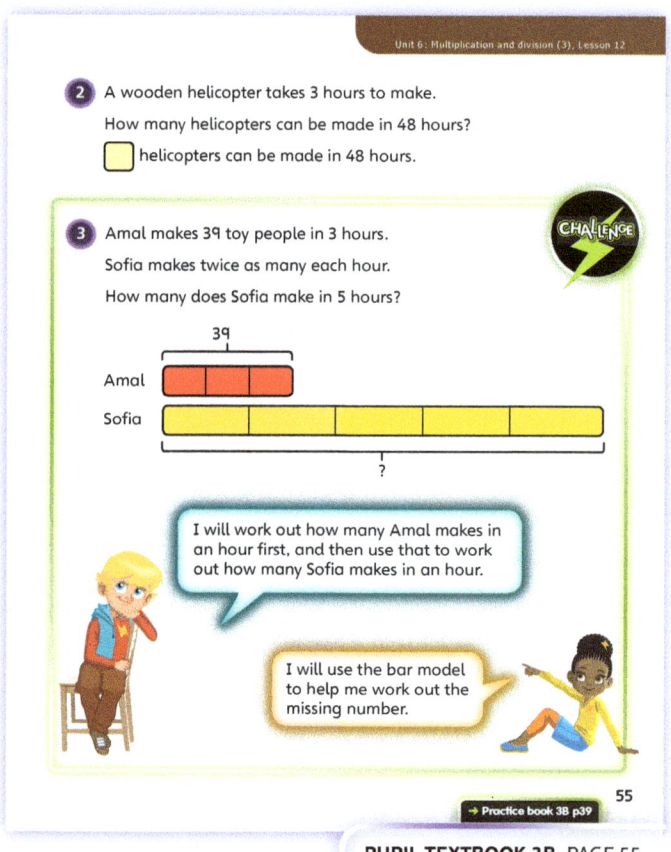

PUPIL TEXTBOOK 3B PAGE 55

Practice

WAYS OF WORKING Independent thinking

IN FOCUS Question ③ introduces a mixed multi-step problem broken into stages, using prompting rather than a bar model to scaffold. This is important since it will assist children with working systematically through a mixed problem to decide what operation they need to use.

STRENGTHEN Questions ① and ② are important because the problems are represented clearly on bar models – the first a multiplication, the second a division. Encourage children to use the bar models by asking: *Why are the bar models useful? What are the differences between the multiplication and division bar models?* Develop this further, if needed, using additional word problems and encourage children to represent the problems on bar models before solving the word problems in written format.

DEEPEN When solving question ⑥, encourage children to explain their strategies by asking: *How can the problem be solved? What step do you need to do first? How can you check your work?* Encourage children to realise that 2 does not go exactly into 5 and their first step is to find the cost of 1 book, before doubling this to find the cost of 2 books.

ASSESSMENT CHECKPOINT Question ③ will give a good opportunity to assess children's ability to interpret a word problem, and if they can use prior knowledge of multiplication and division strategies to find solutions to mixed problems involving 2-digit numbers. Question ⑥ will assess children's ability to interpret multi-step word problems linking to proportional thinking involving multiplication and division. Look for children who can solve this question and explain the reasoning behind their steps, since they are likely to have mastered this topic.

ANSWERS Answers for the **Practice** part of the lesson can be found in the *Power Maths* online subscription.

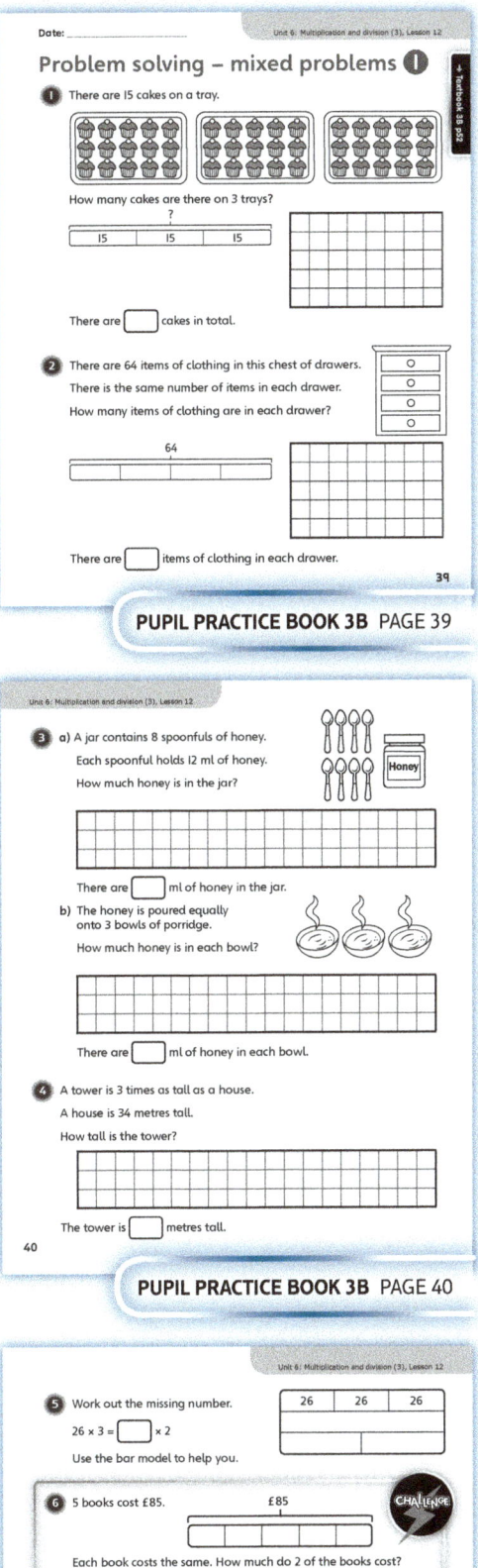

PUPIL PRACTICE BOOK 3B PAGE 39

PUPIL PRACTICE BOOK 3B PAGE 40

Reflect

WAYS OF WORKING Independent thinking

IN FOCUS This question requires children to use their learning to create a problem of their own based on the bar model. Challenge children to develop this into a multi-step problem.

ASSESSMENT CHECKPOINT Children should be able to interpret the bar model as either a multiplication or division problem involving either 4 × 18 = 72 or 72 ÷ 18 = 4. Look carefully at the vocabulary they use and the accuracy of their calculations.

ANSWERS Answers for the **Reflect** part of the lesson can be found in the *Power Maths* online subscription.

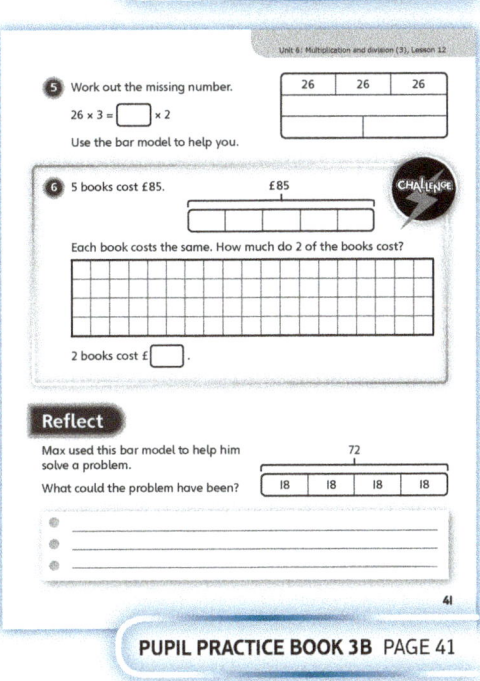

PUPIL PRACTICE BOOK 3B PAGE 41

After the lesson

- Could children interpret the bar models used?
- Could children represent a problem using a bar model?
- Were children confident in understanding the steps in a word problem and whether to use multiplication or division?

Unit 6: Multiplication and division (3), Lesson 13

Problem solving – mixed problems ②

Learning focus

In this lesson, children will use their understanding of all four operations to solve mixed multi-step problems.

Before you teach

- How will you link the bar models to the problems?
- How will you break multi-step problems down?
- Are children confident with using the four operations?

NATIONAL CURRICULUM LINKS

Year 3 Number – multiplication and division

Solve problems, including missing number problems, involving multiplication and division, including positive integer scaling problems and correspondence problems in which *n* objects are connected to *m* objects.

Write and calculate mathematical statements for multiplication and division using the multiplication tables that they know, including for 2-digit numbers times 1-digit numbers, using mental and progressing to formal written methods.

ASSESSING MASTERY

Children can confidently represent and interpret mixed multi-step problems using a bar model and solve them using an efficient method involving all four operations.

COMMON MISCONCEPTIONS

Children may not realise that there are two (or more) steps to a problem. Ask:
- *Have you found the final answer?*

Children may interpret the question incorrectly and confuse the operations. Ask:
- *Can you represent the problem with a bar model? Should you definitely add/subtract/multiply/divide?*

STRENGTHENING UNDERSTANDING

Use base 10 equipment or place value counters and encourage children to use the question language, and the differences and similarities in the bar models, to break down the steps needed to solve each problem.

GOING DEEPER

Challenge children to create their own multi-step problems involving bar models or to complete a 'reverse' question using a completed bar model where they devise a multi-step problem to match it.

KEY LANGUAGE

In lesson: multiplication, division, subtraction, addition

Other language to be used by the teacher: steps, multi-step, number sentence, partition

STRUCTURES AND REPRESENTATIONS

Bar model, column multiplication

RESOURCES

Mandatory: base 10 equipment, place value counters

 In the eTextbook of this lesson, you will find interactive links to a selection of teaching tools.

Quick recap

Ask children how to work out 2-digit by 1-digit divisions such as 48 ÷ 2 and 72 ÷ 3. Children need to be able to use a written method, supported if necessary by place value equipment. A reminder of these methods will help them with the calculations they need to solve the word problems in this lesson.

Discover

WAYS OF WORKING Pair work

ASK

- Question 1 a): *What is the number sentence you need to use?*
- Question 1 a): *How many steps are needed to solve this question?*

IN FOCUS Question 1 a) is a multi-step problem where children will use their prior knowledge to do a multiplication and then a subtraction to solve the problem. This is an opportunity to pre-assess children's ability to break down word problems into the steps required and their comprehension of the vocabulary used in the questions.

PRACTICAL TIPS Children always find big bar models engaging. Use poster paper, and, as a class, create a giant bar model to represent question 1 a). A section could then be cut off to represent subtracting 10.

ANSWERS

Question 1 a): Kate has 72 marbles.
72 − 10 = 62
Zac has 62 marbles.

Question 1 b): There are 35 marbles in one bag and 27 marbles in the other.

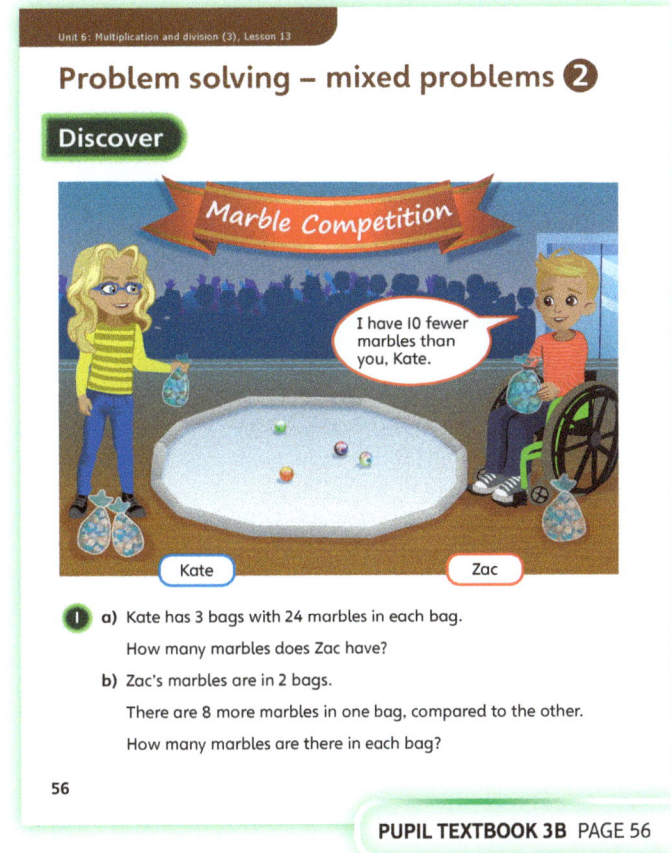

Share

WAYS OF WORKING Whole class teacher led

ASK

- Question 1 a): *What other ways can you use to set this question out?*
- Question 1 a): *How do the columns help with the multiplication?*
- Question 1 b): *How many steps are needed to solve this question? How do you know?*
- Question 1 b): *Could you use a bar model to help you understand the problem?*

IN FOCUS It will be important when focusing on these questions to ensure that children understand that there are two parts to this problem and to identify which part of the word problem tells them this. When solving question 1 b), encourage children to use the bar models to help them understand the question clearly while explaining that first the 8 must be subtracted, and then 54 can be divided by 2 to find one amount. Finally, add 8 back on to give both answers.

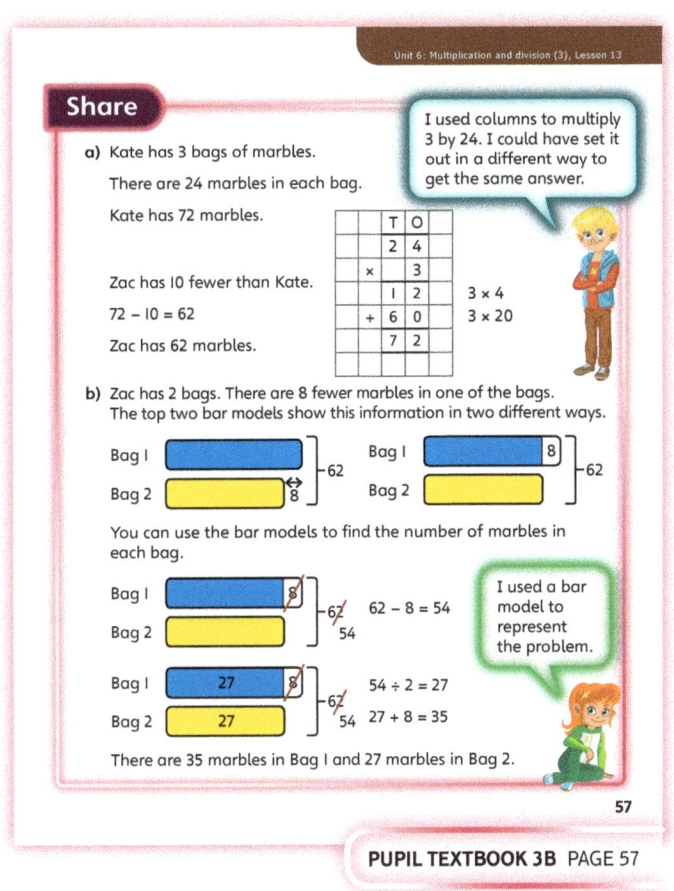

Think together

WAYS OF WORKING Whole class teacher led (I do, We do, You do)

ASK

- Question ❶: *How do the bar models help make the question clear?*
- Question ❶: *What mathematical operations do you need to do? How do you know?*
- Question ❷: *How do you know this is a multiplication and then a subtraction?*
- Question ❷: *What is the difference between a multiplication and a subtraction?*
- Question ❷: *Is there a more efficient way of working this out?*
- Question ❸: *What are the similarities between the two bar models? What is different?*
- Question ❸: *Would subtracting help with this question?*

IN FOCUS Question ❶ supports children in solving problems requiring a combination of operations – in this case, a multiplication and then an addition – while offering the pictorial support of a bar model. Encourage children to use the vocabulary in the question and to explain what it is asking them to do, giving their reasons.

STRENGTHEN In question ❸, children are required to find the cost of a shooter marble before they can find the cost of the tiger marble. Strengthen learning here by explaining the bar models and clarifying the steps needed to solve the problem. Encourage children to recognise the similarities in the bar models and use subtraction to find the cost of 1 shooter model. Some children will give 28 as their answer – remind them that they are trying to find the tiger marble.

DEEPEN Question ❸ deepens children's understanding of solving multi-step problems using bar models since children are required to think carefully about what each marble is worth, to find the solution with minimal prompting. Extend the task by challenging children to find how much 7 shooter marbles cost.

ASSESSMENT CHECKPOINT Question ❸ will give an opportunity to assess children's ability to solve multi-step mixed problems with minimal prompting using an intermediary, implicit step to find the solution using bar models for pictorial support. If children can solve this they are likely to have mastered this lesson.

ANSWERS

Question ❶: The total weight of the marbles is 83 grams.

Question ❷: 12 shooter marbles weigh 36 grams more than 12 tiger marbles.

Question ❸: A tiger marble costs 15p.

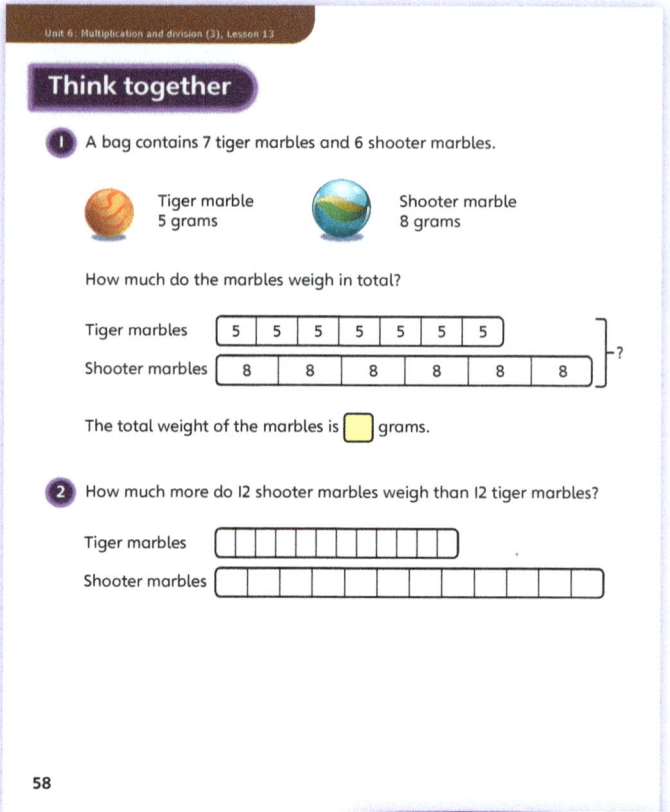

PUPIL TEXTBOOK 3B PAGE 58

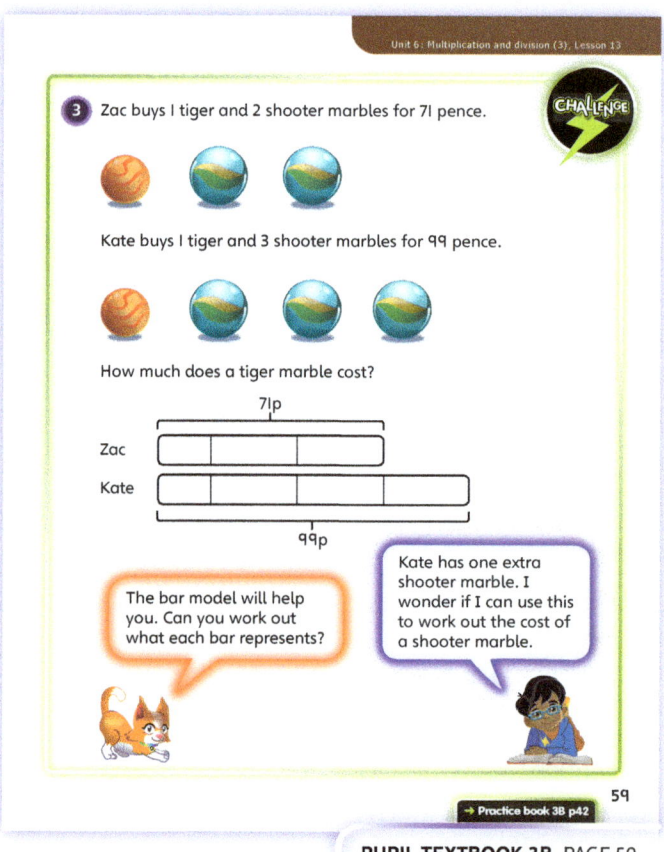

PUPIL TEXTBOOK 3B PAGE 59

Unit 6: Multiplication and division (3), Lesson 13

Practice

WAYS OF WORKING Independent thinking

IN FOCUS Question 3 requires children to solve a multi-step word problem with no pictorial support. Encourage children to break down the questions and translate them into number sentences to understand the mathematical operations required. Questions 4 and 5 develop children's use of bar models to find solutions to missing number problems. It will be beneficial to encourage children to represent each number sentence as a bar model and to develop their fluency with using this important pictorial representation.

STRENGTHEN Key to understanding this topic is the use of bar models; encourage children to use bar models to help break down the questions. In particular with question 5, encourage children to make a bar model for each number sentence. This will support and strengthen their understanding.

DEEPEN Question 5 is an excellent opportunity to deepen learning in this lesson and develop links between multiplication and repeated addition. Encourage children to spot patterns, for example, **4** × 3 + **5** × 3 = **9** × 3, because 4 + 5 = 9.

ASSESSMENT CHECKPOINT Question 6 will assess children's understanding of solving multi-step mixed problems with minimal prompting. Look for children confidently using bar models to translate the word problem into number sentences and using implicit intermediary steps to find the solution.

ANSWERS Answers for the **Practice** part of the lesson can be found in the *Power Maths* online subscription.

Reflect

WAYS OF WORKING Independent thinking

IN FOCUS This question is a good way to end the lesson, bringing with it an opportunity for children to reflect on the strategies they have used and to be able to apply them here. Children should realise that they can find 6 × 5 and 6 × 3, and then add the answers together, or they may add the 5 and 3 and then multiply by 6. Encourage children to discuss the different strategies, focusing particularly on which is the more efficient method.

ASSESSMENT CHECKPOINT Children should be able to find both solutions and explain their reasoning. Observe if children need a bar model to support them.

ANSWERS Answers for the **Reflect** part of the lesson can be found in the *Power Maths* online subscription.

After the lesson

- Did children rely on bar models to help them understand a question?
- Can children solve single-step problems?
- Can children solve multi-step problems?

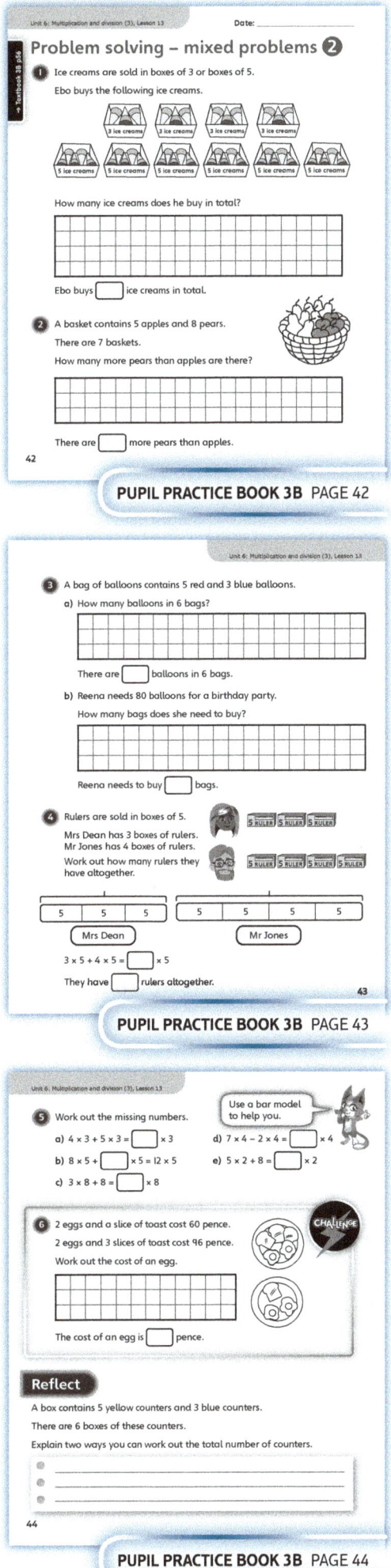

97

Unit 6: Multiplication and division (3)

End of unit check

Don't forget the unit assessment grid in your *Power Maths* online subscription.

WAYS OF WORKING Group work adult led

IN FOCUS

This end of unit check will give an opportunity to evaluate children's understanding of 2-digit multiplication and division and their ability to apply this knowledge to solve multi-step mixed problems.

- Questions ❶ to ❹ offer pictorial support to help children link a number sentence with a pictorial representation. Children should be able to work confidently with a ten frame (question ❶), base 10 equipment (questions ❷ and ❹) and a place value grid (question ❸).
- Look carefully at the answer that is given for question ❼ – this gives ample opportunity to assess children's ability to solve a multi-step mixed problem confidently.

ANSWERS AND COMMENTARY

Children who have mastered this unit will be secure with comparing multiplication and division statements. They should be able to find related multiplication and division facts by using the expanded method to multiply and by partitioning numbers and then dividing. They will also be able to work with remainders. They can also apply their knowledge of bar models to solve mixed problems.

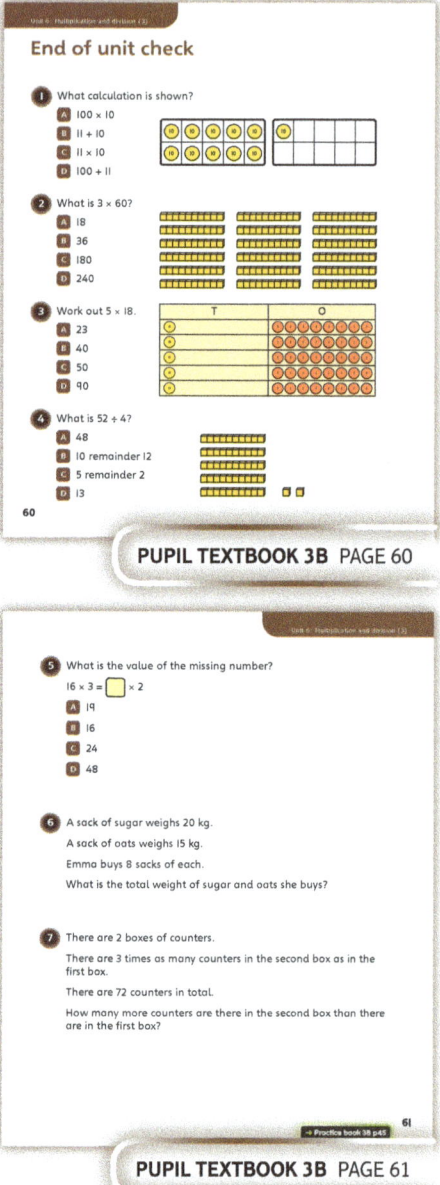

PUPIL TEXTBOOK 3B PAGE 60

PUPIL TEXTBOOK 3B PAGE 61

Q	A	WRONG ANSWERS AND MISCONCEPTIONS	STRENGTHENING UNDERSTANDING
1	C	Any other answer suggests that children need to revise their understanding of how to use place value grids and the value of place value counters.	• Support children with understanding and using the <, > and = signs. • Run intervention sessions in which children practise multiplying 1-digit numbers by multiples of 10, for example, 4 × 5 and 4 × 50. • Throughout the unit, revise the 2, 3, 4, 5 and 10 times-tables (including related multiplication and division facts). • Do some problem solving that helps revise the expanded method of multiplication, and partitioning to divide (including questions with remainders). • Encourage children to explore different ways of solving combined multiplication problems, making a poster to display. • Encourage children to use bar models to solve multi-step mixed problems.
2	C	A indicates that children have worked out the calculation 3 × 6. D suggests that children need to revise their 3 times-table.	
3	D	B and C suggest that children have worked out either 5 × 10 or 5 × 8, but not added the two amounts together.	
4	D	Any other answer suggests children need to work on dividing 2-digits by 1-digit.	
5	C	D suggests that children found the answer for 16 × 3 correctly, but did not apply it to the other side of the equation.	
6	280 kg	Look for children who have added the amounts together.	
7	36 counters	Look for children drawing a bar model to help them work out the answer. If they solve this question successfully they are likely to have mastered the unit.	

98

Unit 6: Multiplication and division (3)

My journal

WAYS OF WORKING Independent thinking

ANSWERS AND COMMENTARY

- Question **1**: First, children should draw a diagram. Then they should write instructions in steps.
- Question **2**: The last digits of the answers in the second column are the same as in the first.

Children can predict what the last digits will be by multiplying the 1s.

Power check

WAYS OF WORKING Independent thinking

ASK

- What new concepts have you learnt in this unit?
- How would you teach someone how to multiply 2-digit numbers by 1-digit numbers?
- Why is partitioning useful when dividing?
- Can you write down the new words you have learnt and what they mean?

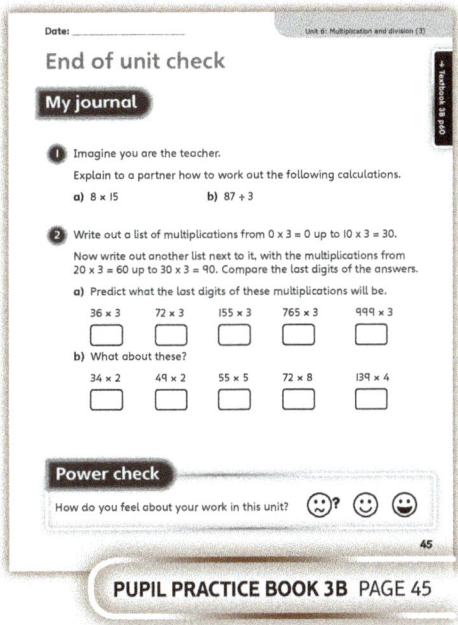

PUPIL PRACTICE BOOK 3B PAGE 45

Power puzzle

WAYS OF WORKING Pair work

IN FOCUS This **Power puzzle** gives an opportunity to assess children's multiplication, division and problem solving skills. Encourage children to explain the methods and strategies used by asking:
- Why did you use that digit?
- What digits did you need to use to get a 3-digit number as your answer?

ANSWERS AND COMMENTARY The **Power puzzle** will assess children's ability to use their knowledge of the strategies introduced in this unit. Listen to the explanations and note any children who may need further strengthening.

a) $30 \times 6 = 180$ or $60 \times 3 = 180$

b) $9 \times 4 + 6 \times 4 = 15 \times 4$

c) Here is one solution:

	H	T	O
		3	2
×			4
			8
+	1	2	0
	1	2	8

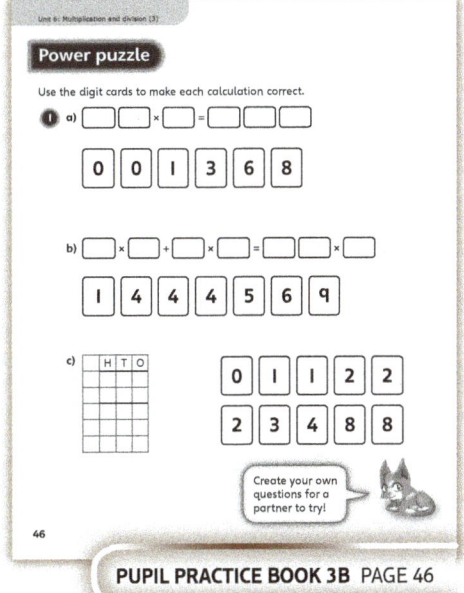

PUPIL PRACTICE BOOK 3B PAGE 46

After the unit

- Did all children understand the key vocabulary of the unit?
- What will you do to support children who struggled during the end of unit check?

Strengthen and **Deepen** activities for this unit can be found in the *Power Maths* online subscription.

Unit 7
Length and perimeter

Mastery Expert tip! 'When teaching this unit, I made sure children had access to as many different measuring tools as I could provide them with. This varied approach really helped to develop their fluency and confidence when measuring in context!'

Don't forget to watch the Unit 7 video!

WHY THIS UNIT IS IMPORTANT

This unit focuses on measurement in millimetres (mm), centimetres (cm) and metres (m). Children will learn how these units of measurement relate to one another and convert between single and mixed units. They will apply their knowledge of number to compare, order, add and subtract measurements of length and calculate the perimeter of 2D shapes.

WHERE THIS UNIT FITS

→ Unit 6: Multiplication and division (3)
→ **Unit 7: Length and perimeter**
→ Unit 8: Fractions (1)

This unit builds on previous units on number, in particular Unit 1: Place value within 1,000 and Units 2 and 3: Addition and subtraction (1) and (2). Children will apply their knowledge of number in the context of length by measuring, comparing and ordering. They will also add, subtract, multiply and divide measurements of length as well as calculating perimeters of 2D shapes. Children will have the opportunity to transfer these skills to other forms of measurement in Unit 9: Mass and Unit 10: Capacity.

Before they start this unit, it is expected that children:
- can count reliably in steps of 1, 2, 5 and 10
- know number bonds to 100 for multiples of 10
- can carry out addition and subtraction for 2- and 3-digit numbers
- can compare and order 2- and 3-digit numbers.

ASSESSING MASTERY

Children who have mastered this unit will be able to confidently measure length in millimetres (mm), centimetres (cm) and metres (m). They will be able to convert between single units and mixed units and apply this when solving measurement problems involving addition, subtraction and finding the perimeter of 2D shapes.

COMMON MISCONCEPTIONS	STRENGTHENING UNDERSTANDING	GOING DEEPER
When measuring length, children may not start at '0' when using measuring equipment. Instead, they may place the object at the start of the ruler or at the '1' division.	Provide plenty of opportunities for practical activities involving measuring. Make the comparison of the scale on a ruler to a number line clear to children by carrying out simple calculations using the scale alongside a number line.	Ask children to use a variety of measuring tools such as tape measures, various rulers, metre sticks and trundle wheels. Ask children to identify the similarities and differences between them.
When converting units or when calculating with units, children may make errors if they are not secure in number bonds to 100 for multiples of 10.	Practise number bonds to 10 alongside number bonds to 100 with multiples of 10, for example, 6 + 4 and 60 + 40. This will enable children to more readily identify the link between the two.	

Unit 7: Length and perimeter

UNIT STARTER PAGES

Use these pages to introduce the focus to children. You can use the characters to explore different ways of working, too.

STRUCTURES AND REPRESENTATIONS

Number line: Number lines will be used to help children make the link between scales, measurement and counting.

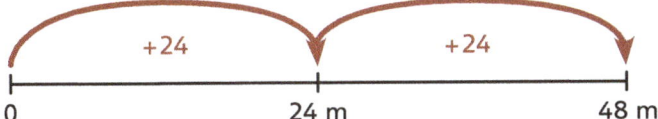

Rulers: Rulers of varying scales and sizes will be used across the unit to develop children's ability to measure accurately.

Column method for addition and subtraction: When adding and subtracting lengths, children should be encouraged to use column methods.

H	T	O
	2	5
	5	0
+ 3	0	0
	7	5

Bar model: This model will be used to support the representation of children's calculations and problem solving.

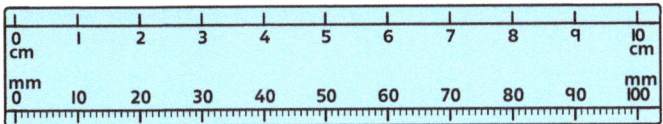

KEY LANGUAGE

There is some key language that children will need to know as part of the learning in this unit:

- millimetres (mm), centimetres (cm), metres (m)
- measure, measurement, unit of measurement
- length, height, width, distance, diagonal
- how long?, how wide?, how tall?, how high?
- ruler, metre stick, metre ruler
- longer, shorter, longest, shortest, furthest
- perimeter
- add, subtract, find the difference, repeated addition, multiply
- greater than (>), less than (<)
- polygon, quadrilateral, triangle, rectangle
- compare, convert, equal, equivalent, ascending, predict, calculate, expression, method

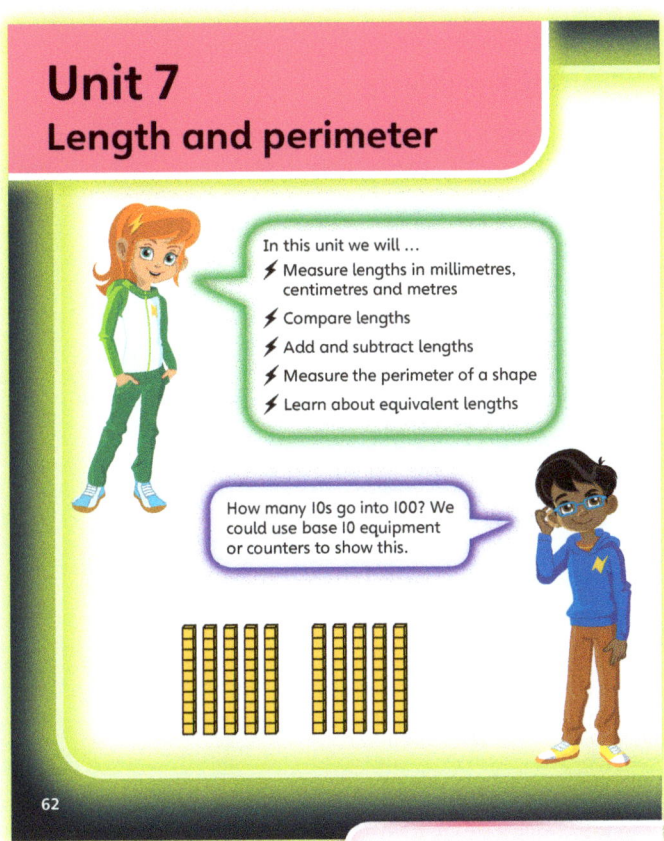

PUPIL TEXTBOOK 3B PAGE 62

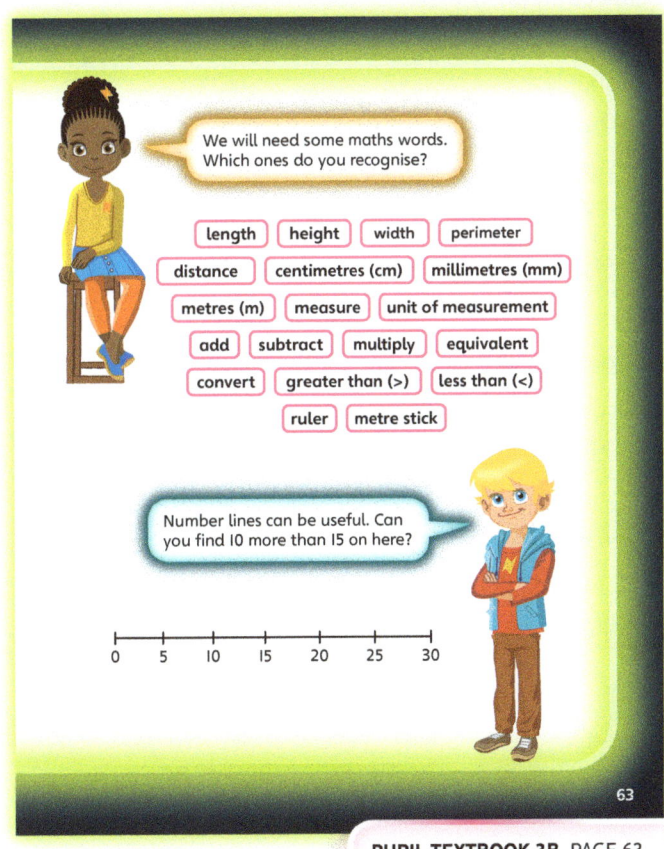

PUPIL TEXTBOOK 3B PAGE 63

Unit 7: Length and perimeter, Lesson 1

Measure in m and cm

Learning focus
In this lesson, children will measure using mixed units for the first time. They will accurately measure and record length using a combination of metres and centimetres.

Before you teach
- Do children understand that length can be recorded using different units, such as metres and centimetres?
- Do children understand that a metre is a larger unit of measure than a centimetre?

NATIONAL CURRICULUM LINKS

Year 3 Measurement

Measure, compare, add and subtract: lengths (m/cm/mm); mass (kg/g); volume/capacity (l/ml).

ASSESSING MASTERY

Children can accurately measure lengths that are over 1 metre, recording their answers using mixed units of metres and centimetres (for example, 1 m 30 cm).

COMMON MISCONCEPTIONS

Children often do not measure from the '0' point of a ruler and instead measure from the end of a ruler. Ask:
- *Where should you actually start measuring from? Where is '0'?*

When using more than one ruler, children often place them end-to-end, rather than lining up the '0' and end points of each ruler. Explore with children the difference between measuring by lining them up end-to-end and lining up the correct start and end points of each ruler. Ask:
- *Which method is more accurate?*

STRENGTHENING UNDERSTANDING

Ensure children are given a range of practical experiences of measuring, using both metres and centimetres. Use a range of representations, including comparing a metre stick to a number line.

GOING DEEPER

Encourage children to explore how they can efficiently and accurately measure a range of lengths, including those that involve lines which are not straight or where they need to use the same ruler more than once.

KEY LANGUAGE

In lesson: length, height, width, metres (m), centimetres (cm), **metre stick**, ruler

Other language to be used by the teacher: distance

STRUCTURES AND REPRESENTATIONS

Number lines

RESOURCES

Mandatory: metre rulers, centimetre rulers, paper to make paper aeroplanes

Optional: feathers

 In the eTextbook of this lesson, you will find interactive links to a selection of teaching tools.

Quick recap

Check that children have seen a metre ruler before. Ask them to use a metre ruler to measure 3 metres on the ground in the classroom. Discuss what the metre ruler shows (for example, measurements every 10 cm).

Discover

WAYS OF WORKING Pair work

ASK

- Question 1 a): *Look at the picture. What might the question be about?*
- Question 1 a): *How can we work out how far Zac's plane has gone? Is it over 2 metres? Is it under 3 metres?*
- Question 1 a): *How can we accurately record the distance Zac's plane has travelled?*
- Question 1 b): *Whose plane has travelled further: Bella's or Emma's?*

IN FOCUS Questions 1 a) and b) introduce children to measuring length using a combination of metres and centimetres for the first time. Whilst this is represented pictorially in the image, children will benefit from practical experiences of measuring using mixed units.

PRACTICAL TIPS Set up a similar scenario outside or in the school hall. Lay out a tape measure or several metre sticks. Ask children to make paper aeroplanes (or other flying machines) and then to take it in turns to throw them and see how far they can fly. Ask children to provide their measurements in mixed units of m and cm. If working on a smaller scale, lay out two metre sticks on the classroom floor and ask children to blow a piece of paper or a feather and measure how far it travels (measure in line with its furthest travelled point).

ANSWERS

Question 1 a): Zac's paper aeroplane has travelled 2 m 50 cm.

Question 1 b): Bella's paper aeroplane has travelled 4 m 30 cm.

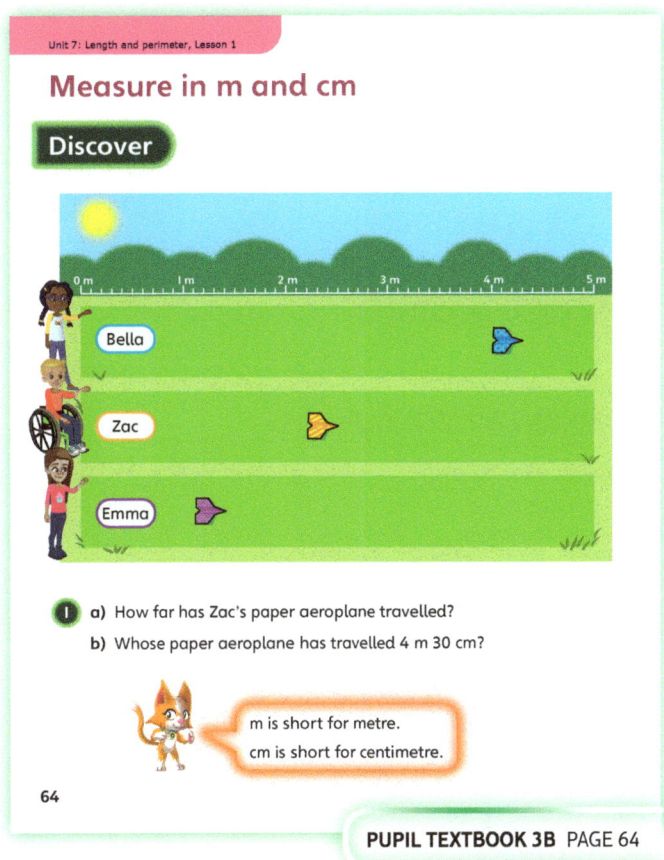

PUPIL TEXTBOOK 3B PAGE 64

Share

WAYS OF WORKING Whole class teacher led

ASK

- Question 1 a): *How did you find an answer to this question? What approaches did you use? Did anyone do anything differently?*
- Question 1 a): *How can you give an exact measurement if the distance is over 2 metres but under 3 metres? Can you use both metres and centimetres together?*
- Question 1 a): *How did you make sure your measurements were accurate?*
- Question 1 b): *Is 4 m 30 cm a larger or smaller length than 2 m 50 cm?*

IN FOCUS In questions 1 a) and b), children are introduced to using mixed measurements in order to accurately record measurements that fall between two larger units; for example, by using metres and centimetres to record a measurement that is between two whole-metre values. Children should consider whether saying that a measurement is between two whole units is always accurate enough.

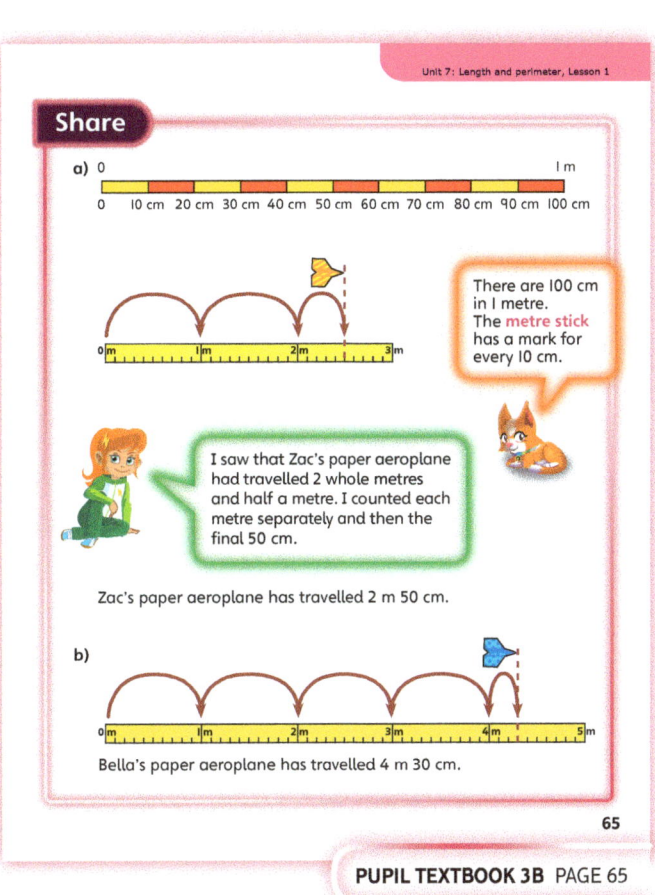

PUPIL TEXTBOOK 3B PAGE 65

Think together

WAYS OF WORKING Whole class teacher led (I do, We do, You do)

ASK

- Question ① a): *How can you work out how far the plain blue ball has rolled? How many whole metres has it travelled? How many centimetres further than the striped ball has it travelled?*
- Question ②: *How can you make sure your measurements are accurate? Do you need to use both metre rulers and 30 cm rulers?*

IN FOCUS Question ② involves children making real-life measurements in metres and centimetres. Depending on their tables, the width and height may be under 1 metre and, therefore, this presents a good opportunity to discuss how children can record measurements under 1 metre. For example, 90 cm should be recorded as 0 m 90 cm. You may want pairs to take turns making only one of the measurements, to ensure a smoother classroom activity.

STRENGTHEN Children who struggle to visualise the measurements of metres when they use the metre stick representation in questions ① a) and b) will benefit from similar situations being represented practically in the classroom and having the opportunity to physically measure for themselves.

If children struggle with measuring accurately in question ②, ensure that you explicitly model measuring from the start point (not the end) of the metre ruler.

DEEPEN Children should begin to explore how they can make their measurements efficient and practical. Question ③ provides a good opportunity to discuss this. Encourage children to consider whether they can measure 3 m 7 cm using a single metre ruler.

ASSESSMENT CHECKPOINT Use question ② to assess whether children can accurately measure using a combination of both metres and centimetres.

ANSWERS

Question ① a): The first ball has rolled 1 m 30 cm. The second ball has rolled 2 m. The third ball has rolled 2 m 75 cm.

Question ① b): Max is 1 m 35 cm tall.

Question ②: Children should measure their own table and write the length, width and height in cm.

Question ③: Children should explain using a metre stick to measure three lengths of 1 m and one length of 7 cm. They should describe lining the metre stick up at 0 cm each time and marking the string to show 1 m each time.

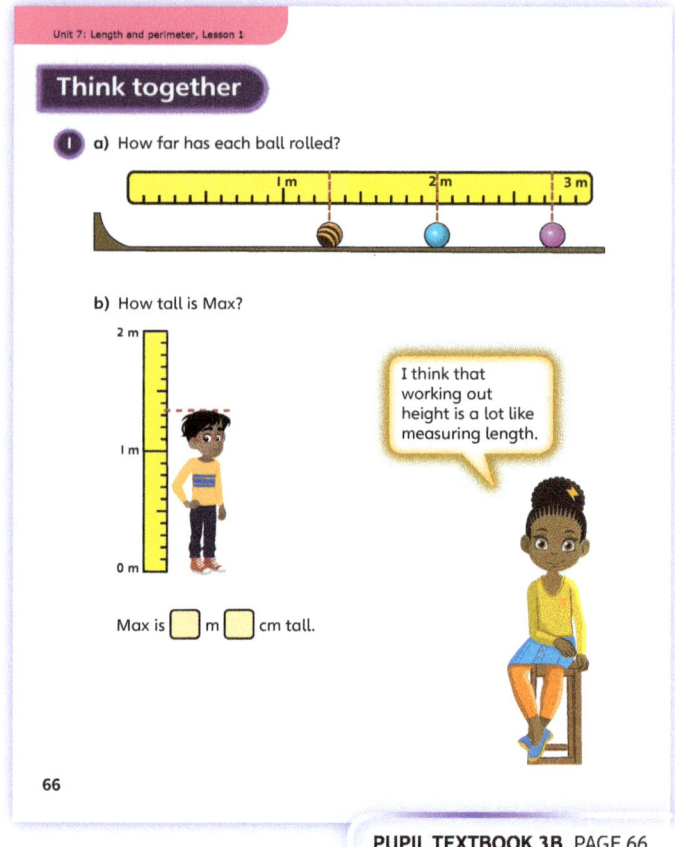

PUPIL TEXTBOOK 3B PAGE 66

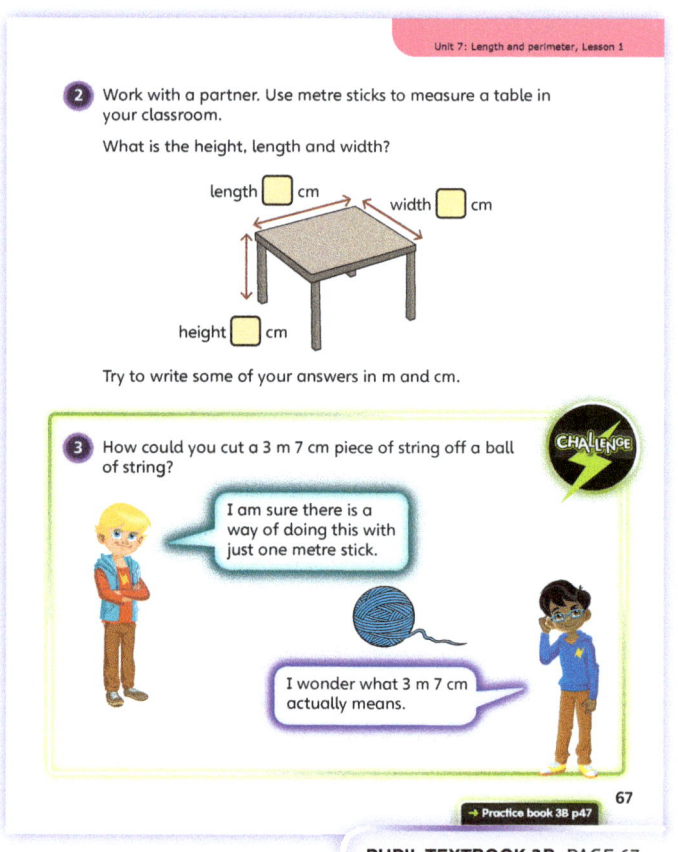

PUPIL TEXTBOOK 3B PAGE 67

Unit 7: Length and perimeter, Lesson 1

Practice

WAYS OF WORKING Independent thinking

IN FOCUS Question 5 encourages children to use their estimation skills to initially identify items that they believe will fulfil the criteria given before going on to accurately measure them.

STRENGTHEN Ensure children have practical experience of measuring length and that the importance of accurate measurements, starting from the '0' point of the ruler, is explicitly modelled. This could be done by comparing two possible measurements: one where the '0' point is used and one where the ends of the rulers are used.

DEEPEN Encourage children to explore different methods they could use to measure the length of something. Question 6 provides an opportunity for children to explore how to measure a line that is not straight. Challenge children to draw their own uneven lines, either on paper or with chalk outside, estimate the line length, and then check their estimates using accurate measuring. They might, for example, use a piece of string to trace the path of the line and then hold the string taut alongside a tape measure or metre stick.

THINK DIFFERENTLY When measuring using multiple rulers, children often place the rulers end to end, without accounting for the 'dead' space that is commonly found at the start and end of a ruler. Question 3 exposes this common error and should be used to help children explore the importance of measuring from the '0' point of each ruler.

ASSESSMENT CHECKPOINT Use questions 2 and 4 to assess whether children are able to accurately measure a range of lengths. Pay particular attention to their ability to correctly record the lengths using mixed units, especially where the length is under 1 metre.

ANSWERS Answers for the **Practice** part of the lesson can be found in the *Power Maths* online subscription.

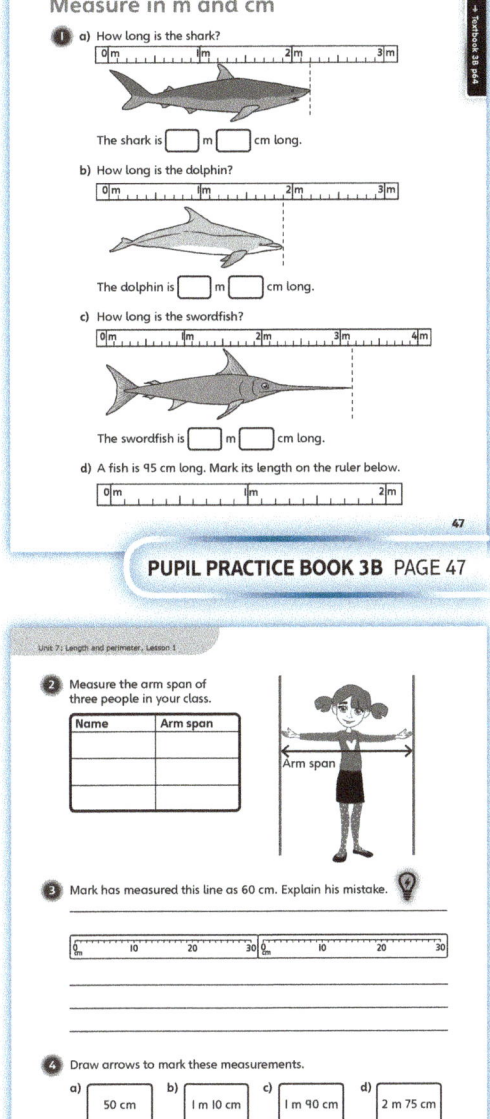

PUPIL PRACTICE BOOK 3B PAGE 47

PUPIL PRACTICE BOOK 3B PAGE 48

Reflect

WAYS OF WORKING Independent thinking

IN FOCUS This **Reflect** question focuses on children's ability to explain their learning from this lesson and whether they can use mixed units to measure and record length and height.

ASSESSMENT CHECKPOINT Use this question to assess if children can explain how to measure using mixed units, including the importance of accurately measuring from the '0' point, when using more than one ruler.

ANSWERS Answers for the **Reflect** part of the lesson can be found in the *Power Maths* online subscription.

After the lesson

- What practical experiences can you give children in measuring and comparing lengths using mixed units?
- Can children now measure accurately?
- Were children comfortable when measuring vertically (for example, measuring height) rather than horizontally (such as measuring length)? Can they see that the same processes apply?

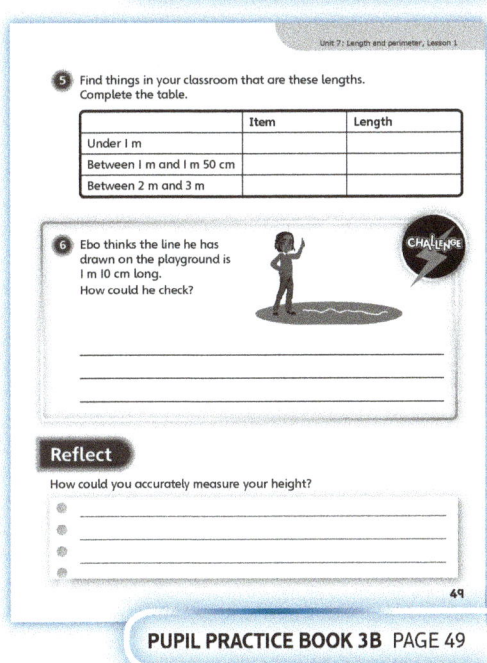

PUPIL PRACTICE BOOK 3B PAGE 49

105

Unit 7: Length and perimeter, Lesson 2

Measure in cm and mm

Learning focus
In this lesson, children will measure accurately, using millimetres and centimetres. They will use a ruler accurately to measure different objects.

Before you teach
- How accurate were children in their use of measuring tools in the previous lesson?
- Were there any misconceptions or weaknesses in children's use of measuring tools that need revisiting?

NATIONAL CURRICULUM LINKS
Year 3 Measurement
Measure, compare, add and subtract: lengths (m/cm/mm); mass (kg/g); volume/capacity (l/ml).

ASSESSING MASTERY
Children can accurately measure a given object using centimetres, millimetres or a combination of both. They can determine which unit of measure is appropriate, explain why and can consistently use a ruler accurately.

COMMON MISCONCEPTIONS
When using a ruler, children may begin measuring from the start of the ruler and not from '0' on the scale. Ask:
- *Where can you see '0' on the ruler? Where should you begin measuring from? Why?*

STRENGTHENING UNDERSTANDING
Provide practical opportunities for children to physically measure objects around them using a ruler. Using a dry wipe marker, highlight the 0 cm mark to remind children where they need to start measuring from.

GOING DEEPER
Ask children to use a 30 cm ruler to measure objects that are longer than the actual ruler. Ask: *How will you use the ruler to measure the object?*

KEY LANGUAGE
In lesson: ruler, centimetres (cm), **millimetres (mm)**, metres (m), measure, measurement, length, width, accurately, exactly

RESOURCES
Mandatory: rulers measuring in cm and mm, metre rulers
Optional: collections of easily measurable objects

 In the eTextbook of this lesson, you will find interactive links to a selection of teaching tools.

Quick recap
Check that children can use a ruler to draw lines in m and cm. For example, ask children to draw a line that is 1 m long and a line that is 1 m and 20 cm long. Their partner should then check that the measurements are accurate. It is important children understand how to measure in mixed units.

Unit 7: Length and perimeter, Lesson 2

Discover

WAYS OF WORKING Pair work

ASK

- Question 1 a): *What is being used to measure the creatures?*
- Question 1 a): *How long is the centipede?*
- Question 1 a): *Is there more than one way of describing the length of the centipede?*
- Question 1 b): *How many mm are there in 1 cm?*

IN FOCUS Discuss with children what they notice about the creatures in the picture. In question 1 a), discuss how the two creatures have been measured, focusing on the appropriate use of mm or cm depending on the size of the item being measured.

PRACTICAL TIPS Encourage children to use a ruler to investigate the length of objects around the classroom. To help assess children's current ability, it may be beneficial to give them objects that have been measured prior to the lesson.

ANSWERS

Question 1 a): The centipede is 5 cm long.
　　　　　　　The ant is 7 mm long.

Question 1 b): The worm is 27 mm long.

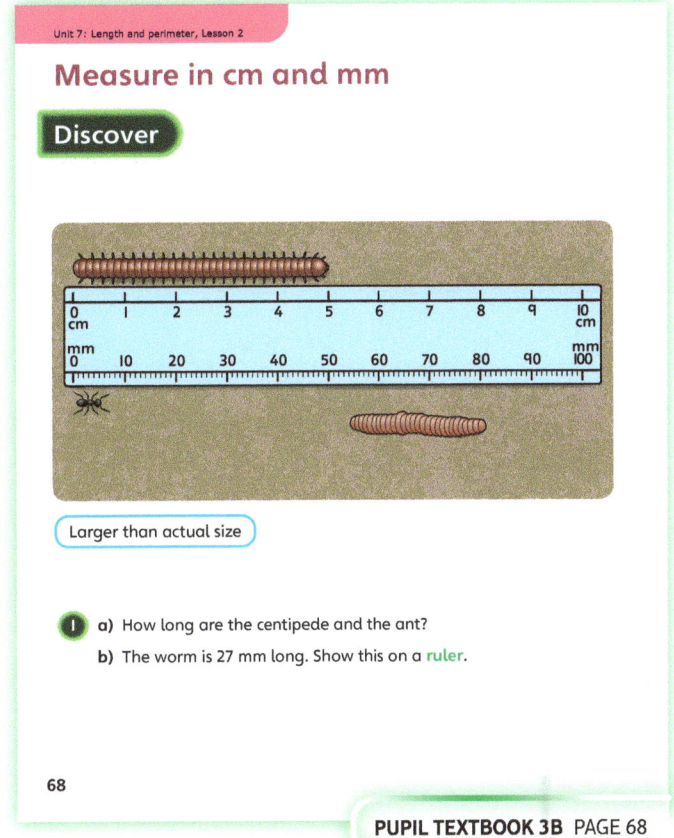

PUPIL TEXTBOOK 3B PAGE 68

Share

WAYS OF WORKING Whole class teacher led

ASK

- Question 1 a): *Can you show me how you would measure the ant using a ruler?*
- Question 1 a): *Could you use mm to measure the centipede?*
- Question 1 a): *Can cm be used to measure the ant? Explain your ideas.*

IN FOCUS The focus of question 1 b) is for children to realise that there are 10 mm in 1 cm and that the cm division on a ruler usually shows the mm divisions too. Encourage children to draw a line without a ruler that is 1 cm long and a line that is 1 mm long. Extend this activity by giving children a number of different lengths, some of which are in cm and some in mm. Can they draw lines of approximately that length? Children could check their estimates using a ruler.

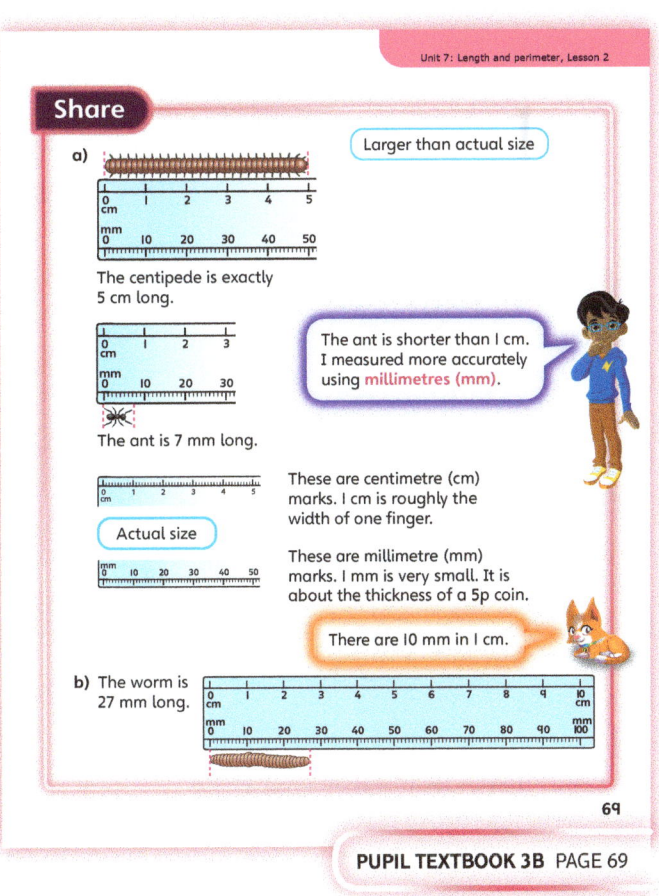

PUPIL TEXTBOOK 3B PAGE 69

107

Unit 7: Length and perimeter, Lesson 2

Think together

WAYS OF WORKING Whole class teacher led (I do, We do, You do)

ASK

- Questions ① and ②: *What unit of measure do the rulers show?*
- Questions ① a) and b): *Are the creatures lined up correctly with the ruler? How do you know?*
- Question ②: *What do you need to remember in order to measure accurately with a ruler?*

IN FOCUS Question ③ requires children to use a ruler to accurately measure the lines. Children will need to remember that the '0' mark on the ruler should be aligned with the beginning of the line as well as which unit of measure they are being asked to measure in.

STRENGTHEN Look out for any children who start measuring from the beginning of the ruler itself, not the '0' mark. Ask: *Where would you normally begin counting on a number line? Can you see this number on the ruler? Where should you begin measuring from?*

DEEPEN Question ② deepens children's awareness and understanding of how to use a ruler accurately by requiring them to spot the misconception in Max's measurements. Extend children's reasoning in the second part of the question by asking: *Where would the line have started if Max had measured 28 mm? What about if he measured 3 cm?*

ASSESSMENT CHECKPOINT At this point in the lesson, children should be able to explain how to accurately use a ruler, recognising the importance of beginning measuring at '0' on the ruler. They should be confident when measuring in cm or mm and more confident at measuring using mixed units of measure.

ANSWERS

Question ① a): The snail is 13 mm long.

Question ① b): The grasshopper is 30 mm long.
The ladybird is 6 mm long.

Question ① c): The children should draw lines that are 13 mm, 30 mm and 4 mm long.

Question ②: No. Max has not aligned his line with the 0 mm mark. He started at 2 mm. To work out the actual length of the line, subtract 2 mm from the measured length.
27 − 2 = 25 mm
Max's line is 25 mm long.

Question ③ a): 26 mm

Question ③ b): 40 mm

Question ③ c): 45 mm

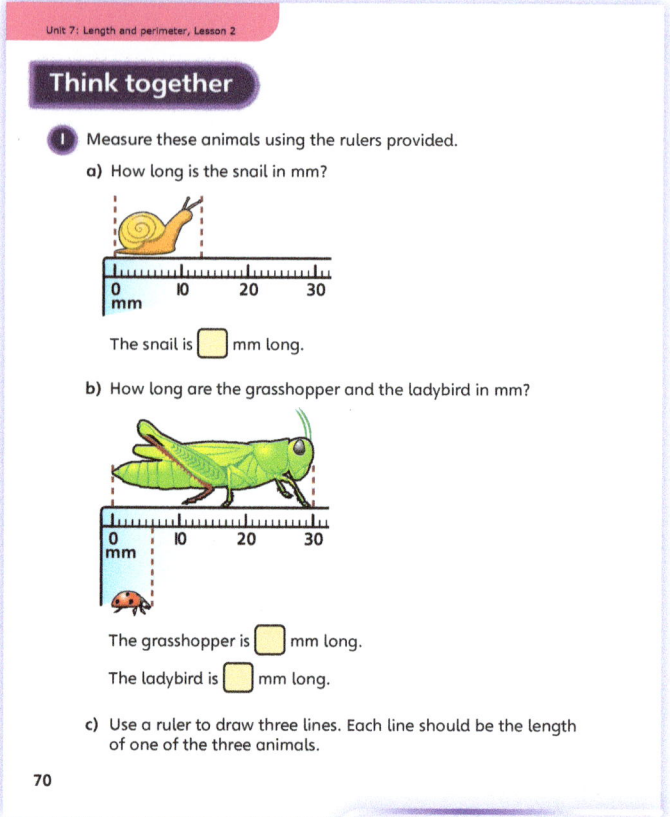

PUPIL TEXTBOOK 3B PAGE 70

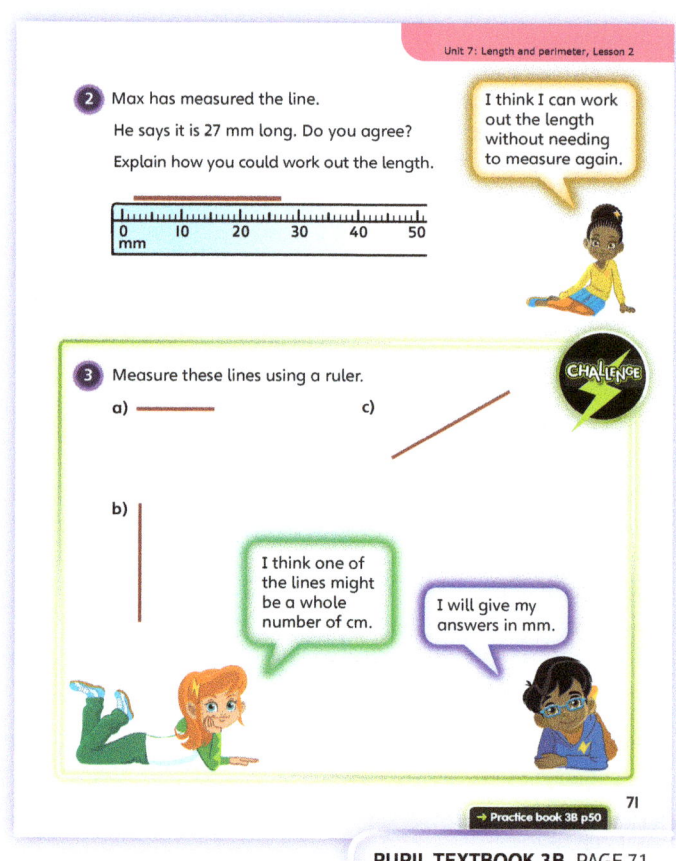

PUPIL TEXTBOOK 3B PAGE 71

Unit 7: Length and perimeter, Lesson 2

Practice

WAYS OF WORKING Independent thinking

IN FOCUS Question 2 requires children to draw lines accurately using a ruler. They should begin their line from the '0' point on the ruler. Question 3 confirms that children are able to accurately use a ruler to measure lines, starting from the '0' point.

STRENGTHEN Provide children with many practical opportunities to measure objects in mm and cm. Reinforce how the object needs to be aligned with the ruler.

DEEPEN Question 5 offers children an opportunity to prove their understanding of the proportionality of each unit of measure. Look for children's reasoning when they give their ideas. Ask: *Why did you choose that unit of measure for the elephant/mouse? Why wasn't the other unit of measure appropriate? Can you prove your thinking using real world objects and measuring tools?*

ASSESSMENT CHECKPOINT At this point in the lesson, children should be measuring objects accurately and fluently, in both cm and mm, using a ruler. They should also be more confident when measuring in mixed units of measure. Finally, children should be able to choose an appropriate unit of measure to use when measuring objects and give their reasoning for their decision.

ANSWERS Answers for the **Practice** part of the lesson can be found in the *Power Maths* online subscription.

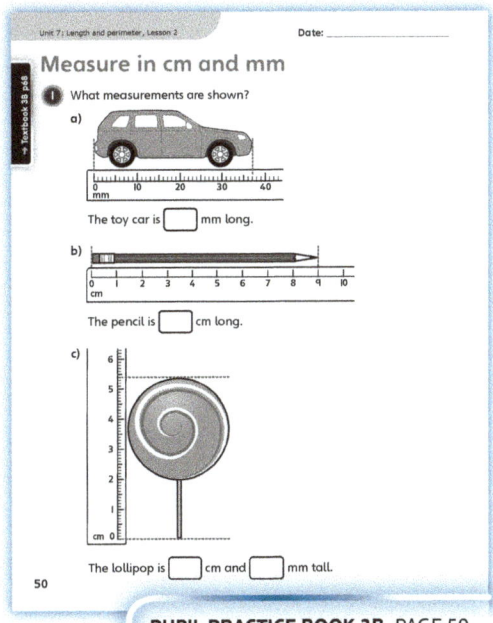

PUPIL PRACTICE BOOK 3B PAGE 50

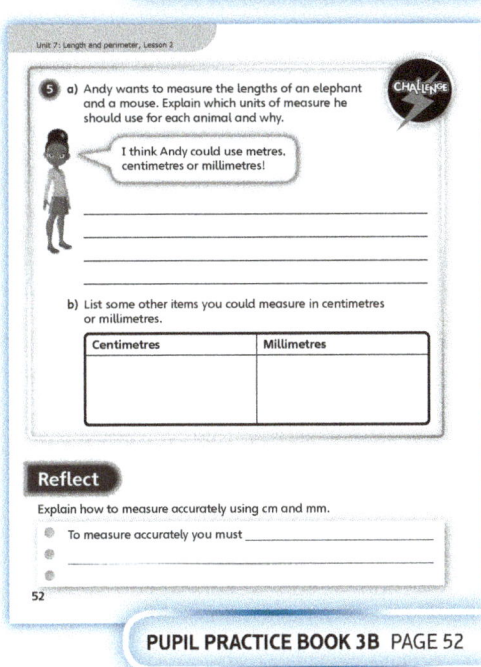

PUPIL PRACTICE BOOK 3B PAGE 51

Reflect

WAYS OF WORKING Independent thinking

IN FOCUS Give children a short amount of time to record their key points on how to measure accurately. Once children have recorded their ideas, ask them to share their ideas with a partner. Ask: *What learning points have you recorded? Have you both given the same advice? Has your partner remembered something you have forgotten?*

ASSESSMENT CHECKPOINT Children should be able to explain the key points of the lesson. Look for children mentioning:
- measuring from '0' on a ruler
- lining the ruler up carefully against the object
- measuring in centimetres and millimetres
- using the large divisions along a ruler to measure efficiently and accurately (1 cm, 2 cm or 10 mm, 20 mm).

ANSWERS Answers for the **Reflect** part of the lesson can be found in the *Power Maths* online subscription.

After the lesson

- Do children require more practical opportunities to measure in cm and mm?
- How could you include the concepts covered in this lesson in other areas of the curriculum, such as Design Technology or Science?
- Were children consistently able to measure accurately?

109

Unit 7: Length and perimeter, Lesson 3

Metres, centimetres and millimetres

Learning focus
In this lesson, children will use their knowledge from the previous lessons in this unit to convert between measurements in mm, cm and m. They will know how many mm are in 1 cm and how many cm are in 1 m. They will use these facts to convert between measurements.

Before you teach
- Are children able to measure a line in cm and mm?
- Can children count in 10s and 100s?
- Can children recognise simple bar models?

NATIONAL CURRICULUM LINKS

Year 3 Measurement
Measure, compare, add and subtract: lengths (m/cm/mm); mass (kg/g); volume/capacity (l/ml).

ASSESSING MASTERY

Children can convert between measurements in mm, cm and m using common/known facts.

COMMON MISCONCEPTIONS

Children can often use incorrect conversion facts. For example, children may think that there are 10 cm in 1 metre and so, when converting, they get an incorrect answer. Show children a metre ruler and 30 cm ruler to help them with the conversions. Ask:
- How many centimetres are in 1 metre? How many millimetres are in 1 centimetre? How can you use the rulers to help you?

Children may not understand that they need to multiply to do conversions. You may, at this stage, just want children to be able to count in 10s or 100s. Ask:
- What do you need to do to convert between cm and m? What do you need to do to convert between mm and cm?

STRENGTHENING UNDERSTANDING

Use different lengths of rulers, 15 cm, 30 cm and 1 m, to help children with their conversions. You may want to ask them to draw lines of a given length to help them convert. For example, to work out how many mm there are in a 7 cm line, you could ask them to draw the line and/or look at the measurements on a ruler.

GOING DEEPER

Ask children to do more complicated conversions and ask them to work out how many mm there are in 1 m. This relies on them converting first to cm and then to mm.

KEY LANGUAGE

In lesson: mm, cm, m, convert

STRUCTURES AND REPRESENTATIONS

Bar models, number lines

RESOURCES

Mandatory: metre rulers, 30 cm rulers, 15 cm rulers

Optional: number lines, a coin

 In the eTextbook of this lesson, you will find interactive links to a selection of teaching tools.

Quick recap
Check that children can count in 10s from 0 to 200 and count in 100s from 0 to 1,000. As a class, count on and back from these numbers.

110

Unit 7: Length and perimeter, Lesson 3

Discover

WAYS OF WORKING Pair work

ASK

- Question 1 a): *What can you see in the image? Do you think the playground has been measured in mm, cm or m? How do you know?*
- Question 1 b): *Do you know any of these answers? How can you use a ruler to help you?*

IN FOCUS This lesson brings together work on mm, cm and m. Question 1 a) encourages children to choose the correct unit of measure. Provide children with rulers to help them. Question 1 b) checks children's knowledge of known conversions. Encourage children to use a metre ruler to help them, if needed.

PRACTICAL TIPS It is useful to have rulers available. You may also want to go outside into the playground to show children that the playground needs to be measured in metres.

ANSWERS

Question 1 a): Ebo measured the playground in metres.

Question 1 b): There are 100 cm in 1 m.
There are 10 mm in 1 cm.

PUPIL TEXTBOOK 3B PAGE 72

Share

WAYS OF WORKING Whole class teacher led

ASK

- Question 1 a): *How do you know the playground is measured in m and not cm or mm?*
- Question 1 b): *How can you use a ruler to help you with the conversions? How do the bar models represent the conversions?*

IN FOCUS In question 1 a), discuss with the class what each of the three characters are saying. You might want to begin with Astrid: if 20 cm is about as long as a paint brush, Ebo's map cannot be measured in cm. Then, use what Ash says: if 20 mm is the smallest of them all, then 20 mm is even smaller than the width of the paint brush. Lastly, look at what Flo says: if 20 m is about 20 strides, then the playground must be measured in m. You may want to ask children to draw lines that are 20 mm and 20 cm long and show that this cannot be the length of the playground. You may also want to choose to go outside and measure 20 metres too to show the difference. In question 1 b), show children how they can use a ruler to show that 10 mm = 1 cm and that 100 cm = 1 m. Explain that the bar models help show the conversions and they will help when converting other measures later on.

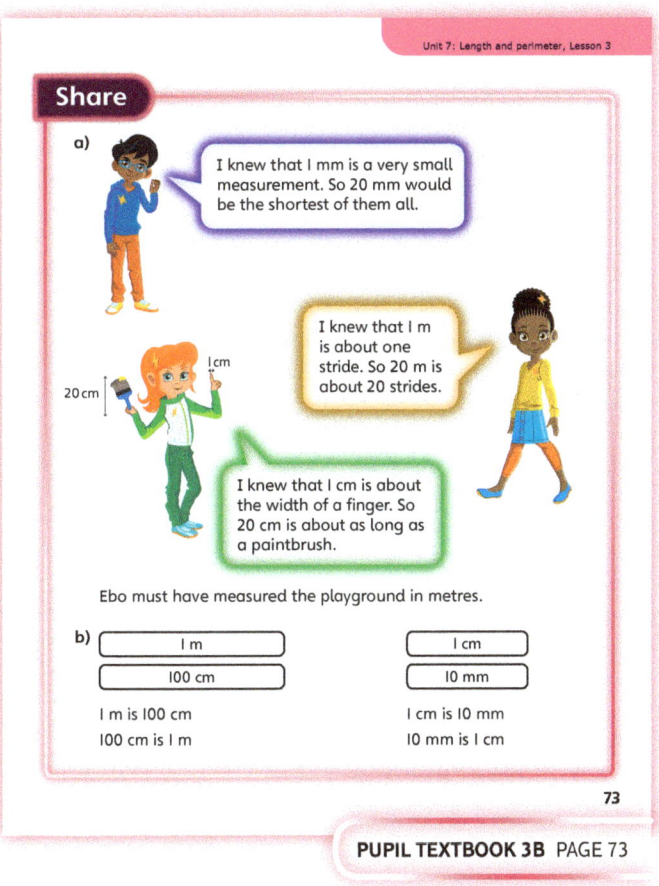

PUPIL TEXTBOOK 3B PAGE 73

111

Think together

WAYS OF WORKING Whole class teacher led (I do, We do, You do)

ASK

- Question ❶: *What units of measurement do you know? Which is the biggest unit of measurement? Which is the smallest? Which unit is the correct one for each of the lengths in the diagrams?*
- Question ❷: *What do the bar models show?*
- Question ❸: *How do you convert 1 m into cm? How can you then work out how many mm are in 1 m?*

IN FOCUS In question ❶, children are required to choose the correct unit of measure. They can do this by considering which of the mm, cm and m units of measure are the shortest and longest. Use practical objects, such as a coin, to help children see this. Question ❷ provides some bar models to help children make simple conversions from mm to cm and from cm to m. You may want to add some additional quick examples to convert both ways between these units of measure. For larger numbers, see if children can convert between units without having to draw bar models.

STRENGTHEN Use bar models to support children with their understanding of conversion. Encourage children to find the answers by counting in 10s and 100s. Children may need a number line to help then.

DEEPEN Ask children if they can measure the same object with different units. Ask: *Why might this not always be a good idea? For example, why is it better to say and use 8 m rather than 800 cm?*

ASSESSMENT CHECKPOINT Check that children can convert between cm and mm and cm and m using simple numbers.

ANSWERS

Question ❶: Book: 25 cm
Coin: 20 mm
Speed boat: 12 m

Question ❷ a): i) 1m = 100 cm   ii) 2m = 200 cm
iii) 3m = 300 cm

Question ❷ b): i) 1 cm = 10 mm   ii) 2 cm = 20 mm
iii) 3 cm = 30 mm   iv) 4 cm = 40 mm

Question ❸: There are 100 mm in 10 cm, so there will be 1,000 mm in 100 cm or 1 m.

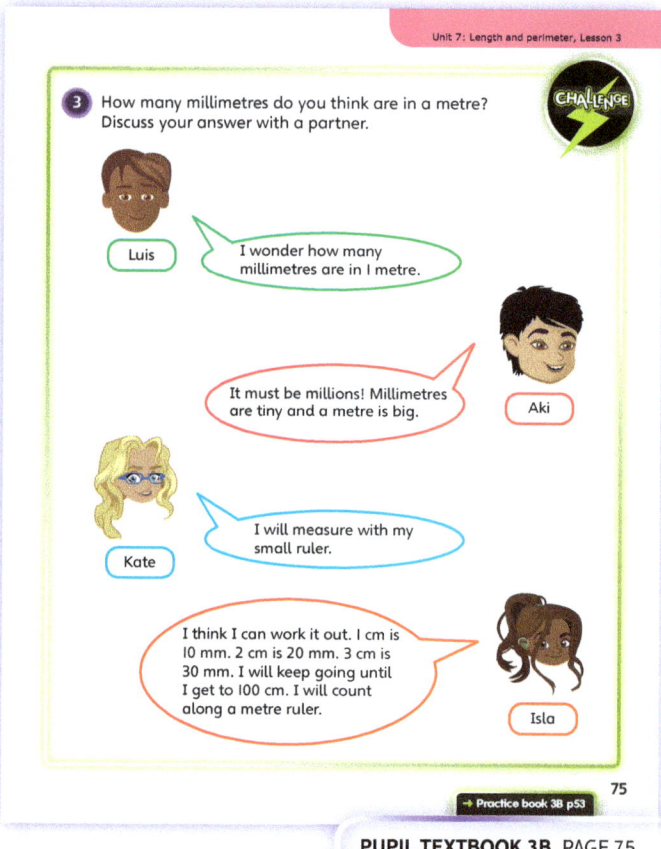

PUPIL TEXTBOOK 3B PAGE 74

PUPIL TEXTBOOK 3B PAGE 75

Unit 7: Length and perimeter, Lesson 3

Practice

WAYS OF WORKING Independent thinking

IN FOCUS In questions ❶ and ❷, children have to select the correct unit of measurement that has been used to measure common objects. In questions ❸ and ❹, children are asked to complete the bar models to help them work out the unit conversions. Children may just know the answers, use multiplication, or count in 10s and 100s to work out the answers. Look out for children using the correct conversion value, for example using 10 mm = 1 cm. Children can often get the conversions mixed up. Questions ❺ and ❻ ask children to complete conversions, this time without having bar models to support them. Children may draw their own bar models to help them.

STRENGTHEN Use different lengths of rulers to help children with the conversions. You may want to ask them to draw lines of a given length to help them convert. For example, to work out how many mm there are in a 7 cm line, you could ask them to draw the line and/or look at the measurements on a ruler.

DEEPEN In question ❻, children are required to fluently convert between mm, cm and m. They should know the conversion factors.

ASSESSMENT CHECKPOINT Questions ❸ and ❹ can be used to check that children can convert between cm and mm and cm and m.

ANSWERS Answers for the **Practice** part of the lesson can be found in the *Power Maths* online subscription.

Reflect

WAYS OF WORKING Whole class

IN FOCUS Ask children to think about objects that are 7 mm, 7 cm and 7 m long. Encourage them to use rulers to check their measurements. This should help children consolidate their knowledge of different units of measure and which ones to use.

ASSESSMENT CHECKPOINT Check that children can suggest different objects that can be measured in mm, cm and m.

ANSWERS Answers for the **Reflect** part of the lesson can be found in the *Power Maths* online subscription.

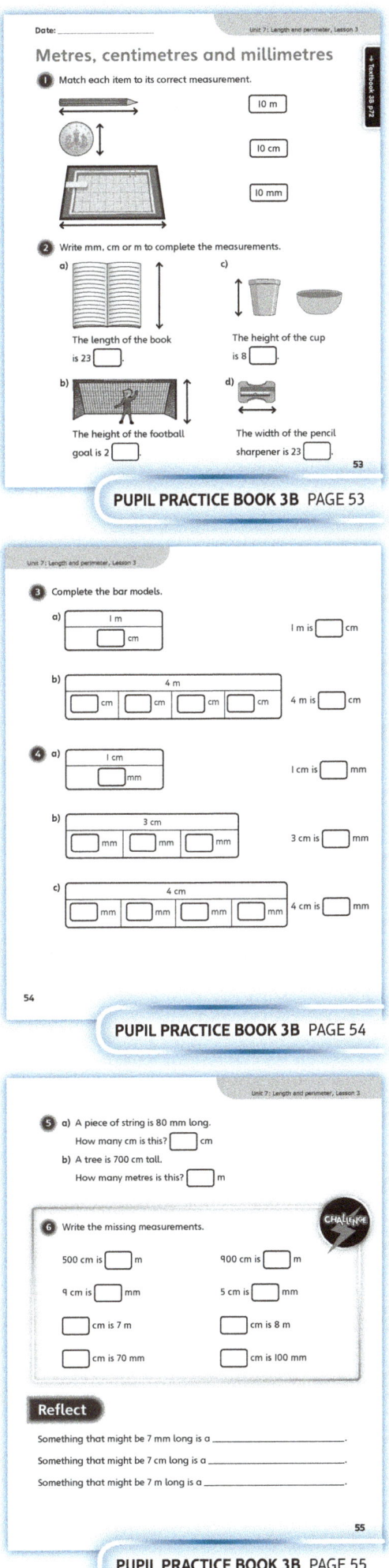

After the lesson
- Can children convert between mm and cm and cm and m?

113

Unit 7: Length and perimeter, Lesson 4

Equivalent lengths (m and cm)

Learning focus
In this lesson, children will explore the equivalence between measurements given in centimetres and measurements given in metres and centimetres.

Before you teach
- Are children secure recording measurements using the mixed units of m and cm?
- What opportunities can you provide for children to physically measure and demonstrate the equivalence between metres and centimetres?

NATIONAL CURRICULUM LINKS

Year 3 Measurement

Measure, compare, add and subtract: lengths (m/cm/mm); mass (kg/g); volume/capacity (l/ml).

ASSESSING MASTERY

Children can explain that 100 cm is equivalent to 1 m and use this to record measurements in both centimetres and in mixed units of metres and centimetres. Children can solve related problems and explain their methods.

COMMON MISCONCEPTIONS

Children may incorrectly partition measurements over 100 cm, sometimes partitioning into 100s, 10s and 1s, thus offering inaccurate equivalences. Ask:
- *How many centimetres are there in 1 metre? Can you use a bar model and place value counters to help you separate measurements over 100 cm into 100s and then 10s with 1s?*

STRENGTHENING UNDERSTANDING

Provide children with plenty of practical experiences, such as identifying things in their classroom or school environment, that will help them see the equivalence between 100 cm and 1 m. Using bar models, initially with place value counters, will also support children as they learn to record the same measurement in both centimetres and mixed units.

GOING DEEPER

Encourage children to say and record a wide range of measurements in both centimetres and mixed units, including those which are under 1 metre. Challenge children to explore measurements over 9 m 99 cm.

KEY LANGUAGE

In lesson: metre, centimetre, equivalent, length, width, total, same as, high

Other language to be used by the teacher: size, distance, equal, measure, equivalence

STRUCTURES AND REPRESENTATIONS

Number line, bar model

RESOURCES

Mandatory: rulers, metre stick(s), place value counters

Optional: number lines, blank bar models, base 10 equipment

 In the eTextbook of this lesson, you will find interactive links to a selection of teaching tools.

Quick recap

On the board, write down a 3-digit number. Ask children to partition the number into 100s, 10s and 1s. Now ask children if they can partition the number into two parts only, for example, 236 into 200 and 36. Repeat for other 3-digit numbers. They may use a part-whole model to support them if needed.

Discover

WAYS OF WORKING Pair work

ASK
- Question 1 a): *What does the picture show?*
- Question 1 a): *How can you work out which of the windows will fit?*
- Question 1 a): *Are the measurements Toshi gives the same as on the windows?*
- Question 1 b): *What do you need to do to change 1 m 21 cm into centimetres?*

IN FOCUS Question 1 a) introduces children to the idea that measurements that are given in metres and centimetres can also be given in just centimetres (or vice versa), for example, 2 m 13 cm is the same as 213 cm.

PRACTICAL TIPS Offer children the opportunity to model a similar scenario in their classroom. Measure the height and length of one of the windows in metres. Ask children to convert their measurement into centimetres.

ANSWERS

Question 1 a): The 213 cm window will fit.

Question 1 b): 1 m 21 cm is 121 cm.

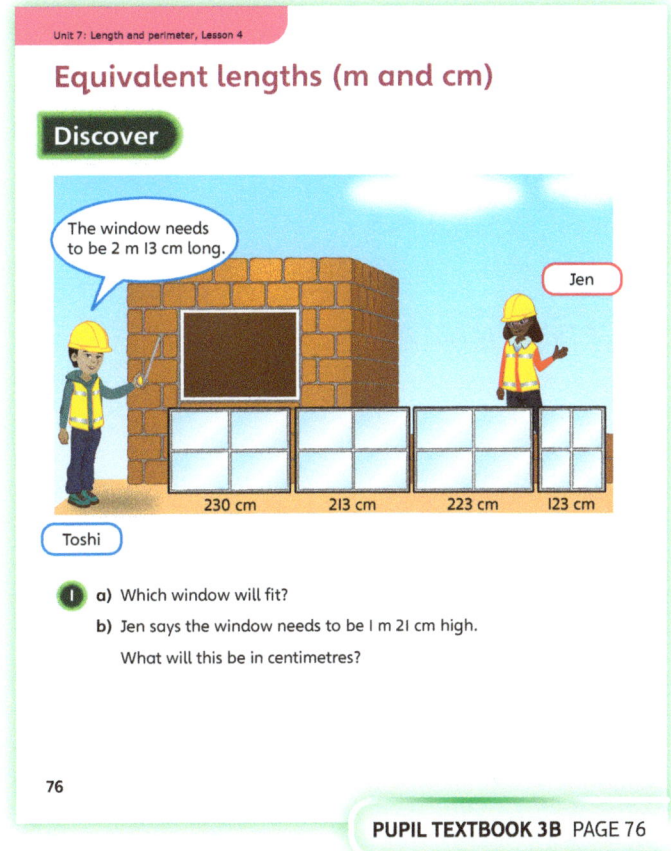

PUPIL TEXTBOOK 3B PAGE 76

Share

WAYS OF WORKING Whole class teacher led

ASK
- Question 1 a): *How did you work out which window will fit?*
- Question 1 a): *What representations did you use to work out which of the measurements were the same?*
- Question 1 b): *How does the bar model help you with this question?*
- Question 1 b): *Is there another word you could use instead of 'the same as'?*

IN FOCUS In questions 1 a) and b), the bar model is introduced to help children convert from measurements given in metres and centimetres and vice versa. It is important to first represent this visually, using place value counters or base 10 equipment, so that children can see that they 'unitise' 100 cm 'chunks' into 1 m (or vice versa).

The comparison between 1 m and 100 cm on a number line is also used as an alternative representation. Use Ash's comment to remind children that the word 'equivalent' means 'the same as'.

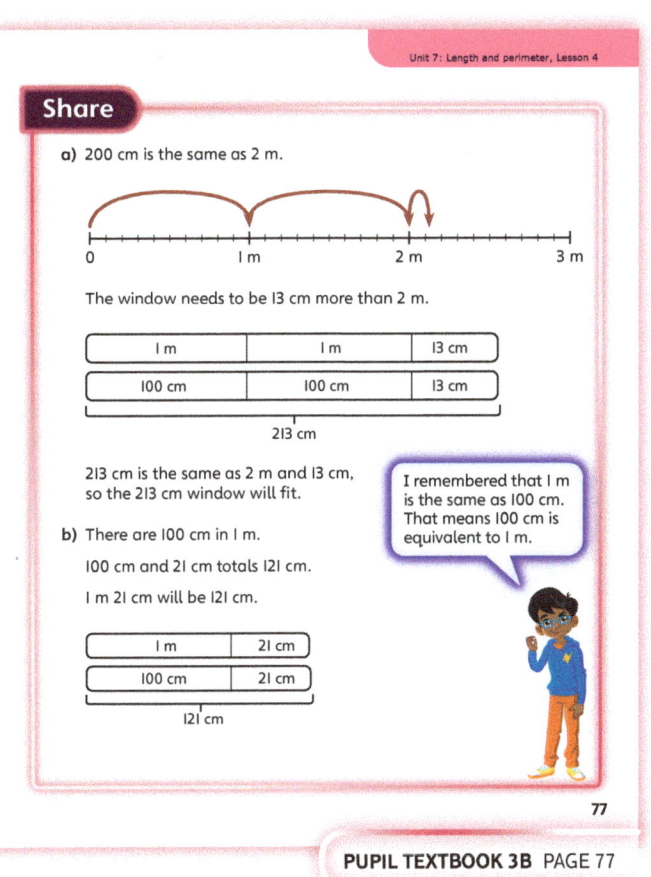

PUPIL TEXTBOOK 3B PAGE 77

115

Unit 7: Length and perimeter, Lesson 4

Think together

WAYS OF WORKING Whole class teacher led (I do, We do, You do)

ASK

- Question 1 a): *How many metres are there in 314 cm?*
- Question 1 b): *How could you use the bar model to help you?*
- Question 2: *What do you do when there are less than 100 cm when converting into m and cm?*
- Question 3: *Can you write down each crocodile's length in cm?*
- Question 3: *How can you use a bar model to help you convert the lengths? Will you convert from m and cm to cm, or from cm to m and cm?*

IN FOCUS This section begins to introduce children to measurements where an understanding of place value is essential. For example, in question 2, children are asked to convert 93 cm into a measurement in m and cm. To do this, they need to realise that, because 93 cm is under 100 cm, there are 0 m and, therefore, it can be expressed as 0 m 93 cm.

STRENGTHEN If children are struggling to convert, ensure that you use practical resources such as place value counters or base 10 equipment and a bar model to help children unitise 100 cm into 1 m and vice versa. The model of a number line could also be used.

DEEPEN Explore with children when they may use a measurement given in cm and when it may be more efficient to use a measurement given in m and cm. Ask them to think about longer distances, such as 15 m, and consider whether converting a distance works in the same way as converting a length.

ASSESSMENT CHECKPOINT Use question 3 to help assess if children can recognise equivalence between a range of measures both in m and cm and in cm.

ANSWERS

Question 1 a): The window is 107 cm high.

Question 1 b): The window is 3 m and 14 cm wide.

Question 2:

2 m 34 cm	234 cm
3 m 17 cm	317 cm
4 m 63 cm	463 cm
0 m 93 cm	93 cm

Question 3 a): 4 m 4 cm is 404 cm.

Question 3 b): 4 m 44 cm is 444 cm.

Question 3 c): 4 m 40 cm is 440 cm.

Question 3 d): 4 m is 400 cm.

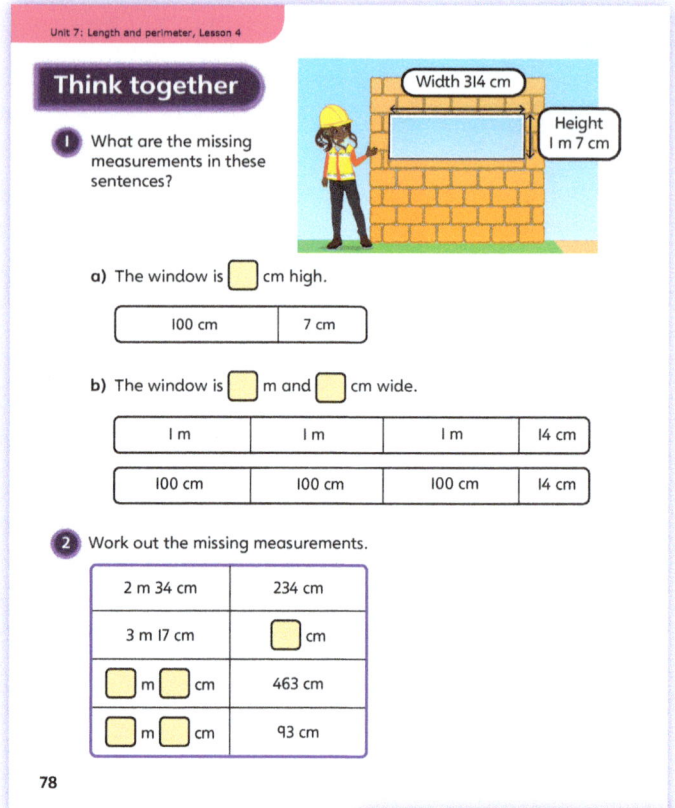

PUPIL TEXTBOOK 3B PAGE 78

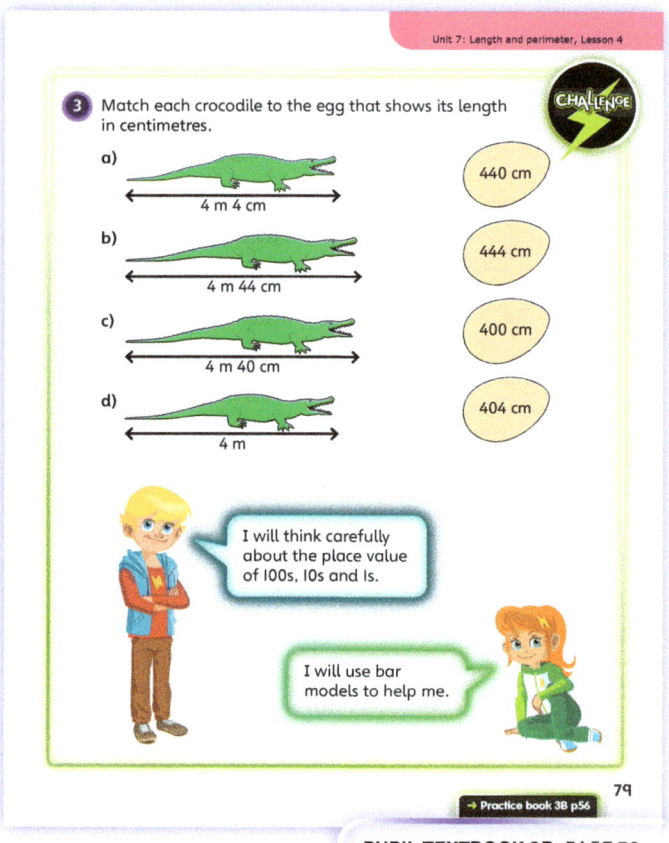

PUPIL TEXTBOOK 3B PAGE 79

Unit 7: Length and perimeter, Lesson 4

Practice

WAYS OF WORKING Independent thinking

IN FOCUS Questions ③ and ④ focus on children exploring common misconceptions that could occur with equivalence. Misconceptions might include thinking that, for example, 4 m 4 cm is 440 cm or that 0 m 76 cm is 76 m. Support children with identifying the misconceptions in questions ③ and ④, followed by accurate reasoning.

STRENGTHEN Encourage children to draw bar models to help them with their conversions.

DEEPEN Encourage children to find the equivalence for numbers up to 9 m 99 cm. Question ⑤ will challenge children to work mentally and verbalise their conversions from m and cm to cm. Extend this challenge question by encouraging children to make up their own measurements and say them aloud to a partner, who responds with the equivalent in cm (or vice versa from cm to m and cm).

THINK DIFFERENTLY Question ④ requires children to use reasoning to explain the mistake that has been made. They should be able to recognise that the place value within the conversion is incorrect.

ASSESSMENT CHECKPOINT Use question ① to determine whether children can use number lines and bar models respectively to convert between m and cm. Use question ③ to assess whether children can convert between m and cm when no pictorial support is given. Use question ④ to assess if children can explain why 2 m 4 cm is not 240 cm and to see if they are secure at converting when place value misconceptions may be a factor.

ANSWERS Answers for the **Practice** part of the lesson can be found in the *Power Maths* online subscription.

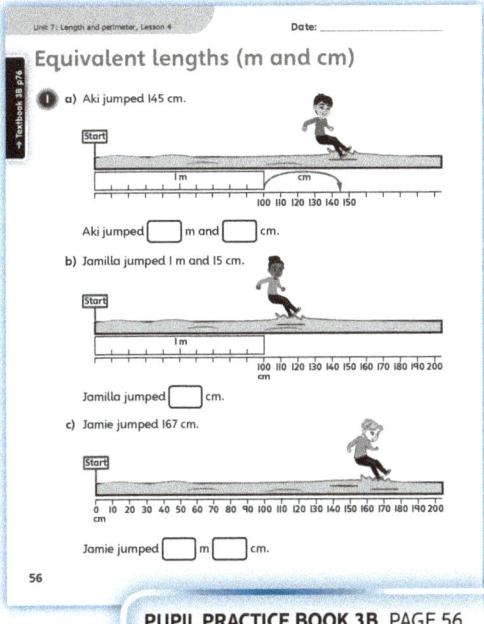

PUPIL PRACTICE BOOK 3B PAGE 56

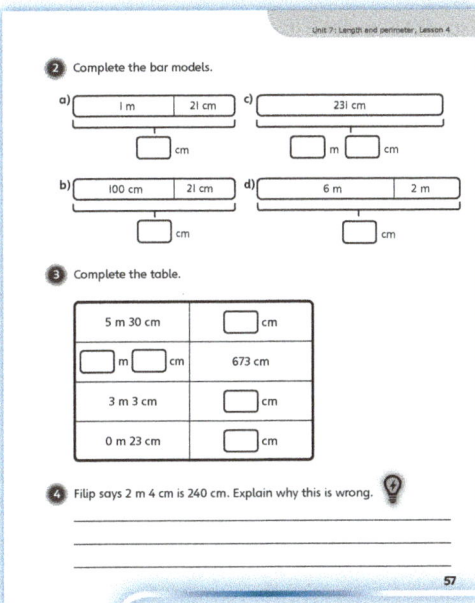

PUPIL PRACTICE BOOK 3B PAGE 57

Reflect

WAYS OF WORKING Pair work

IN FOCUS This section allows children to reflect on how to convert to equivalent measurements to ensure they have mastered the lesson.

ASSESSMENT CHECKPOINT Use the discussion from this activity to assess if children are accurately converting between cm and measurements given in m and cm (and vice versa) and if they can reason with these measurements.

ANSWERS Answers for the **Reflect** part of the lesson can be found in the *Power Maths* online subscription.

After the lesson

- Are children struggling to convert measurements?
- Would a classroom display support learning?

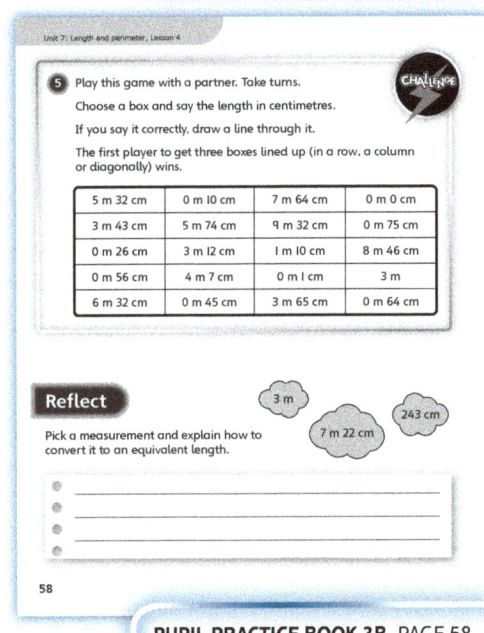

PUPIL PRACTICE BOOK 3B PAGE 58

Unit 7: Length and perimeter, Lesson 5

Equivalent lengths (mm and cm)

Learning focus
In this lesson, children will read lengths in centimetres and millimetres – converting between these units.

Before you teach
- Are children secure with converting from metres to centimetres?
- Could you put some conversions on your working wall to support children?

NATIONAL CURRICULUM LINKS

Year 3 Measurement

Measure, compare, add and subtract: lengths (m/cm/mm); mass (kg/g); volume/capacity (l/ml).

ASSESSING MASTERY

Children can quickly convert between centimetres and millimetres. They can effectively explain the methods they used.

COMMON MISCONCEPTIONS

If children are given rulers that have one side in centimetres and the other in millimetres, they may use the wrong side of the ruler. Ask:
- How do you know which side shows centimetres?

Sometimes children make errors with place value when converting. For example, they might say that 8 cm 4 mm is 804 mm. Ask:
- What could you do to check your answer?

STRENGTHENING UNDERSTANDING

Knowledge of multiplying by 10 will be important in this lesson. Run intervention sessions in which numbers are multiples of 10. Apply this learning to cm to mm conversion, for example, by looking at how many mm there are in 5 cm.

GOING DEEPER

To deepen learning in this lesson, challenge children to think of when centimetres and millimetres would be used to measure things in real-life situations.

KEY LANGUAGE

In lesson: centimetres (cm), millimetres (mm), convert, equal, equivalent, same, length, long, measure, measurement

Other language to be used by the teacher: conversion

STRUCTURES AND REPRESENTATIONS

Bar models, number lines

RESOURCES

Mandatory: number lines, rulers (a variety of different types)

Optional: place value counters, base 10 equipment, flashcards of key vocabulary, blank pieces of card, thread, small items that could be measured in millimetres such as beads, pegs, paperclips

 In the eTextbook of this lesson, you will find interactive links to a selection of teaching tools.

Quick recap
Ask children to convert 3 m, 5 m and 7 m to cm and then ask them to convert 80 cm, 90 cm and 120 cm into m. Discuss with children how they got their answers.

Discover

WAYS OF WORKING Pair work

ASK

- Question 1 a): *How are the rulers different? Why are they different?*
- Question 1 a): *Can you use both rulers to measure the length of the thread?*
- Question 1 b): *Will you measure the thread in mm or in cm and mm? Do you find one way easier than the other?*

IN FOCUS For question 1 a), encourage children to discuss their answers and explain why they made their choice. This question focuses on children's understanding of how centimetres and millimetres compare and emphasises the multiplicative relationship between the two units of measurement. It may also highlight any issues with place value when reading the mm ruler, for example, do children recognise that the thread is 24 mm or do they think it is 204 mm?

PRACTICAL TIPS Show children different rulers which use cm and mm or both. Ask them to line the rulers up. What do they notice? Replicate Luis's activity and cut some lengths of thread. Ask children to measure them with both the cm ruler and the mm ruler. Encourage children to draw lines that are 24 mm and 2 cm 4 mm long using different types of ruler. This should help them recognise that they are working with equivalent measurements and are drawing the same length of line.

ANSWERS

Question 1 a): Luis could use either ruler to measure his thread.

Question 1 b): The thread is 24 mm or 2 cm 4 mm.

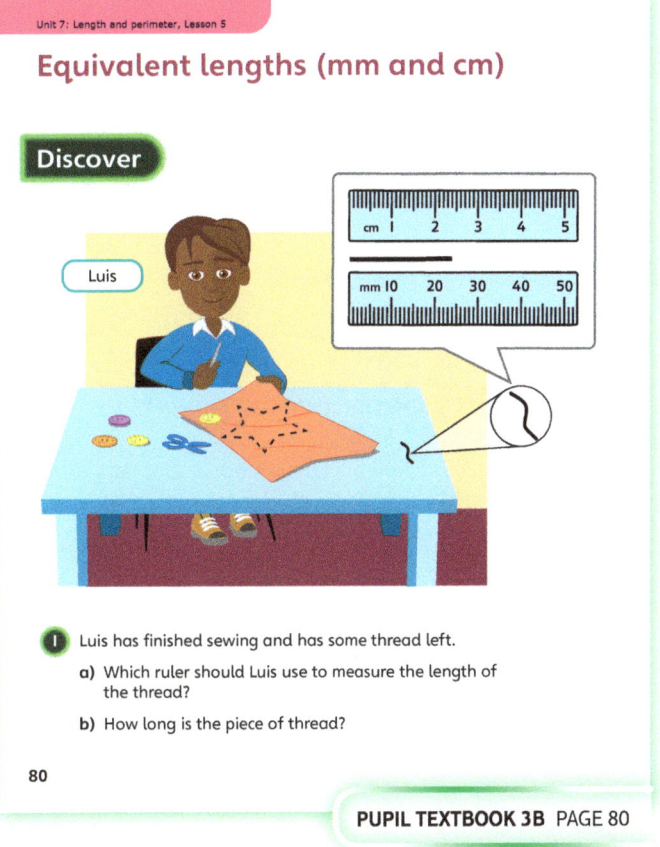

PUPIL TEXTBOOK 3B PAGE 80

Share

WAYS OF WORKING Whole class teacher led

ASK

- Question 1 a): *Do you remember how many millimetres are equal to 1 centimetre?*
- Question 1 b): *How does the bar model help you to convert measurements?*
- Question 1 b): *Can you explain to a partner how to convert measurements from millimetres to centimetres?*
- Question 1 b): *What other things might be best measured in millimetres?*

IN FOCUS For question 1 b), encourage children to convert to both millimetres and centimetres and millimetres.

STRENGTHEN Where children are finding it challenging to relate to the units of measurement, have them find small items around the classroom (such as beads, pegs, paperclips) that would be most accurately measured in millimetres. They could then find some larger items that might be more efficiently measured in centimetres.

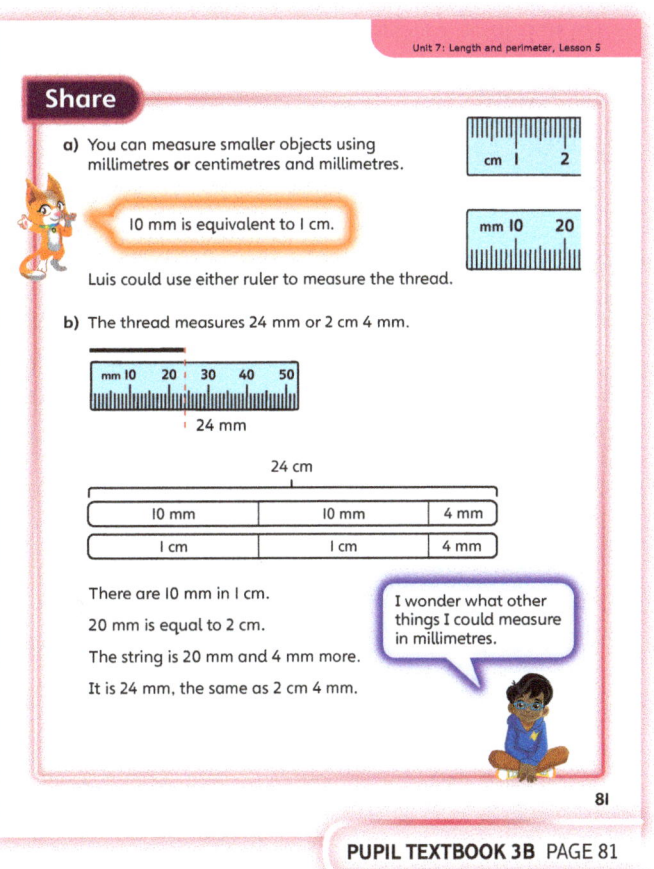

PUPIL TEXTBOOK 3B PAGE 81

Unit 7: Length and perimeter, Lesson 5

Think together

WAYS OF WORKING Whole class teacher led (I do, We do, You do)

ASK

- Question ① a): *Which piece of string is the longest? How do you know? How would you work this out without looking at the picture?*
- Question ① c): *If there are only 9 mm, are there any whole centimetres?*
- Question ②: *Can you draw lines that have a variety of different lengths? Are there some that are easier to draw using a cm ruler or a mm ruler?*
- Question ③: *How can you measure a line that isn't straight? What could you use to help you?*

IN FOCUS Question ② focuses on children's understanding of how to accurately convert between measurements. Some children will need support with place value. For example, discuss that 13 mm is 1 cm 3 mm and **not** 10 cm 3 mm. In question ③, children should be very careful when tracing the shape of the curve with the string. They should:

- keep the string on the paper
- use a finger to always keep at least one part of the string pressed down onto the curved line
- start with the end of the string at one end of the line
- hold tightly to the point on the string that corresponds to the other end of the line
- stretch the string taut to measure against the ruler.

STRENGTHEN Strengthen learning through playing matching pairs games. Write down lengths in mm on half the cards and then their equivalent in cm and mm on the other half. Mix them up and ask children to match up the pairs.

DEEPEN As an extension of question ②, challenge children to draw lines of varying lengths, including some that cross the 100 millimetre mark, for example, 11 cm 4 mm = 114 mm. Children should write the measurements for each line in cm and mm and just in mm.

ASSESSMENT CHECKPOINT Question ① c) will give a good indication of children who have mastered the lesson. They will be able to accurately complete the equivalent lengths and explain why there are not any whole cm in 9 mm.

ANSWERS

Question ① a): The shorter piece of string is 37 mm long.

Question ① b): The longer piece of string is 6 cm 6 mm long.

Question ① c):

2 cm 9 mm	29 mm
8 cm 4 mm	84 mm
6 cm 5 mm	65 mm
0 cm 9 mm	9 mm

Question ②: Children should draw lines of the correct lengths:
  a) 5 cm 8 mm   b) 102 mm
  c) 10 cm 3 mm   d) 13 mm

Question ③: Children should place string along the curves of the lines. Then they should straighten the string and hold it against a ruler to measure the lengths.

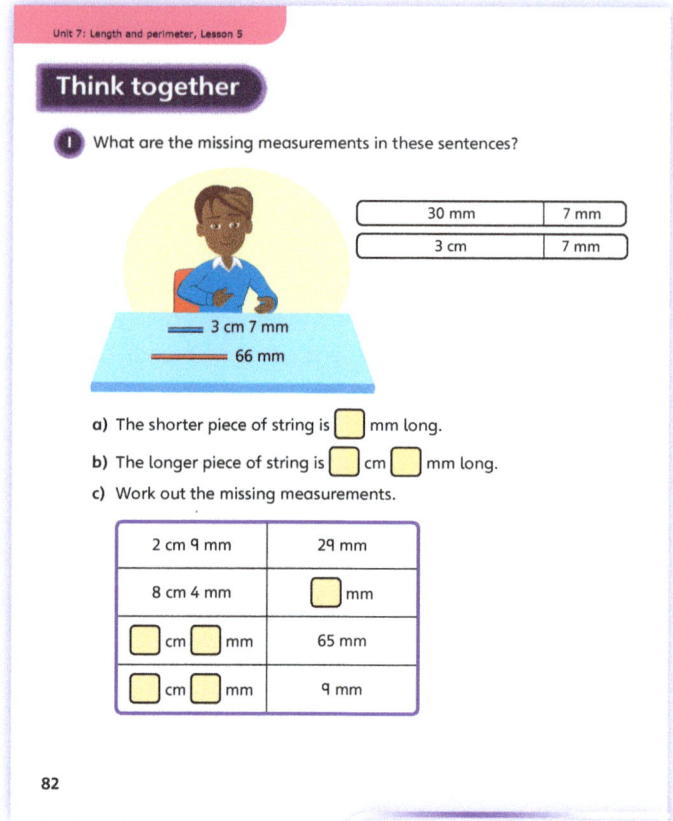

PUPIL TEXTBOOK 3B PAGE 82

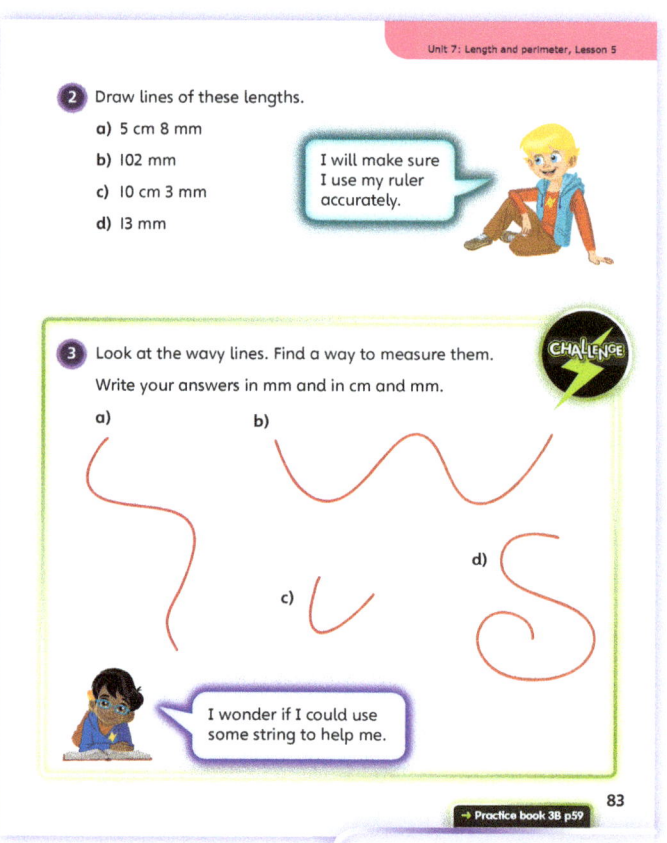

PUPIL TEXTBOOK 3B PAGE 83

Unit 7: Length and perimeter, Lesson 5

Practice

WAYS OF WORKING Independent thinking

IN FOCUS The focus of question ❶ is on children recognising the different units of measurement on the different rulers. This reinforces their understanding of converting between cm and mm. Some children may want to use rulers that have centimetres on one side and millimetres marked on the other to support them when converting lengths.

STRENGTHEN The practical nature of question ❺ makes it perfect for strengthening understanding. Where children need further support, challenge them to cut out additional strips of paper (or draw lines) of specified lengths. To start with, suggest lengths that are slightly simpler to work with, for example, that do not bridge 100 mm and are not likely to lead to place value errors.

DEEPEN Challenge children to create some word problems for converting measurements. You could also make links between converting from millimetres to centimetres and converting from centimetres to metres or from millimetres to metres.

THINK DIFFERENTLY Question ❹ asks children to consider and then explain why there are no whole cm in 5 mm. This should elicit their understanding that a mm is a smaller unit of measurement than a cm and that anything less than 10 mm is not a whole cm.

ASSESSMENT CHECKPOINT Question ❷ will allow you to assess children's knowledge of partitioning in order to support them when converting.

ANSWERS Answers for the **Practice** part of the lesson can be found in the *Power Maths* online subscription.

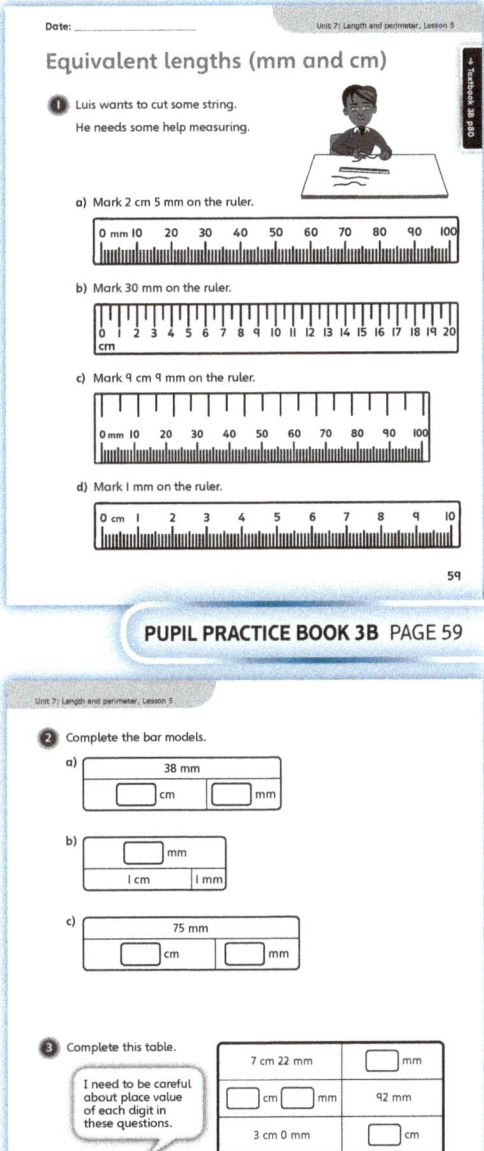

PUPIL PRACTICE BOOK 3B PAGE 59

PUPIL PRACTICE BOOK 3B PAGE 60

Reflect

WAYS OF WORKING Pair work

IN FOCUS This question will encourage rich discussion about why centimetres and millimetres are used and which might be more appropriate in certain contexts. You may want to put some of the key language from this lesson on the board to support the maths talk.

ASSESSMENT CHECKPOINT This activity will offer an opportunity to assess children's use of mathematical vocabulary. It should also clarify whether children understand why converting is useful when measuring.

ANSWERS Answers for the **Reflect** part of the lesson can be found in the *Power Maths* online subscription.

After the lesson

- Do children understand the key vocabulary?
- Are children confident converting between millimetres and centimetres?

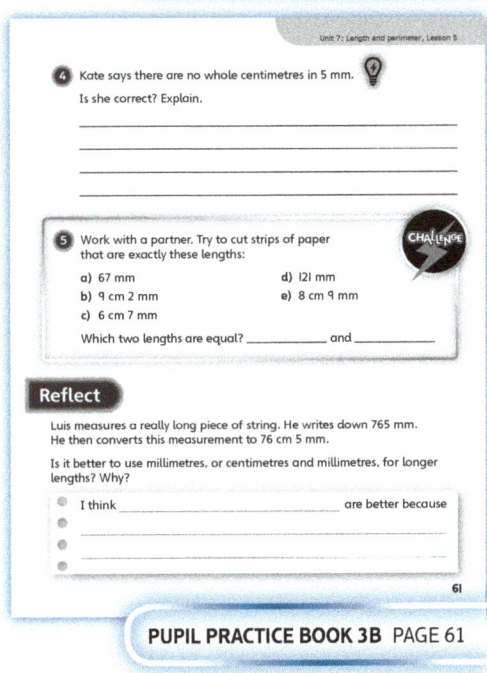

PUPIL PRACTICE BOOK 3B PAGE 61

Unit 7: Length and perimeter, Lesson 6

Compare lengths

Learning focus
In this lesson, children will compare and order measurements given in millimetres, centimetres and metres.

Before you teach
- Are children confident converting lengths to a common unit of measure?
- Are children confident using the < and > signs?

NATIONAL CURRICULUM LINKS

Year 3 Measurement

Measure, compare, add and subtract: lengths (m/cm/mm); mass (kg/g); volume/capacity (l/ml).

ASSESSING MASTERY

Children can compare lengths in mixed units, place lengths in ascending or descending order and complete expressions using appropriate comparison signs.

COMMON MISCONCEPTIONS

Children may ignore the unit of measurement and compare lengths based on the size of the numbers. Ask:
- *Are all of the lengths in the same unit of measurement? Can you use your understanding of equivalence and convert so that all lengths are in the same unit of measurement?*

Children may also make calculation errors when converting lengths to a common unit of measurement. Ask:
- *Does your conversion look accurate? What lengths are equivalent to each other?*

STRENGTHENING UNDERSTANDING

Continue to provide children with the opportunity to measure as many different objects as possible. Encourage children to use different units of measurement and to investigate the similarities and differences between the values. Physically placing objects next to each other will increase understanding of the > and < signs.

GOING DEEPER

Encourage children to write expressions that use more than one greater than or less than sign. For example, 120 cm < 140 cm < 150 cm. They could also practise recording lengths using different units of measurement.

KEY LANGUAGE

In lesson: compare, measurement, millimetres (mm), centimetres (cm), metres (m), **convert**, equivalent, length, height, greater than (>), less than (<), first (1st), second (2nd), third (3rd), furthest, ascending

Other language to be used by the teacher: known fact, equal to, descending, unit of measurement

STRUCTURES AND REPRESENTATIONS

Number line

RESOURCES

Mandatory: metre sticks, rulers

Optional: chalk, real-life objects to measure, paper to make paper aeroplanes, balls or bean bags

 In the eTextbook of this lesson, you will find interactive links to a selection of teaching tools.

Quick recap
Ask children to compare two numbers to check they know how to compare. Compare two 2-digit numbers first, then two 3-digit numbers and finally a 2-digit number and a 3-digit number. For each pair of numbers, ask: *Which number is bigger? Which is smaller?*

Unit 7: Length and perimeter, Lesson 6

Discover

WAYS OF WORKING Pair work

ASK

• Question 1 a): *Can you convert all the lengths to metres and centimetres?*
• Question 1 a): *What is 100 cm equivalent to?*

IN FOCUS In questions 1 a) and 1 b), children use their understanding of equivalence to compare different heights to a set value. The focus is on developing an appreciation that lengths can be recorded using different units of measurement.

PRACTICAL TIPS Provide children with the opportunity to measure each other's heights and then order themselves based on their relative heights. Children should also practise measuring the lengths of objects using different units of measurement to help them understand how the units relate to each other.

ANSWERS

Question 1 a): Olivia and Lee are tall enough to go on the ride.

Question 1 b): Lee's height is closest to 1 m 30 cm.

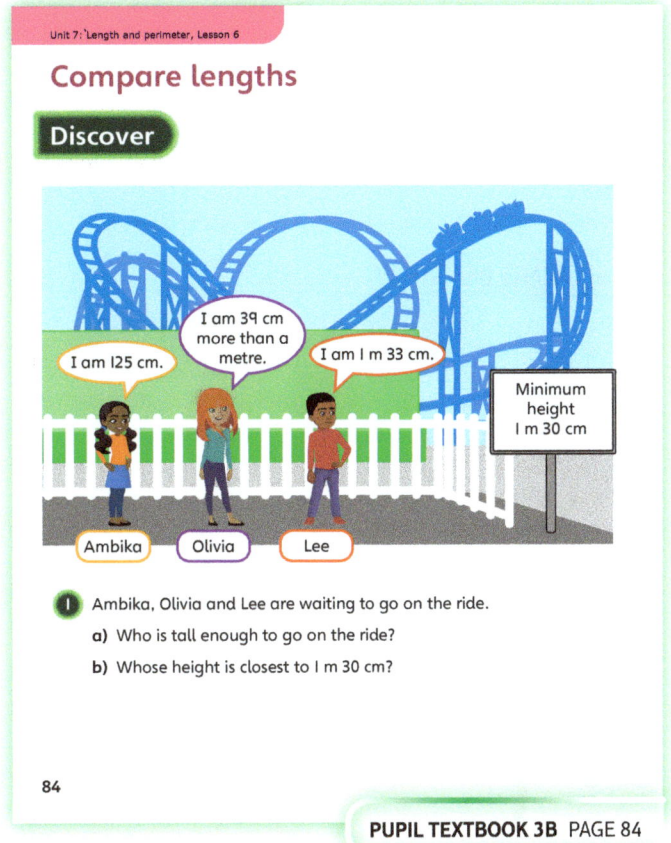

PUPIL TEXTBOOK 3B PAGE 84

Share

WAYS OF WORKING Whole class teacher led

ASK

• Question 1 a): *What is 125 cm equivalent to?*
• Question 1 a): *What is more significant, the size of the number or the unit of measurement?*
• Question 1 a): *Is Astrid correct when she says that heights with a 3-digit number must be tall enough to go on the ride? Why not? Can you give an example of a 3-digit number that is less than 130 cm?*
• Question 1 b): *What height is closest to 1 m 30 cm? What height is furthest away from 1 m 30 cm?*

IN FOCUS In question 1 a), encourage children to think of the height measurement sticks as number lines. This should help them examine equivalent facts and make comparisons.

STRENGTHEN Encourage children to relate the context to their own lives by asking them to plot their own heights on to a number line and see who is closest to 1 m 30 cm.

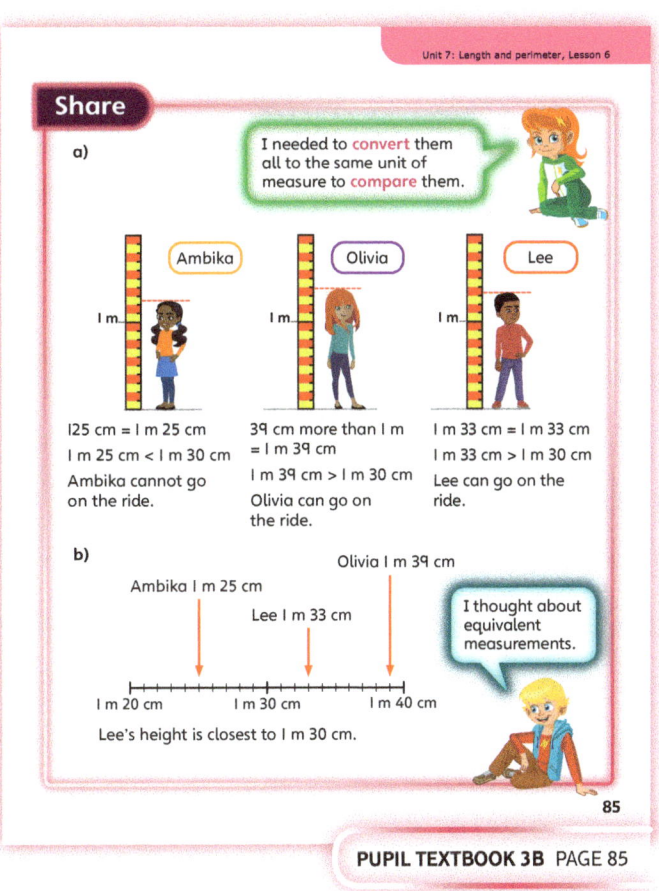

PUPIL TEXTBOOK 3B PAGE 85

123

Think together

WAYS OF WORKING Whole class teacher led (I do, We do, You do)

ASK

- Question ❶: *Is it easiest to convert all measurements to centimetres, metres or a combination of different measurements?*
- Question ❶: *Is it necessary to convert all lengths that are given?*

IN FOCUS In this section, lengths are the focus as opposed to heights. The lengths are presented horizontally rather than vertically. Children should continue to make conversions based on their understanding of equivalence, rather than using their place value understanding of dividing or multiplying by 10 or 100.

STRENGTHEN Children could complete a similar activity to that presented in question ❶. They could throw a ball or bean bag and measure the distances thrown using different units of measurement. This will help them understand the equivalent lengths.

DEEPEN Encourage children to convert known lengths into different units of measurement. Once the lengths have been converted, children can create expressions comparing known quantities, for example 194 mm < 1 m 21 cm.

ASSESSMENT CHECKPOINT In question ❸, children should be able to justify the way they have compared different lengths by using the language of equivalence. They should be able to explain the facts that they have used within their mental calculations and be able to present lengths in many different ways.

ANSWERS

Question ❶: Lexi is in first place (363 cm = 3 m 63 cm), Mo is in second place (3 m 59 cm) and Danny is in third place (2 m 99 cm).

Question ❷: Box B has the shortest ribbons.

Question ❸: The longer measurement in each pair is:
a) 30 cm (30 cm is 300 mm, so is longer than 294 mm)
b) 490 cm (4 m 9 cm is 409 cm, so 490 cm is longer)
c) 3 m 1 cm (3 m 1 cm is 301 cm, so is longer than 199 cm).

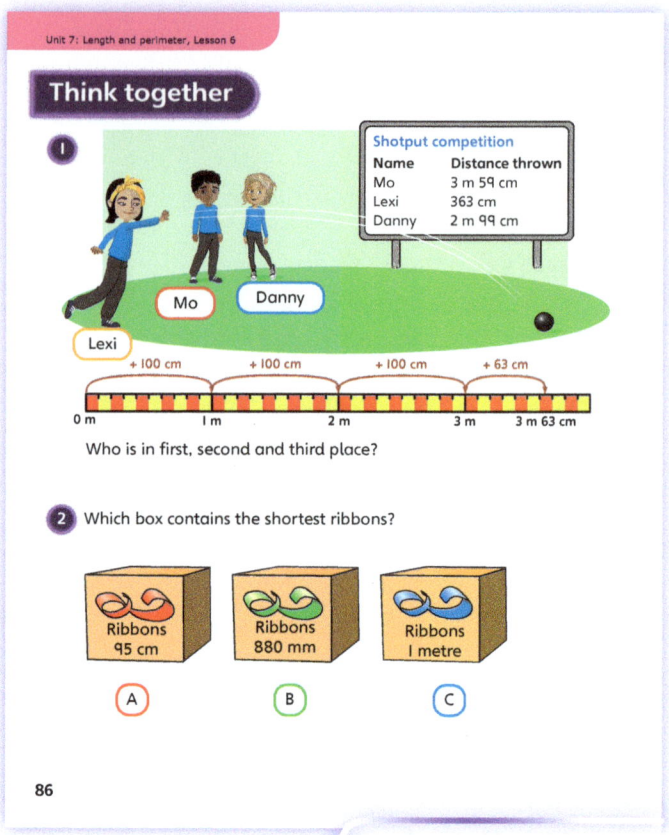

PUPIL TEXTBOOK 3B PAGE 86

PUPIL TEXTBOOK 3B PAGE 87

Unit 7: Length and perimeter, Lesson 6

Practice

WAYS OF WORKING Independent thinking

IN FOCUS Question ① provides children with lengths in different units of measurement. They are required to compare these lengths in order to demonstrate their understanding of equivalence.

STRENGTHEN Children could recreate some of the fun practical activities demonstrated in these questions. They could make their own paper aeroplanes and measure the dimensions of their planes. They could then hold a paper aeroplane competition and measure the distances flown by the aeroplanes using different units of measurement. Ordering these distances would benefit children that find the concept more challenging.

Children could also be given a list of key equivalent facts to help them with their conversions, such as 10 mm = 1 cm, 100 mm = 10 cm, and so on.

DEEPEN Challenge children to complete expressions with more than one possible answer. For example, 125 cm ☐ 1 m 50 cm. Further restrictions could be made to extend the challenge, such as 'all unknown values are multiples of 5'.

THINK DIFFERENTLY In question ⑥, children must apply the rules that they have learnt in previous lessons, in order to calculate the length of each item before making comparisons. Children may be inclined to think that the box will have the greater length because it measures to the greater point on the ruler, ignoring the starting point of each object. Note that there is only 0·5 mm difference in the lengths, so answers may differ. Discuss how such differences in children's answers may result from how accurately they measured.

ASSESSMENT CHECKPOINT To assess children's understanding of the greater than and less than signs in relation to comparing lengths, ask them to write two expressions for each comparison. For example, 152 cm > 125 cm so 125 cm < 152 cm. Encourage children to explain how they know both expressions are true.

ANSWERS Answers for the **Practice** part of the lesson can be found in the *Power Maths* online subscription.

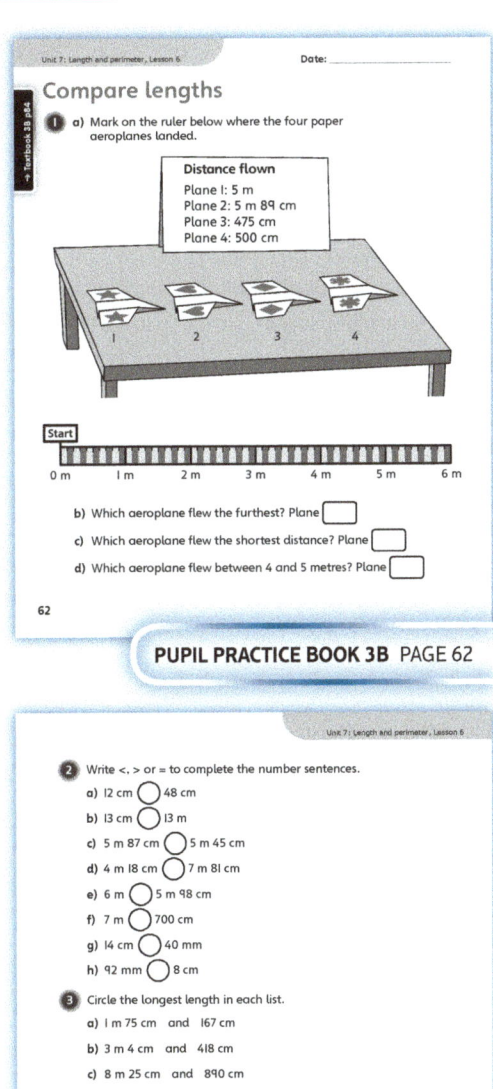

PUPIL PRACTICE BOOK 3B PAGE 62

PUPIL PRACTICE BOOK 3B PAGE 63

Reflect

WAYS OF WORKING Independent thinking

IN FOCUS Encourage children to work independently to explain how the different lengths that have been given can be converted to a common unit of measurement. Once these conversions have been made, the different choices that children have made should be discussed to compare efficiency and ease.

ASSESSMENT CHECKPOINT The examples given in children's independent reflection plus their ability to explain known equivalent facts should allow an assessment of their true understanding to be made.

ANSWERS Answers for the **Reflect** part of the lesson can be found in the *Power Maths* online subscription.

After the lesson ⏸

- Are children secure with using known equivalent lengths within mental calculations?
- Do children understand the importance of the unit of measurement for each length, rather than just the size of each number?

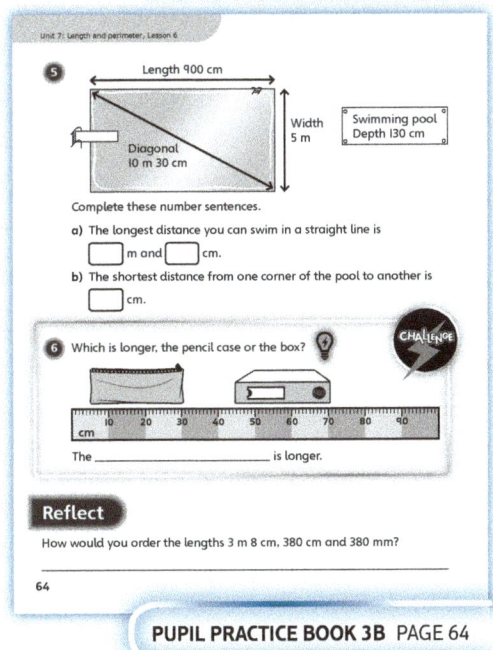

PUPIL PRACTICE BOOK 3B PAGE 64

125

Unit 7: Length and perimeter, Lesson 7

Add lengths

Learning focus
In this lesson, children will find the totals of two or more lengths given in centimetres, metres or simple combinations of both units. They will convert answers into millimetres, centimetres or metres as appropriate.

Before you teach
- Do children have a good practical understanding of the relative size of metres, centimetres and millimetres?
- Do children know the relationships 1 m = 100 cm, and 1 cm = 10 mm? Can they convert between the units where necessary?

NATIONAL CURRICULUM LINKS

Year 3 Measurement

Measure, compare, add and subtract: lengths (m/cm/mm); mass (kg/g); volume/capacity (l/ml).

ASSESSING MASTERY

Children can add pairs of lengths given in millimetres, centimetres, metres or mixed units using appropriate strategies such as the column method for addition.

COMMON MISCONCEPTIONS

Children may find conversions between centimetres and metres challenging. The examples in this lesson are quite simple, mainly using lengths that are multiples of 10 cm or 5 cm. This should help ensure that children can see the connections between the units – that 150 cm is the same as 1 m 50 cm, and so on – and build their understanding of the various conversions and equivalences. Ask:
- *What is 100 cm equivalent to? So, what would 150 cm be if you converted it into metres and centimetres?*

STRENGTHENING UNDERSTANDING

Use the number line as a model for addition, as it provides a direct picture of the process of combining lengths by addition. Ensure children use the number line as an aid to understanding the process involved in each calculation, particularly the idea of 'completing the larger unit' and then finding the remaining length. Children can also use the number line as an aid to calculation but this should not stop them from using known number facts and mental calculation where appropriate.

GOING DEEPER

More confident learners may be ready to tackle more complex calculations. For example, ask children to find the total of 1 m 74 cm and 38 cm, or they could create a problem involving adding three or more lengths.

KEY LANGUAGE

In lesson: metre, centimetre, convert, converted, add, total, addition, height, length, method, combinations

STRUCTURES AND REPRESENTATIONS

Number line

RESOURCES

Optional: tape, tape measure, metre ruler, ribbon, counters, paper strips and similar materials for modelling calculations

 In the eTextbook of this lesson, you will find interactive links to a selection of teaching tools.

Quick recap
Ask children to add two 2-digit numbers where one exchange is needed. They should use column methods. Ask children to then add two 3-digit numbers together, and a 2-digit number and 3-digit number together.

Discover

WAYS OF WORKING Pair work

ASK

- Question 1 a): *What units have been used to measure the bunting?*
- Question 1 a): *Which piece of bunting is the longest? Which piece is the shortest?*
- Question 1 a): *Do you need to do a conversion? Which numbers will you need to add?*
- Question 1 b): *Is there a quick way to do this? What units will you use in your answer?*

IN FOCUS Question 1 a) sets the scene by introducing a practical situation where you might need to add lengths to solve a problem. It is important to be able to apply learning in a real-life context.

PRACTICAL TIPS In an open space, ask children to place a 'start' marker such as a counter, then measure 25 cm and place another marker. They should then measure a further 450 cm and place another marker, then measure 3 m from that marker and place a final marker. This should provide some visual and concrete context to the lengths they are dealing with. If they measure from the start to the end marker, do they get 7 m 75 cm? Note whether they choose to measure in metres and centimetres or just in centimetres.

ANSWERS

Question 1 a): The new piece of bunting is 7 m 75 cm or 775 cm.

Question 1 b): They would have 7 m 50 cm or 750 cm of bunting.

Share

WAYS OF WORKING Whole class teacher led

ASK

- Question 1 a): *Let's look at how Flo found the total. What did she do first? Why do you think she did that?*
- Question 1 a): *Dexter tackled the problem in a different way. What did he do?*

IN FOCUS In question 1 a), discuss whether both methods give the same answer in order to establish the idea that 7 m 75 cm and 775 cm are the same length. Do children prefer one method over another? Why?

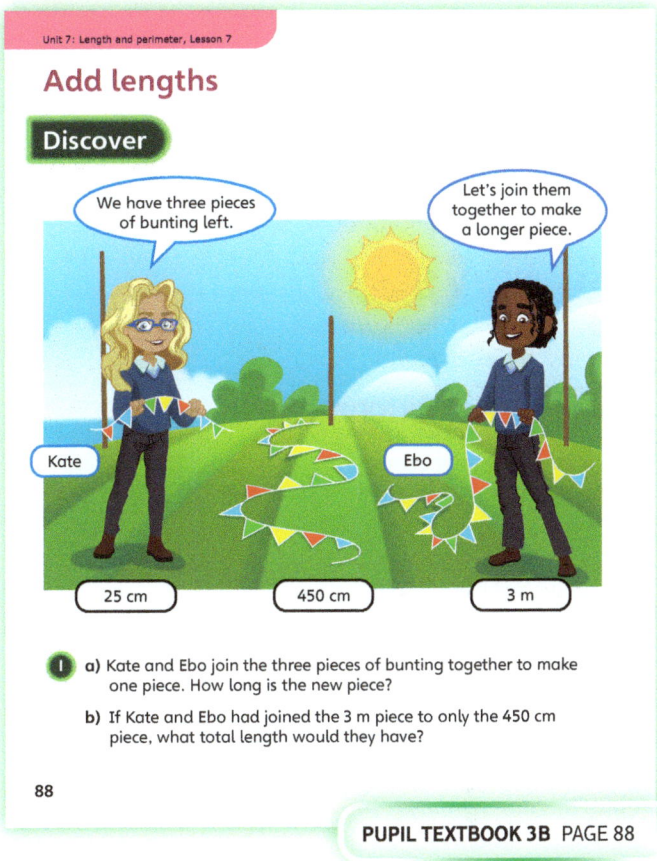

PUPIL TEXTBOOK 3B PAGE 88

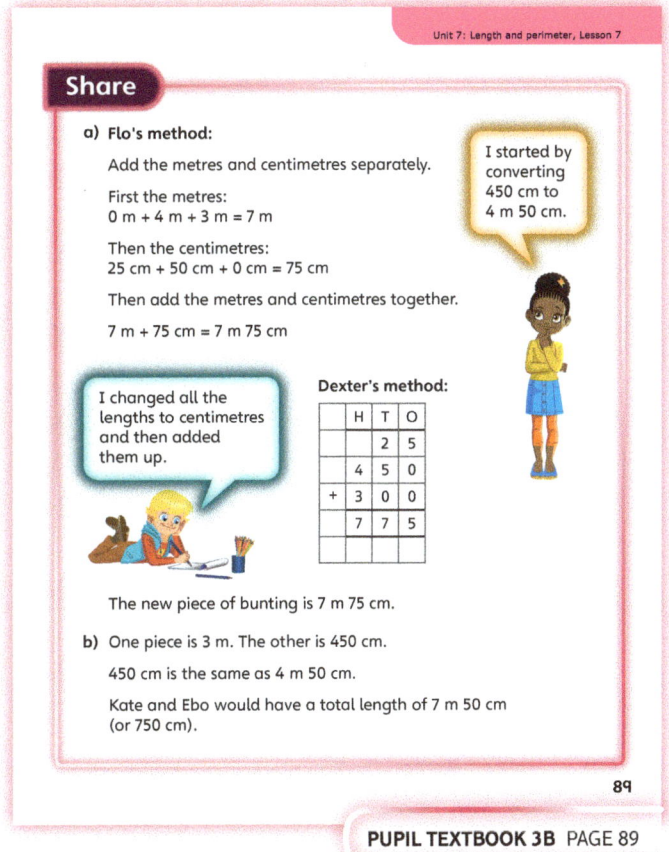

PUPIL TEXTBOOK 3B PAGE 89

Unit 7: Length and perimeter, Lesson 7

Think together

WAYS OF WORKING Whole class teacher led (I do, We do, You do)

ASK

- Question 1 b): *The body length is given as 12 mm. Flo's method would involve changing 12 mm into centimetres and millimetres. How would she do the calculation?*
- Question 1 b): *What unit would Dexter change everything to? How would he do the calculation?*

IN FOCUS Question 2 provides an opportunity to discuss conversions and relative sizes of units. You could model all of these calculations on a number line (or ruler). Use the number line representation alongside known number bonds, for example, ask: *Where would 12 mm be marked? How much do you need to add on to 12 mm to get to 20 mm (or 2 cm)?*

STRENGTHEN Children who are finding the conversion between centimetres and metres difficult may benefit from additional practical experience. Ask them to use a ruler to cut paper strips to the sizes shown in question 3, join them together and then measure them to find the total length.

DEEPEN Make sure that children know all the different ways in which the lengths from question 3 could be written and recorded. For example, adding all the lengths together gives a total of 295 cm, which could also be given as 2 m 95 cm. Challenge them to give this length in millimetres.

ASSESSMENT CHECKPOINT Use question 3 to check that children understand the conversion between centimetres and metres. Ask: *What if I joined the 50 cm and 60 cm pieces together? What would that make? Would that be more or less than 1 metre? How do you know?*

ANSWERS

Question 1 a): The bunting is 3 m 75 cm or 375 cm.

Question 1 b): The total height of the model is 3 cm 8 mm or 38 mm.

Question 2 a): 50 cm + 50 cm = 1 m

Question 2 b): 4 cm + 60 mm = 10 cm

Question 2 c): 12 mm + 8 mm = 2 cm

Question 2 d): 1 m 40 cm + 160 cm = 3 m

Question 3 : Flo's method:
  1 m 85 cm + 50 cm = 2 m + 35 cm = 2 m 35 cm
  1 m 85 cm + 60 cm = 2 m + 45 cm = 2 m 45 cm
  50 cm + 60 cm = 1 m + 10 cm = 1 m 10 cm
  1 m 85 cm + 50 cm + 60 cm =
  2 m + 35 cm + 60 cm = 295 cm

  Dexter's method:
  1 m 85 cm = 185 cm
  185 + 50 = 235 cm
  185 + 60 = 245 cm
  50 + 60 = 110 cm
  185 + 50 + 60 = 295 cm

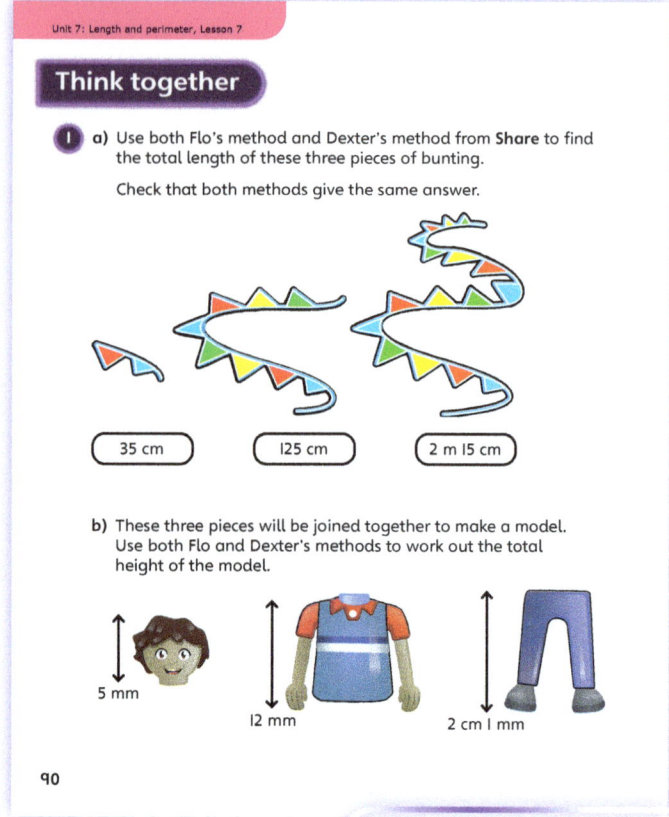

PUPIL TEXTBOOK 3B PAGE 90

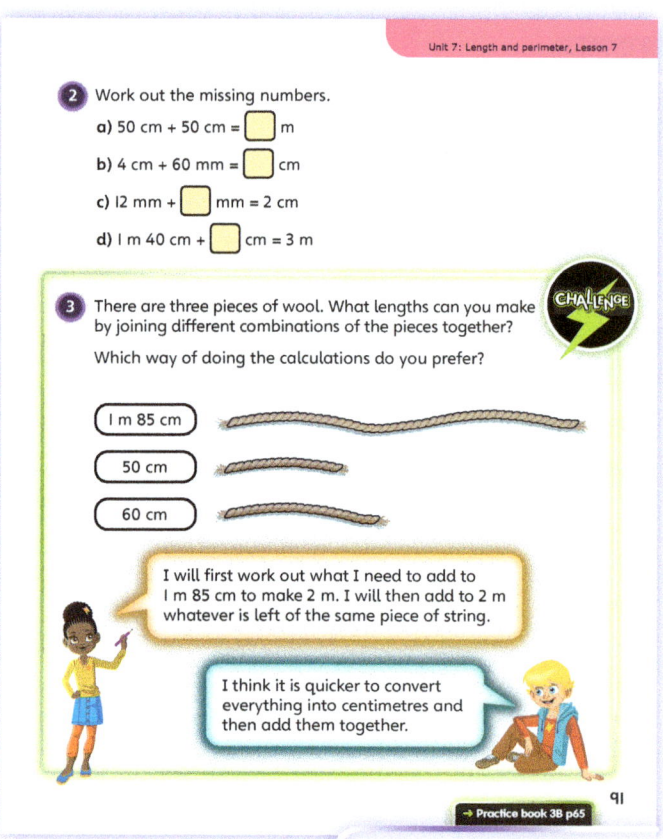

PUPIL TEXTBOOK 3B PAGE 91

Unit 7: Length and perimeter, Lesson 7

Practice

WAYS OF WORKING Independent thinking

IN FOCUS Question ④ involves mixed units and conversions. Children will need to adopt suitable strategies for dealing with this, for example, by changing everything to centimetres in order to carry out the calculations (perhaps with the aid of a number line) and then converting the answer to metres, if appropriate.

STRENGTHEN You could model question ⑥ by cutting some tape to represent each stage of the hop, skip and jump for each of the competitors. Ask children to stick the strips end-to-end and then use a tape measure or metre ruler to find the total for Jamilla and then for Andy.

DEEPEN Challenge children to make up (and solve in pairs) more complex versions of some of the existing questions. For example, for question ⑦, they could ask: *What are the different heights that you could make by stacking different combinations of these books? Could you make a stack that is 54 mm high? What about a stack that is 1 m 15 cm high?*

ASSESSMENT CHECKPOINT Use question ⑥ to check that children can successfully perform additions using mixed units. Ask: *Who did the longer 'skip'? Who did the longer 'jump'? How are you going to work out the total distance for Jamilla and Andy?*

ANSWERS Answers for the **Practice** part of the lesson can be found in the *Power Maths* online subscription.

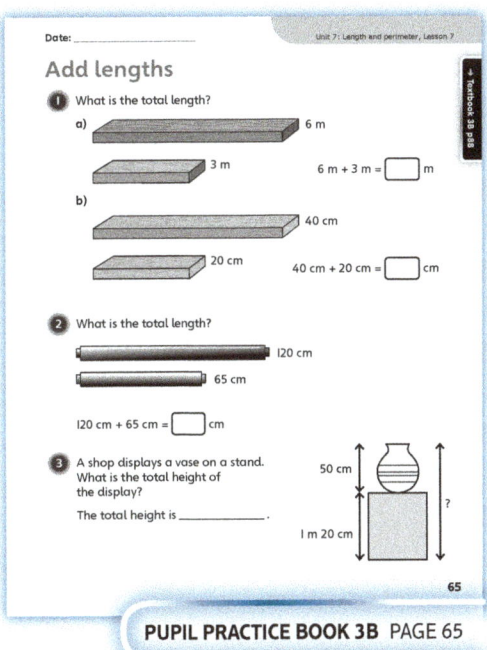

PUPIL PRACTICE BOOK 3B PAGE 65

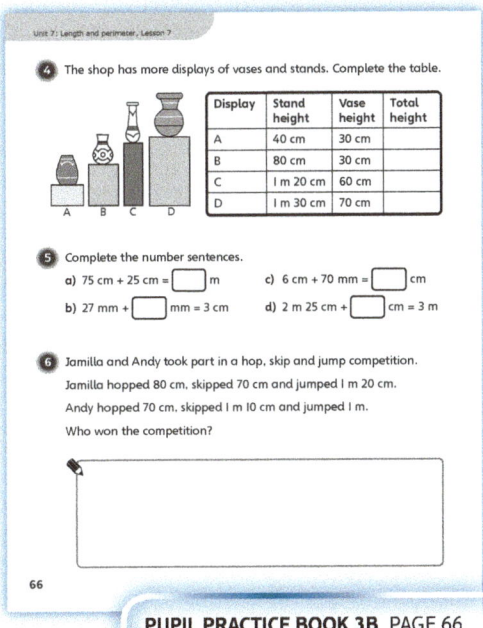

PUPIL PRACTICE BOOK 3B PAGE 66

Reflect

WAYS OF WORKING Independent thinking

IN FOCUS This question involves two of the main learning points from the lesson: the need to convert units and the need to deal with an answer that 'crosses the unit boundary'. In this case, it is necessary to work out 1 m 70 cm + 60 cm = 170 cm + 60 cm = 230 cm = 2 m 30 cm.

ASSESSMENT CHECKPOINT Check each of the required stages of the solution and consider:
- Do children recognise the need to add in this situation?
- Have they converted to a common unit (centimetres) before adding?
- Have they carried out the addition correctly and recorded the answer appropriately?
- Are children able to explain the mistake that Zac has made?

Note that some children may already be able to compress these steps and carry out the whole process mentally. In this case, encourage them to explain their steps to show the process that they used.

ANSWERS Answers for the **Reflect** part of the lesson can be found in the *Power Maths* online subscription.

After the lesson ⏸

- What approaches to calculation did you see in the lesson?
- Are children confident in converting between centimetres and metres?

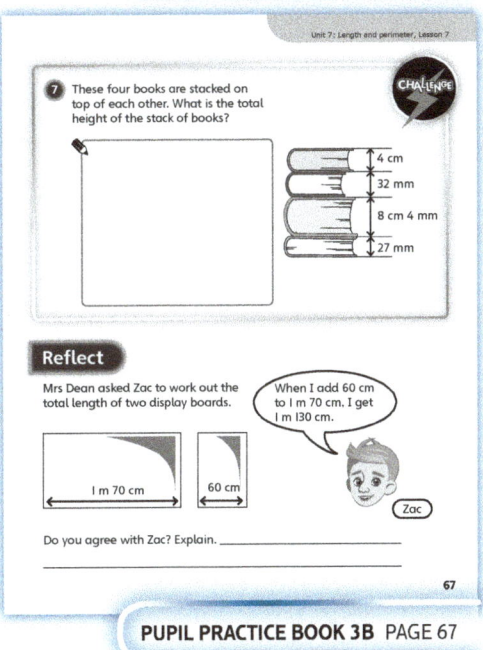

PUPIL PRACTICE BOOK 3B PAGE 67

129

Unit 7: Length and perimeter, Lesson 8

Subtract lengths

Learning focus
In this lesson, children will use subtraction to find the difference between two lengths given in centimetres, metres or simple combinations of both units. They will convert answers into either centimetres or metres as appropriate.

Before you teach
- What strategies do you expect children to use for subtraction?
- Are children confident in converting between centimetres and metres?

NATIONAL CURRICULUM LINKS

Year 3 Measurement

Measure, compare, add and subtract: lengths (m/cm/mm); mass (kg/g); volume/capacity (l/ml).

ASSESSING MASTERY

Children can find the difference between pairs of lengths given in metres, centimetres or simple combinations of both units and will select the appropriate methods depending on the question.

COMMON MISCONCEPTIONS

As with the previous lesson, a likely source of errors here arises from converting between units. Check that children understand that, for example, 1 m 50 cm and 150 cm are equivalent. Ask:
- *How many centimetres are in 1 m 50 cm? How many whole metres are in 1 m 50 cm?*

STRENGTHENING UNDERSTANDING

As with addition in the previous lesson, the number line provides a very direct and clear means of modelling the various calculations that are encountered in this lesson. Make sure that this is used appropriately. In many cases, the number line will only be needed to set up or justify the choice of calculation. The actual subtraction may be carried out mentally.

GOING DEEPER

More confident learners may be ready to go on to more complex conversions, for example, calculating the difference between 1 m 42 cm and 68 cm.

KEY LANGUAGE

In lesson: metre, centimetre, convert, difference, subtract, solve

STRUCTURES AND REPRESENTATIONS

Number line

RESOURCES

Optional: number lines, ribbon or string or paper strips to model some of the calculations, multilink cubes, toy animal

 In the eTextbook of this lesson, you will find interactive links to a selection of teaching tools.

Quick recap
Write the following numbers on the board:

46, 95, 139, 375.

Ask children to find the difference between two of the numbers. How many difference can they find?

Unit 7: Length and perimeter, Lesson 8

Discover

WAYS OF WORKING Pair work

ASK

- Question 1 a): *How long is the whole board? Could you fit it in the classroom?*
- Question 1 a): *Where has Holly drawn a line to help her cut the board?*

IN FOCUS Question 1 a) uses a simple practical context to model the process of subtraction as 'taking away'. There is an opportunity here for children to work with lengths expressed as centimetres or in mixed units as metres and centimetres.

PRACTICAL TIPS Ask children to use equipment such as multilink cubes to make a long line that is 2 m 50 cm long. Encourage them to break the line around the 1 m position and again at the 2 m position. What length do they have left? Ask them to create four 1 m lengths. They can arrange them into a square on the floor and then put a toy animal into the middle to represent the guinea pig run.

ANSWERS

Question 1 a): There will be 1 m 50 cm or 150 cm left.

Question 1 b): Yes. There is 1 m 50 cm left, so another 1 m can be cut from this and still leave 50 cm.

PUPIL TEXTBOOK 3B PAGE 92

Share

WAYS OF WORKING Whole class teacher led

ASK

- Question 1 a): *Flo used a number line to get the same answer as Ash. What kind of calculation has she done?*
- Question 1 b): *If you use a column subtraction, what is the whole number you start with? What is the number you are subtracting from the whole?*

IN FOCUS Although the calculations involved in questions 1 a) and b) should be quite straightforward, it is important to make sure that children are confident in moving between the different representations of the lengths that are involved. Focus on ensuring that they are comfortable with the equivalence between 1 m 50 cm, 150 cm, and one and a half metres.

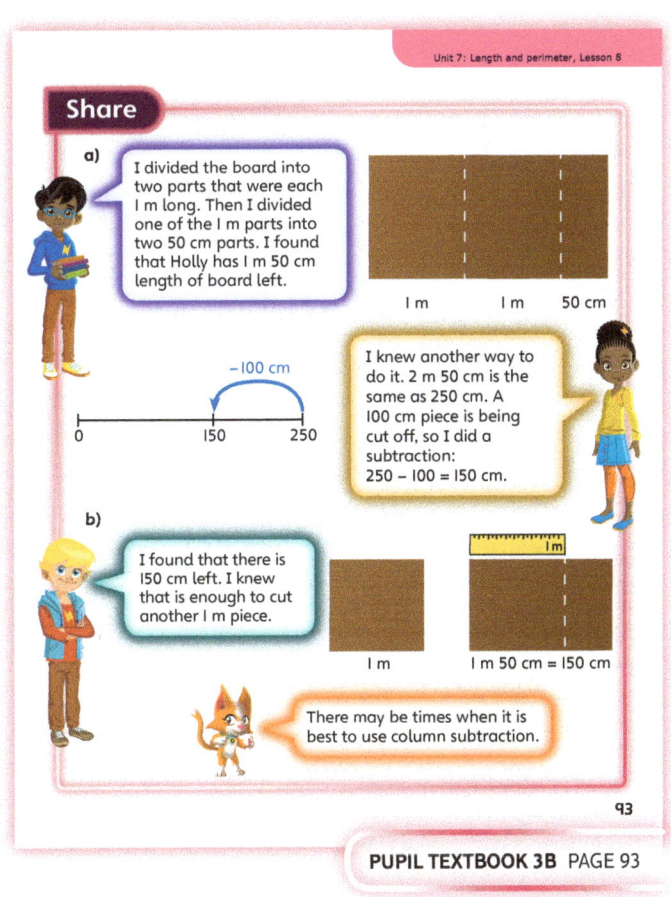

PUPIL TEXTBOOK 3B PAGE 93

131

Think together

WAYS OF WORKING Whole class teacher led (I do, We do, You do)

ASK

- Question 1 b): *What calculation will you need to do to work out how long the picture will be after Emma has cut off the end piece?*
- Question 2: *This is a bit different from the other questions. Why? How will you do it?*
- Question 3: *What other methods could you use to solve this? Would it be easier to use a written method?*

IN FOCUS Question 2 introduces the idea of subtraction as 'finding the difference'. This lends itself to different methods, for example, calculating the difference 85 – 65, or just looking at what you would need to add on to the 65 to make 85.

STRENGTHEN Check that children understand the scenario in question 3. This could be modelled in real life for children who would benefit from a more practical approach. Provide some ribbon or string and adapt the question so that children start with a more manageable amount, such as 3 m. Ask pairs to cut a 3 m length (check their measurements before they cut) and then ask them to cut off pieces of the required lengths (again, check their measurements first). Ask them to calculate how much should be left. Measure to check they are correct.

DEEPEN Deeper thinking can be elicited through discussion surrounding equivalence, including different representations of the same lengths and different methods for tackling the same calculations. More confident learners can be challenged throughout this lesson. Ask: *Can you write that in a different way? How many other ways could you do that calculation?*

ASSESSMENT CHECKPOINT Question 3 involves conversion of units and some more complicated calculations. Take note of the methods that children use here, as there are several steps and alternative methods available. Encourage the use of column subtraction or part-whole models, if needed. If children decide to change everything into centimetres, check they understand that 10 m = 1,000 cm.

ANSWERS

Question 1 a): 300 cm – 50 cm = 250 cm

Question 1 b): The picture is 75 cm now.

Question 2: Lee has cut off 2 cm or 20 mm.

Question 3: Children should suggest converting to centimetres: 1,000 cm – 150 cm – 77 cm = 773 cm or 7 m 73 cm or 1,000 cm – (150 cm + 77 cm) = 773 cm or 7 m 73 cm.

They could also suggest subtracting the metres first and then the centimetres: 150 cm + 77 cm = 227 cm, 10 m – 2 m = 8 m, 8 m – 27 cm = 7 m 73 cm.

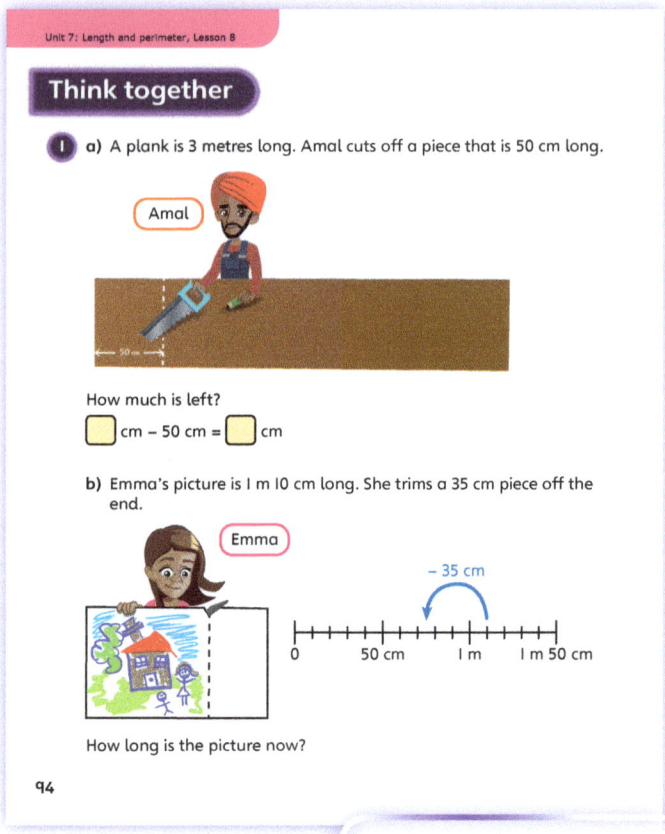

PUPIL TEXTBOOK 3B PAGE 94

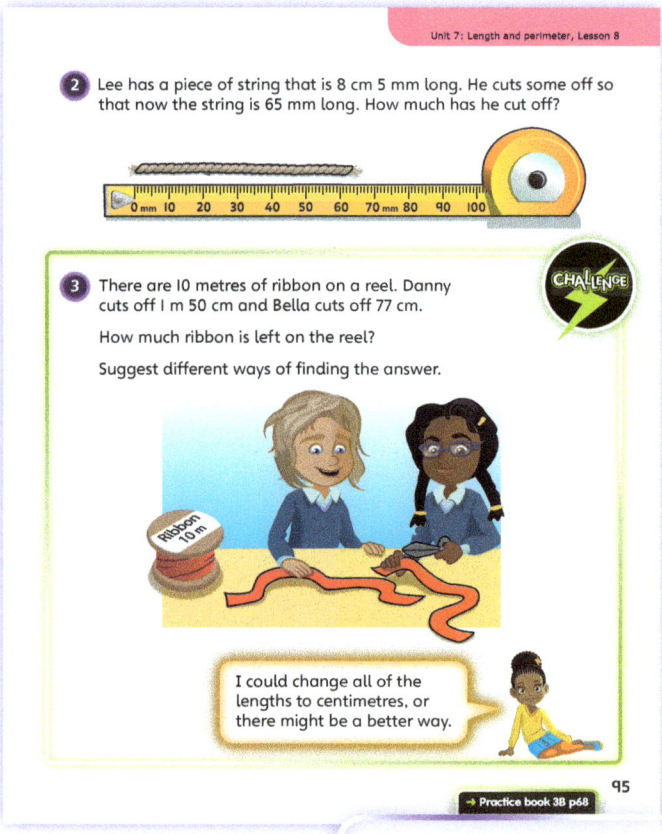

PUPIL TEXTBOOK 3B PAGE 95

132

Unit 7: Length and perimeter, Lesson 8

Practice

WAYS OF WORKING Independent thinking

IN FOCUS Question ④ involves conversion between units, as well as some 'working backwards' to find required answers. The emphasis here (and in question ③) is on subtraction as difference. This provides a useful contrast to the idea of subtraction as 'taking away' as in question ①.

STRENGTHEN Questions ③ and ④ could be modelled using paper strips or other suitable materials. You could also encourage children to convert mixed units to centimetres where this is helpful and to use a number line to help with the calculations.

DEEPEN More confident learners could make up further examples for question ③, possibly including more complicated calculations such as a flower of length 1 m 23 cm in a 78 cm vase.

ASSESSMENT CHECKPOINT Check that children can answer all of the parts of question ④. Support children in deciding what conversion to do in each case, and what would be the best unit to use for their answer.

ANSWERS Answers for the **Practice** part of the lesson can be found in the *Power Maths* online subscription.

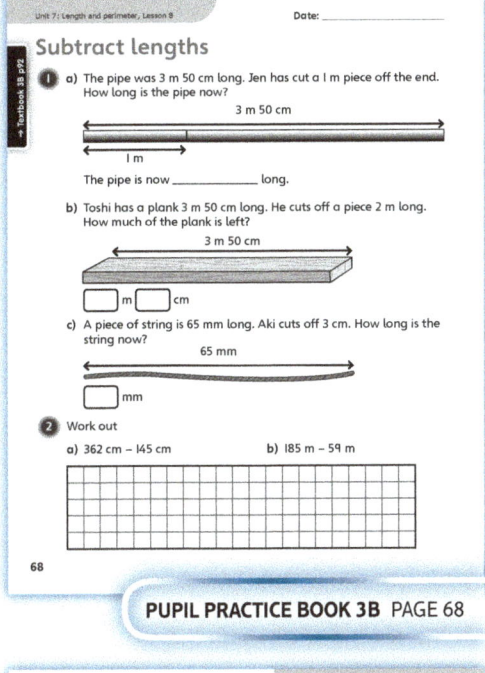

PUPIL PRACTICE BOOK 3B PAGE 68

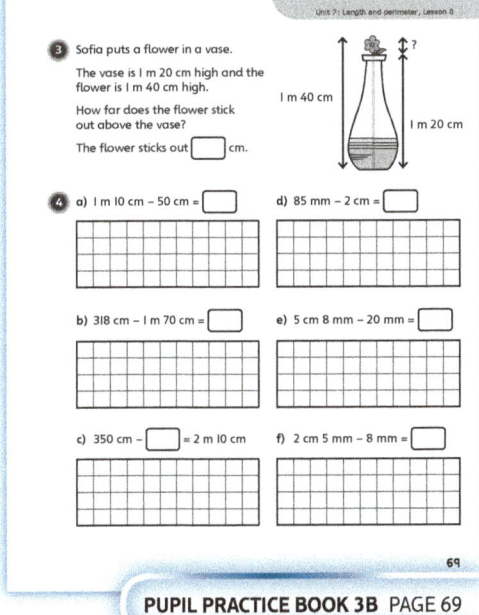

PUPIL PRACTICE BOOK 3B PAGE 69

Reflect

WAYS OF WORKING Independent thinking

IN FOCUS The aim here is to get children to articulate and reflect on the process of using subtraction to solve a problem involving lengths with different units where some exchange may be required. Different methods may be appropriate for the two examples: the first may be best tackled as a column subtraction, while the second could be done mentally.

ASSESSMENT CHECKPOINT Check that children have explained how they would handle the conversion of units, and that they have chosen appropriate methods for each part of the question.

ANSWERS Answers for the **Reflect** part of the lesson can be found in the *Power Maths* online subscription.

After the lesson

- Did children understand the need for subtraction when tackling the questions in this lesson?
- Were some of the models of subtraction (take away/find the difference) used more than others? Could children select these models as needed in particular questions?
- How confident were children in converting between units (metres and centimetres) where required?

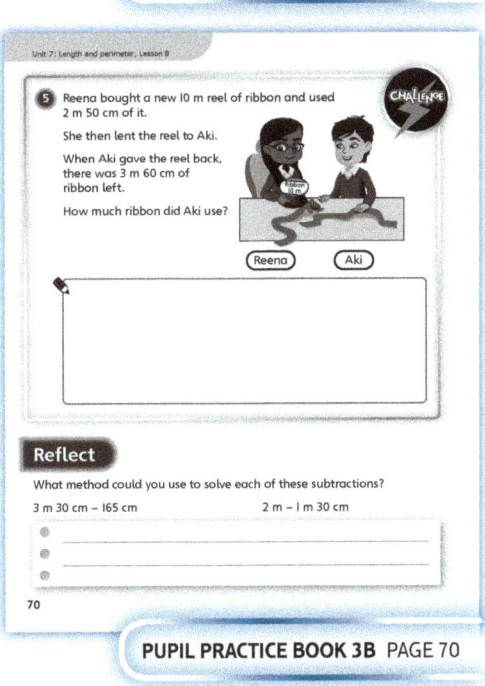

PUPIL PRACTICE BOOK 3B PAGE 70

Unit 7: Length and perimeter, Lesson 9

Measure perimeter

Learning focus
In this lesson, children will measure the perimeters of a range of shapes in both centimetres and millimetres.

Before you teach
- Are children secure in their addition and subtraction calculation skills from Lessons 7 and 8 of this unit?
- Are all children able to accurately measure the lengths of lines in cm and mm?

NATIONAL CURRICULUM LINKS

Year 3 Measurement
Measure the perimeter of simple 2D shapes.

ASSESSING MASTERY

Children can explain that the perimeter is the total measurement of the distance around a shape and can be found by totalling all the side lengths of a shape. They can accurately measure the perimeters of shapes.

COMMON MISCONCEPTIONS

Children sometimes measure and add the same side length more than once when calculating the perimeter. In order to help them avoid this, encourage them to work systematically from a starting point moving in one direction around the shape, recording each side length as they go. Ask:
- *Where did you start measuring? How many measurements have you worked out? Which ones do you have left to do? Is there a way you can keep track of which sides you have already measured?*

STRENGTHENING UNDERSTANDING

Ensure children are given a range of practical experiences of measuring a perimeter, including using real-life objects and shapes they have constructed. Use a range of representations, including the number line, in order to support children with the calculations involved when adding multiple numbers together to calculate the perimeter.

GOING DEEPER

Children should be able to apply their knowledge of perimeter and their subtraction skills to solve problems where the perimeter and a side length are known and they have to complete the shape. For example, ask: *This shape has a perimeter of 20 cm and this side is 10 cm long. Can you complete the shape? Is there more than one way to do it?*

KEY LANGUAGE

In lesson: perimeter, distance, length, metres (m), centimetres (cm), millimetres (mm), polygon, diagonal
Other language to be used by the teacher: total, addition, subtraction

STRUCTURES AND REPRESENTATIONS

Number lines

RESOURCES

Optional: number lines, rulers, geoboards, wool, card, string, art straws, scissors

 In the eTextbook of this lesson, you will find interactive links to a selection of teaching tools.

Quick recap
Ask children to draw a line that is 5 cm long. Ask them to get a partner to check that the line is the correct length. Ask children to draw other lines and get their partner to check them. Use a mix of lines in cm and mm.

Unit 7: Length and perimeter, Lesson 9

Discover

WAYS OF WORKING Pair work

ASK
- Question 1 a): *How can you work out how much red wool has been used to make the triangle?*
- Question 1 a): *How are you going to make sure that your measurements are accurate?*
- Question 1 b): *Does the blue polygon look like it has used more wool or less wool than the red triangle? How can you check?*

IN FOCUS Questions 1 a) and 1 b) introduce children to the concept of perimeter for the first time. The term 'perimeter' is deliberately not used at this stage. Instead, through your questioning and discussion with children, begin to introduce the term 'perimeter' as the distance around the sides of a shape.

PRACTICAL TIPS This **Discover** activity would be a good one to replicate practically in your classroom using either wool and cards, string and geoboards, or other construction materials such as art straws.

ANSWERS

Question 1 a): Alex used 12 cm of wool to make the red triangle.

Question 1 b): Alex used 10 cm of wool to make the blue polygon.
She used more wool to make the red triangle.

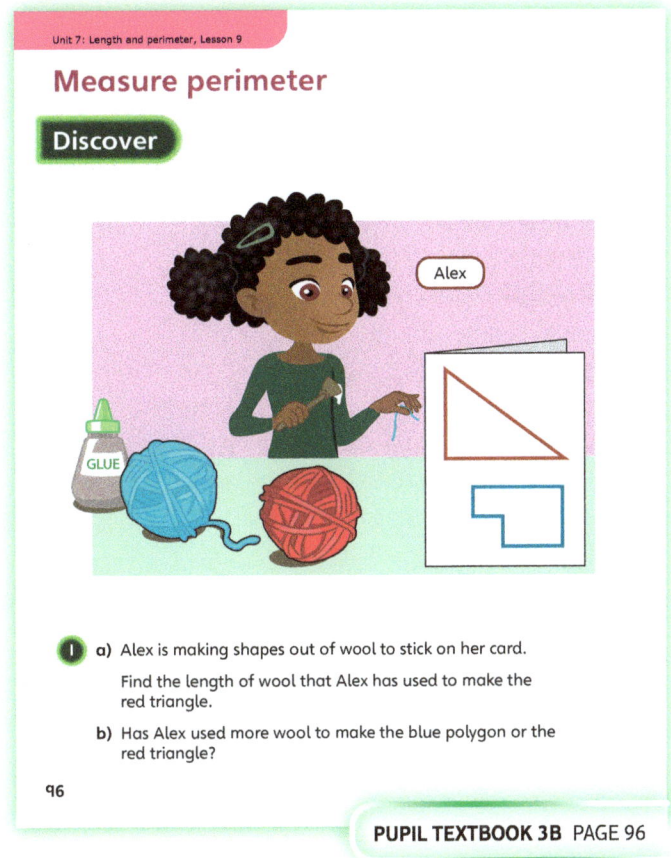

PUPIL TEXTBOOK 3B PAGE 96

Share

WAYS OF WORKING Whole class teacher led

ASK
- Question 1 a): *How did you solve this problem? What approaches did you use? Did anyone do anything differently?*
- Question 1 a): *If you knew the length of each side, how could you work out the amount of wool used for the triangle?*
- Question 1 a): *How did you make sure your measurements were accurate?*
- Question 1 b): *Did you use any models to help you add up the side lengths?*
- Question 1 b): *Did anyone add the side lengths in any other way?*

IN FOCUS In this part of the lesson, through the characters and class discussion, introduce children to the term 'perimeter' for the measurement of the distance around the outside of a shape. In questions 1 a) and 1 b), children are introduced to the idea that they can calculate the perimeter by adding the lengths of all the sides of a shape together.

Encourage children to use their addition skills and models or representations they have developed and applied previously in Year 3 to help them add the side lengths together.

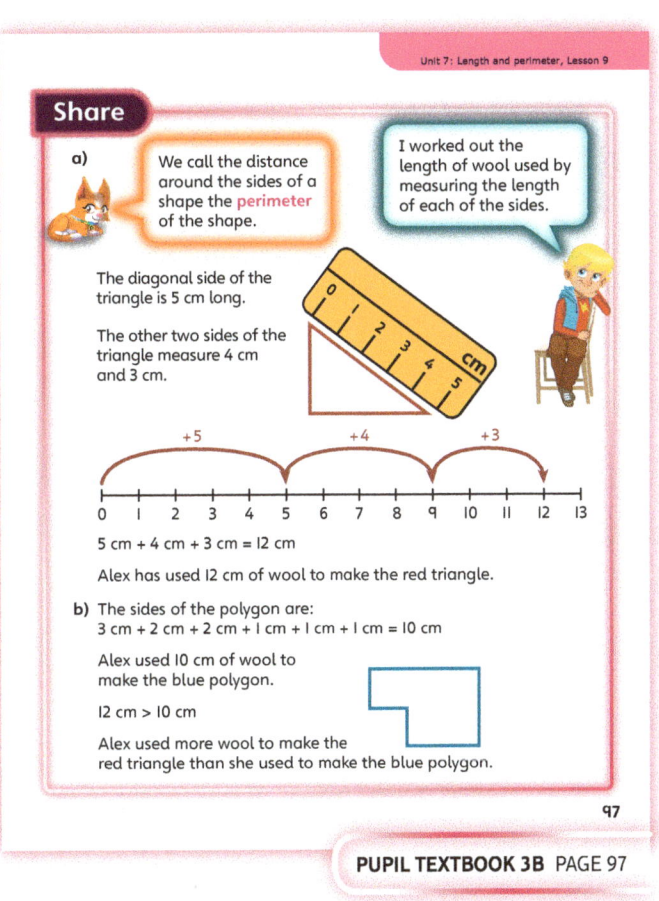

PUPIL TEXTBOOK 3B PAGE 97

135

Think together

WAYS OF WORKING Whole class teacher led (I do, We do, You do)

ASK

- Question ①: *How can you work out how much wool you would need?*
- Question ①: *What is the mathematical name for what you are calculating?*
- Question ②: *How can you make sure you add up all of the side lengths? What strategies can you use?*
- Question ②: *Do you need to measure all of the sides or do you know that some of them are the same length?*
- Question ②: *How will you convert the measurements so you are working with the same units?*

IN FOCUS Question ② introduces children to calculating the perimeter of shapes with mixed units of measurement. It is a good opportunity to revisit the discussions around choosing an appropriate unit of measurement from earlier on in this unit. Ask children to consider whether any of the sides are the same length. Can they use this information to reduce the number of measurements they need to do?

STRENGTHEN When exploring question ②, encourage children to discuss how many millimetres are in a centimetre. Continue to support children in using a range of calculation strategies for adding the side lengths together.

DEEPEN Children should begin to understand that there is often more than one possible shape that has a set perimeter. Question ③ helps them to develop this understanding by asking them to draw a shape with a set perimeter, rather than measuring a given shape. Draw children's attention to Ash's comment. Encourage children to investigate if they can draw any more shapes with a perimeter of 14 cm. Consider providing string or wool and some scissors, so that children can form shapes and measure them.

ASSESSMENT CHECKPOINT Use question ② to assess if children can calculate the perimeter based on the measurements they take. Do they measure all the sides and record them accurately? Can they convert between units of measurement successfully?

ANSWERS

Question ①: The blue rectangle would need 10 cm of wool. The red polygon would need 8 cm of wool.

Question ②: The hexagon is 3 cm × 6 = 18 cm or 180 mm. The square is 25 mm × 4 = 100 mm or 10 cm. The rectangle is 15 mm × 2 + 50 mm × 2 = 30 mm + 100 mm = 130 mm or 13 cm.

Question ③: Children should draw a shape with a perimeter of 14 cm. For example, a rectangle with 2 sides of 2 cm each and 2 sides of 5 cm each.

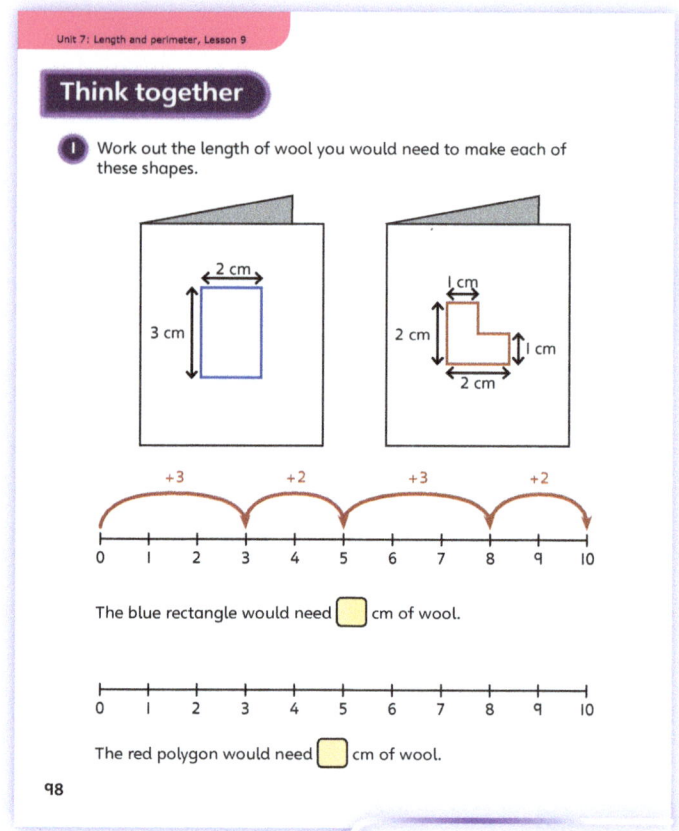

PUPIL TEXTBOOK 3B PAGE 98

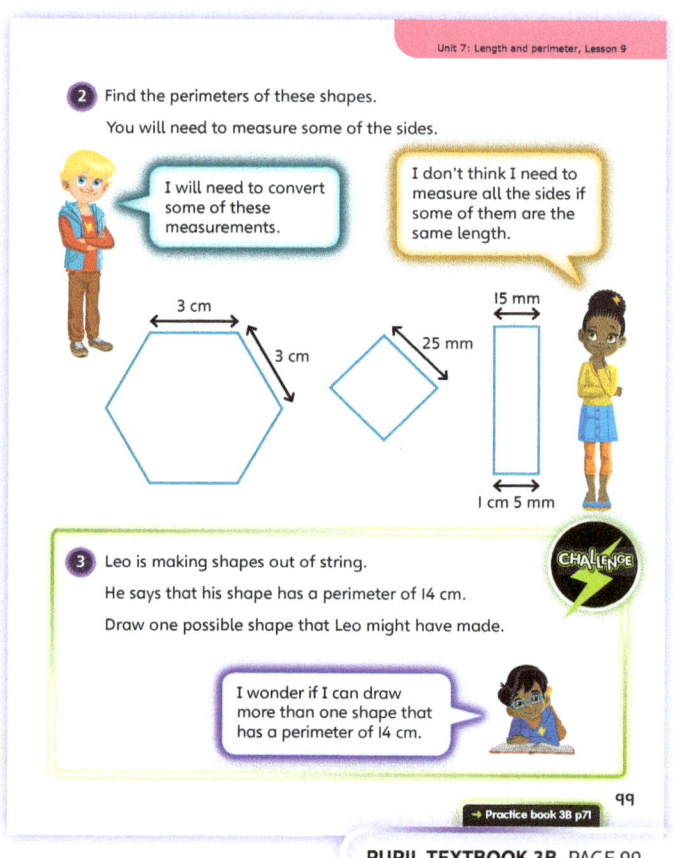

PUPIL TEXTBOOK 3B PAGE 99

Unit 7: Length and perimeter, Lesson 9

Practice

WAYS OF WORKING Independent thinking

IN FOCUS Question ③ asks children to draw a shape with a set perimeter, rather than measuring an existing shape. Children should use their knowledge of how a perimeter is calculated in order to achieve this. Their drawings will vary, so this question may provide a good discussion opportunity around how different shapes can share the same perimeter.

Question ④ asks children to order shapes based on the lengths of their perimeters and should provide some useful discussion points, as the shape that may appear largest at first glance does not always have the largest perimeter.

STRENGTHEN The perimeter questions in this lesson involve the addition of three or more numbers together. Children should be supported with their calculations using a range of models, including the number line. They should be reminded and encouraged to use their learning from earlier in Year 3 in order to complete the calculations.

DEEPEN Children should be able to apply their knowledge of perimeter to solve problems where they are given a perimeter measurement plus one or more side lengths. They then use this information to work out the lengths of the remaining sides.

ASSESSMENT CHECKPOINT Use questions ① and ② to assess if children are able to independently measure and calculate the perimeter of a given shape in cm (as in question ①) and in mm (as in question ②).

ANSWERS Answers for the **Practice** part of the lesson can be found in the *Power Maths* online subscription.

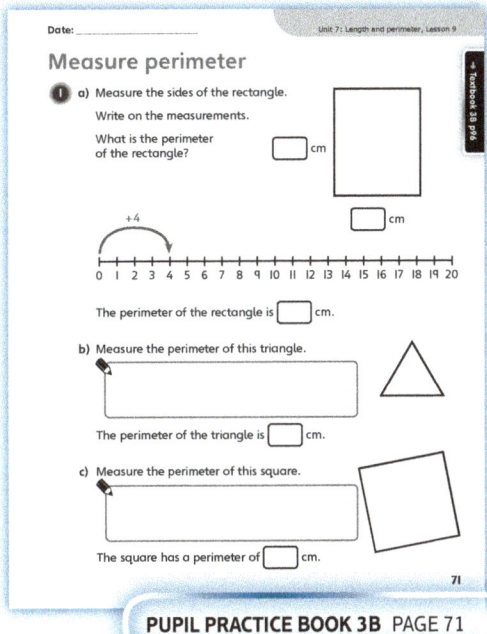

PUPIL PRACTICE BOOK 3B PAGE 71

Reflect

WAYS OF WORKING Independent thinking

IN FOCUS Children reflect on what they have done in this lesson to explain how to find the perimeter of a shape. Encourage them to collaborate with a partner if needed. Discuss methods as a whole class.

ASSESSMENT CHECKPOINT Look for children who say that they would measure the sides of the shape and then add up the measurements to find the distance all the way around.

ANSWERS Answers for the **Reflect** part of the lesson can be found in the *Power Maths* online subscription.

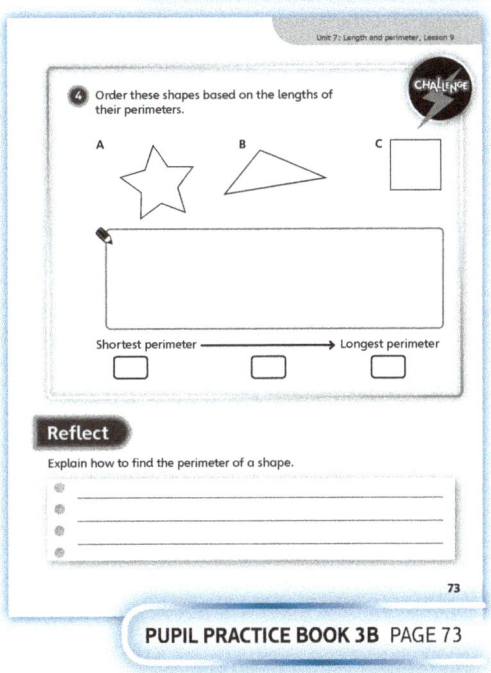

PUPIL PRACTICE BOOK 3B PAGE 73

After the lesson

- What practical experiences can you give children in measuring and calculating the perimeter of a shape?
- Are children secure with how to calculate a perimeter?

137

Unit 7: Length and perimeter, Lesson 10

Calculate perimeter

Learning focus
In this lesson, children will calculate perimeter in situations where side lengths are given but they cannot physically measure them for themselves.

Before you teach
- Were there any common misconceptions from Lesson 8? If so, how will you adapt this lesson so that you are able to address these?

NATIONAL CURRICULUM LINKS

Year 3 Measurement
Measure the perimeter of simple 2D shapes.

ASSESSING MASTERY

Children can explain that the perimeter can be found by totalling all the side lengths of a shape and can accurately calculate the perimeters of shapes when given the side lengths. They can also calculate missing side length(s) based on a given perimeter.

COMMON MISCONCEPTIONS

Children sometimes add the same side length more than once when calculating the perimeter. In order to avoid this, encourage them to work systematically when adding the side lengths, working from a starting point and moving in one direction around the shape, recording each side length as they go. Ask:
- *How are you going to add your lengths to make sure you do not miss any or add any twice?*

Children may believe they cannot calculate the perimeter without measuring. Encourage children to make the connection between measuring the side lengths and adding them together, and using side lengths that have already been given and adding those together. Ask:
- *Do you need to measure any of the sides or have you already been given all the information you need to calculate the perimeter?*

STRENGTHENING UNDERSTANDING

Use a range of representations, including the number line, to support children when adding multiple numbers together in order to calculate the perimeter. When a diagram is not provided, encourage children to sketch their own diagram to help them visualise the problem and the question being asked.

GOING DEEPER

Encourage children to find different ways of creating the same perimeter. For example, ask: *I have a quadrilateral with a perimeter of 30 cm. One of the side lengths is 12 cm; what could the other side lengths be?*

KEY LANGUAGE

In lesson: perimeter, distance, metres (m), centimetres (cm), millimetres (mm), quadrilateral, triangle, rectangle, measure

Other language to be used by the teacher: total, addition, multiplication, subtraction, repeated addition

STRUCTURES AND REPRESENTATIONS

Number lines

RESOURCES

Optional: number lines, rulers, birds-eye-view of school playground (taken from the internet)

 In the eTextbook of this lesson, you will find interactive links to a selection of teaching tools.

Quick recap
Provide children with a 6 cm × 4 cm rectangle. Ask children to find the perimeter of the rectangle and ask them to explain how they worked it out. Discuss what the perimeter of the shape is.

Unit 7: Length and perimeter, Lesson 10

Discover

WAYS OF WORKING Pair work

ASK

• Question 1 a): *What word can you use to describe the distance around the outside of a shape or area?*
• Question 1 a): *Do you need to measure the sides? How else could you work out the perimeter?*

IN FOCUS Question 1 a) introduces children to the concept of calculating a perimeter by using the given information rather than physically measuring the sides of the shape. Children should be encouraged to use their knowledge of perimeter and its definition from Lesson 9 to help them calculate the perimeter of the playground and the amount of fence needed.

Question 1 b) requires children to complete a calculation based on the value of the perimeter they have found in question 1 a).

PRACTICAL TIPS The activity could be made immediately relevant to children by asking them to recreate a similar scenario in their own environment. Ask: *Suppose we wanted to paint a mural around the walls of our classroom/school hall. Can you work out the perimeter of the school hall/classroom?*

Alternatively, use birds-eye-view images of your playground (for example, taken from the internet) and approximate measurements to replicate question 1 a).

ANSWERS

Question 1 a): The school will need to buy 24 metres of fence.

Question 1 b): Amelia runs 48 metres.

Share

WAYS OF WORKING Whole class teacher led

ASK

• Question 1 a): *Could you measure the side lengths to work out the perimeter? How did you calculate the perimeter?*
• Question 1 a): *Did you use any representations to help you add up the side lengths?*
• Question 1 a): *Did anyone add the side lengths in any other way?*
• Question 1 b): *How did you work out how far Amelia had run? Did anyone calculate this in any other way?*

IN FOCUS In question 1 a), encourage children to systematically add the given side lengths in order to calculate the perimeter of the playground. Watch that they do not add the same side length more than once; suggesting they work in one direction around the shape can help with this. Children should also be urged to use and apply the efficient calculation skills and strategies they learnt earlier in Year 3 in order to calculate the value of the perimeter. For example, they could look for bonds to 10 (7 m + 3 m and 5 m + 5 m) within the calculation.

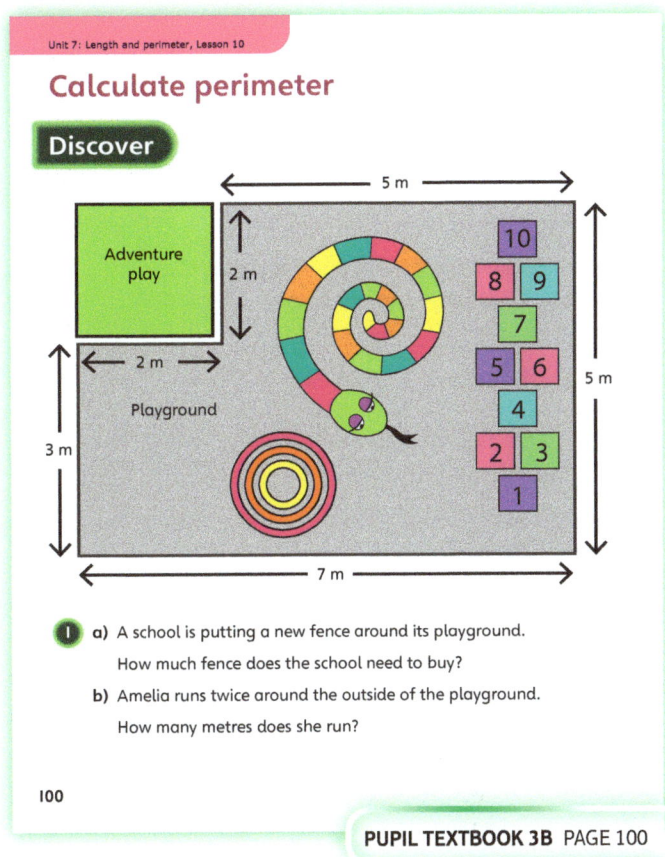

PUPIL TEXTBOOK 3B PAGE 100

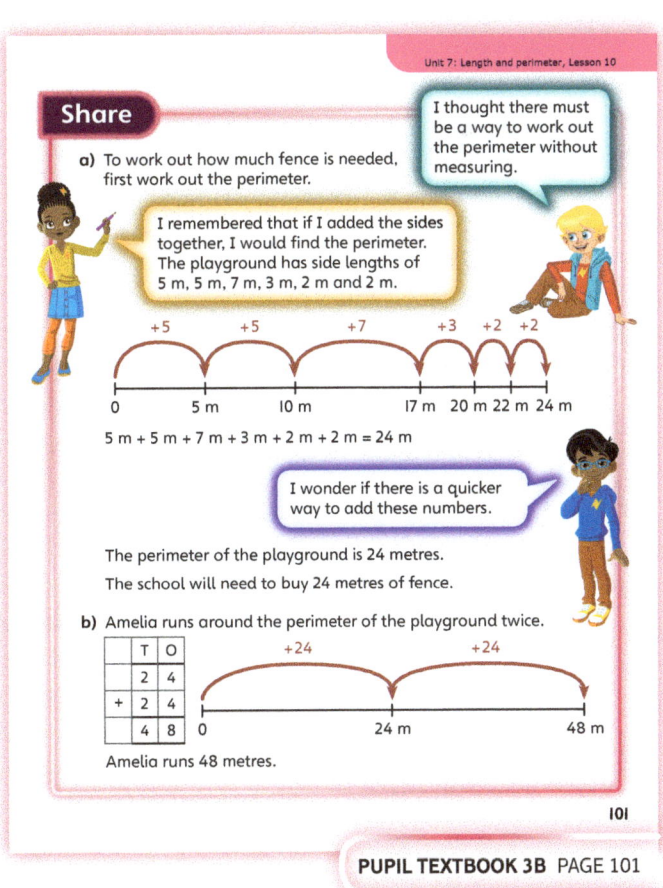

PUPIL TEXTBOOK 3B PAGE 101

139

Unit 7: Length and perimeter, Lesson 10

Think together

WAYS OF WORKING Whole class teacher led (I do, We do, You do)

ASK

- Question ❶: *How can you work out the perimeter? What are the lengths of each of the sides of the adventure play area?*
- Question ❶ b): *How can you make sure you add up all of the side lengths and do not miss any?*
- Question ❶ b): *What strategies or methods could you use to help you add the side lengths?*
- Question ❶ b): *How can you work out three times the perimeter?*
- Question ❸: *If you know the perimeter and two of the side lengths, how can you work out the missing side lengths?*
- Question ❸: *Do you think there is more than one possible answer?*

IN FOCUS In question ❷, children will need to recognise that the unit of measurement has changed from metres to centimetres but that they can still calculate the perimeter from given side lengths – working in centimetres does not always mean the lengths can be physically measured.

STRENGTHEN You can support children in question ❶ b) by encouraging them to sketch each shape, including labelling the side lengths. Work methodically around the shape. In questions ❶ b) and ❷, continue to support children with a range of calculation strategies in order to add the side lengths together and to multiply the perimeter in question ❶ b).

DEEPEN Children should understand that there is often more than one possible shape with the same perimeter. Question ❸ helps them develop this understanding by presenting two side lengths: children have to use this information to work out possible lengths for the missing sides. Children should be encouraged to consider if there is more than one possible answer to this question. They should then work systemically to find all the possible combinations of whole number side lengths.

ASSESSMENT CHECKPOINT Use questions ❶ and ❷ to assess if children can calculate the perimeter based on given measurements. Note whether children notice the change of unit between questions ❶ and ❷.

ANSWERS

Question ❶ a): The perimeter of the adventure play area is 8 metres.

Question ❶ b): The perimeter of the playing field is 110 metres. The children run 330 metres.

Question ❷: The perimeter of the shape is 65 cm.

Question ❸: 6 m + 4 m = 10 m
18 m – 10 m = 8 m
The other two sides added together must be 8 m. They could be: 7 m and 1 m, 6 m and 2 m, 5 m and 3 m or 4 m and 4 m.

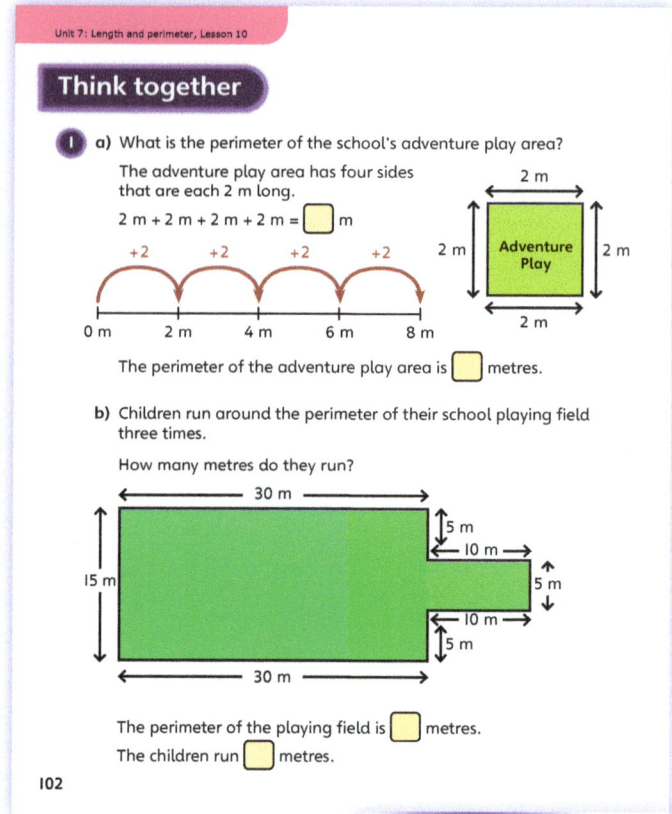

PUPIL TEXTBOOK 3B PAGE 102

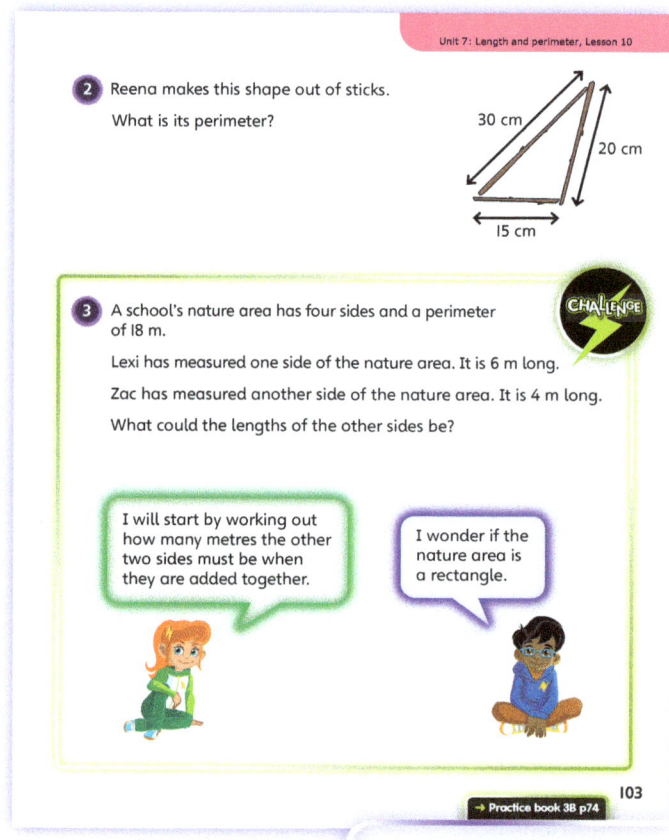

PUPIL TEXTBOOK 3B PAGE 103

Unit 7: Length and perimeter, Lesson 10

Practice

WAYS OF WORKING Independent thinking

IN FOCUS In question 2, children are asked to work out the missing side lengths of shapes based on a given perimeter. Encourage them to do this through subtraction, using their understanding that the perimeter is the total length of all of the sides (the whole).

STRENGTHEN Support children with the calculation elements in these questions using practical resources and strategies from the addition, subtraction, multiplication and division units from Year 3.

DEEPEN Children should have a secure understanding that there can be many different shapes and dimensions that share the same perimeter. Present children with irregular polygons that have a known perimeter and at least two side lengths not labelled. Encourage them to work systematically to find all possible values for the missing side lengths.

THINK DIFFERENTLY Question 4 focuses on children using their estimation skills to estimate the perimeter of certain everyday objects. Encourage children to consider the units of measurement, using their skills and knowledge from earlier in this unit.

ASSESSMENT CHECKPOINT Use question 2 to help you assess if children have a full understanding of the meaning of perimeter. This will be indicated if they are able to accurately find the missing lengths using subtraction.

ANSWERS Answers for the **Practice** part of the lesson can be found in the *Power Maths* online subscription.

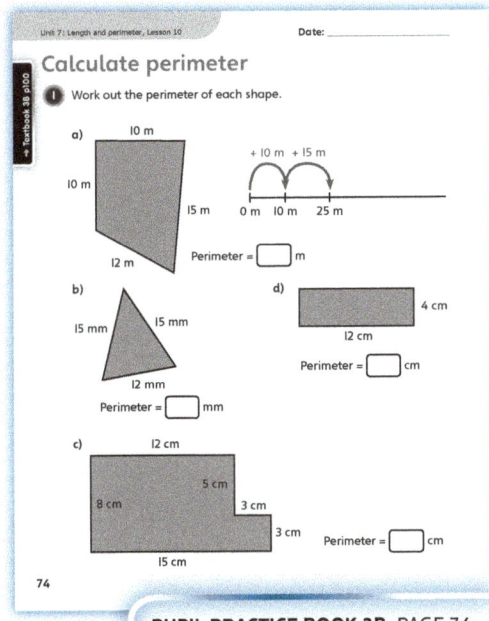

PUPIL PRACTICE BOOK 3B PAGE 74

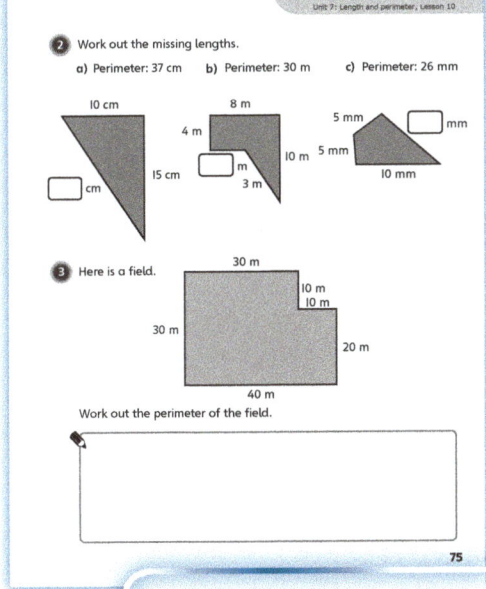

PUPIL PRACTICE BOOK 3B PAGE 75

Reflect

WAYS OF WORKING Independent thinking

IN FOCUS This question helps children reflect on the generalisations they have made throughout Lessons 9 and 10 of this unit.

ASSESSMENT CHECKPOINT Use this question to assess if children can explain how to work out the perimeter of any shape by adding the side lengths together.

ANSWERS Answers for the **Reflect** part of the lesson can be found in the *Power Maths* online subscription.

After the lesson

- Are children secure with the concept of perimeter?
- How can you continue to use the term 'perimeter' in your day-to-day classroom activities?

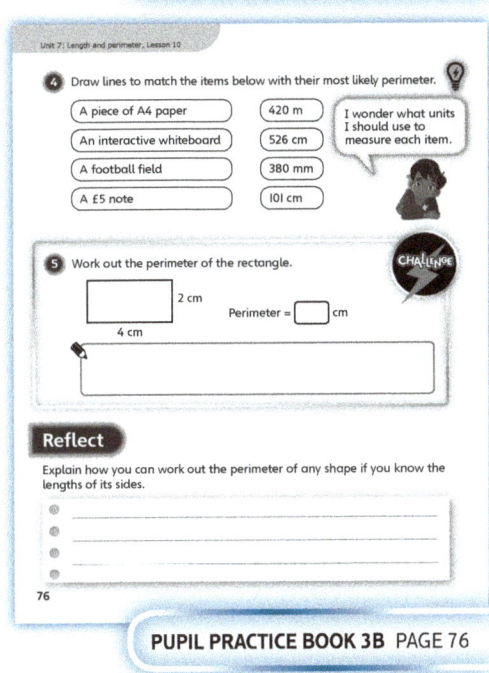

PUPIL PRACTICE BOOK 3B PAGE 76

Unit 7: Length and perimeter, Lesson 11

Problem solving – length

Learning focus
In this lesson, children will solve one-step problems involving length. They will apply their learning from all previous lessons in this unit.

Before you teach
- How secure are children in the concepts taught up to now in this unit?
- How will you support children who find reading challenging?

NATIONAL CURRICULUM LINKS

Year 3 Measurement

Measure the perimeter of simple 2D shapes.

ASSESSING MASTERY

Children can apply their knowledge, understanding and skills in measuring length in order to solve one-step problems. They can identify the operations needed, using pictures and diagrams to support their thinking.

COMMON MISCONCEPTIONS

Children may incorrectly identify the operation needed. Ask:
- *What calculation do you need to do to solve this question? Can you model or draw the problem to make it clearer? Do you recognise any words that link to calculations you have done before?*

STRENGTHENING UNDERSTANDING

Read the problems to children or ask them to read them to each other. Give children time to discuss the problem to ensure that they understand it. Children could then draw a picture to represent the problem. If necessary, support children with drawing the picture.

GOING DEEPER

Ask children to generate their own word problems and then try them out on each other. They could try to come up with a problem related to any of the other lessons in this unit, such as perimeter or converting from cm to m.

KEY LANGUAGE

In lesson: length, height, centimetres (cm), millimetres (mm), metres (m), how long, compare, subtraction, calculation, method

Other language to be used by the teacher: measure, how wide, perimeter

STRUCTURES AND REPRESENTATIONS

Bar model

RESOURCES

Optional: ruler, metre stick, string, modelling clay, multilink cubes, scissors

 In the eTextbook of this lesson, you will find interactive links to a selection of teaching tools.

Quick recap
Write pairs of different lengths on the board and ask children to add and subtract them. This will remind them of key methods they need for this problem-solving lesson. Include an example where children may need to convert first.

Unit 7: Length and perimeter, Lesson 11

Discover

WAYS OF WORKING Pair work

ASK

- Question 1 a): *How long is the snake?*
- Question 1 a): *What operation do you need to do to find out if the snake can grow longer than 2 metres?*
- Question 1 a): *How could you write this as a calculation?*
- Question 1 b): *Is this a 'find the difference' question? What two values are you finding the difference between?*
- Question 1 b): *What information is important in order to solve the problem?*

IN FOCUS Question 1 a) focuses on children's ability to apply a range of skills learnt through the previous lessons, namely taking a measurement and repeated addition or multiplication. In order to find out how much the snake could grow, children need to look at the ruler to find the current length of the snake. They then multiply this value by 3.

PRACTICAL TIPS Model this question using a length of string, modelling clay or multilink cubes to form the snake. Use a metre stick and ask children to mark the start and end point of the snake. Repeat it three times to find the final length. Pose different scenarios with snakes of varying lengths. You could extend this to measuring other items in the classroom and then asking children to find twice their length or three times their length, and so on.

ANSWERS

Question 1 a): Andy is correct.

Question 1 b): The snake could grow another 65 cm.

PUPIL TEXTBOOK 3B PAGE 104

Share

WAYS OF WORKING Whole class teacher led

ASK

- Question 1 a): *Can you use a bar model to show the problem? How would you do this?*
- Question 1 a): *What word(s) in the picture gave you a clue about which operation to use?*
- Question 1 b): *Can you think of a similar problem?*

IN FOCUS Questions 1 a) and 1 b) highlight the importance of using representations to illustrate a problem. From the bar models, children can easily see what is required of them in order to solve the problem. Children need to identify the key information from the text and understand how it relates to the images and how they should then apply it to find the answers.

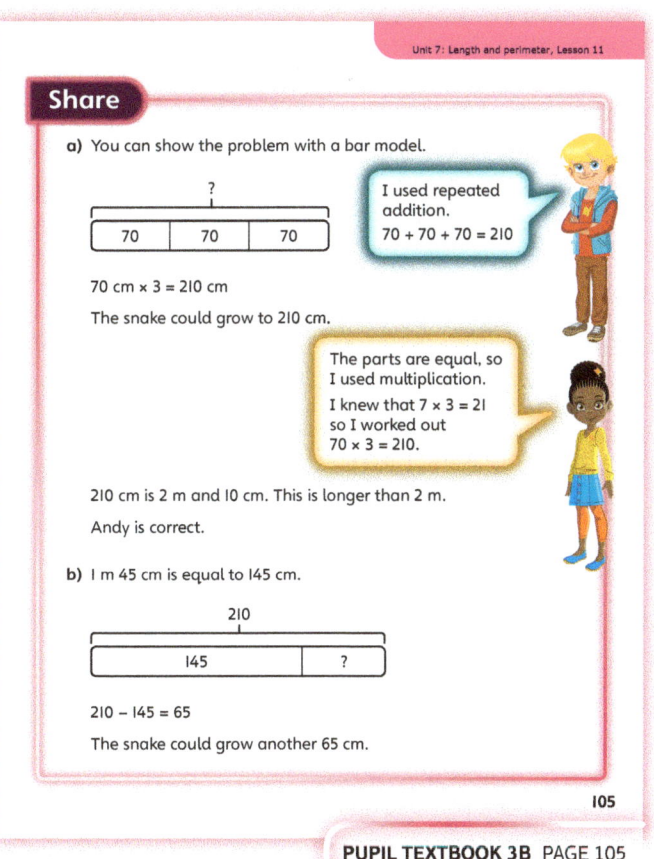

PUPIL TEXTBOOK 3B PAGE 105

143

Unit 7: Length and perimeter, Lesson 11

Think together

WAYS OF WORKING Whole class teacher led (I do, We do, You do)

ASK

- Question ❶: *Can you draw a picture to demonstrate the problem?*
- Question ❶: *Is there more than one way to solve this problem?*
- Question ❶: *Can you solve it using a different operation?*
- Question ❷: *What words were important for you to solve the problem? Why do you think Max's snail is pictured twice?*
- Question ❸: *How tall is the parrot? How can you find out the height of the budgie?*

IN FOCUS Question ❶ is represented as a multiplication problem but could also be solved by using repeated addition. Talk to children about how they would find the total length of the adult snake and ask them to discuss their ideas with a partner. Use this as an opportunity to reinforce the connection between repeated addition and multiplication. Do children recognise that they need to convert the 200 cm to 2 m or do they give the answer as 4 m 200 cm? If the latter, ask: *Is there a better way to show this? What could you convert 200 cm to?*

STRENGTHEN Where children are challenged by certain questions, read the problem to them and then ask them to tell you the problem in their own words. Allow children time to draw a picture to represent each problem and prompt them to explain what they need to do in order to solve it.

DEEPEN Ask children to come up with their own problems that use the same numbers and operations but in a different context.

ASSESSMENT CHECKPOINT At this point in the lesson, children should be becoming more confident in identifying the calculation needed to solve a word problem. They may begin to see that, for some problems, there may be more than one calculation that can solve them.

ANSWERS

Question ❶: The snake could grow to 6 m long.

Question ❷: Max's snail is 4 cm and 1 mm long.

Question ❸: The budgie is 7 cm and 5 mm tall.

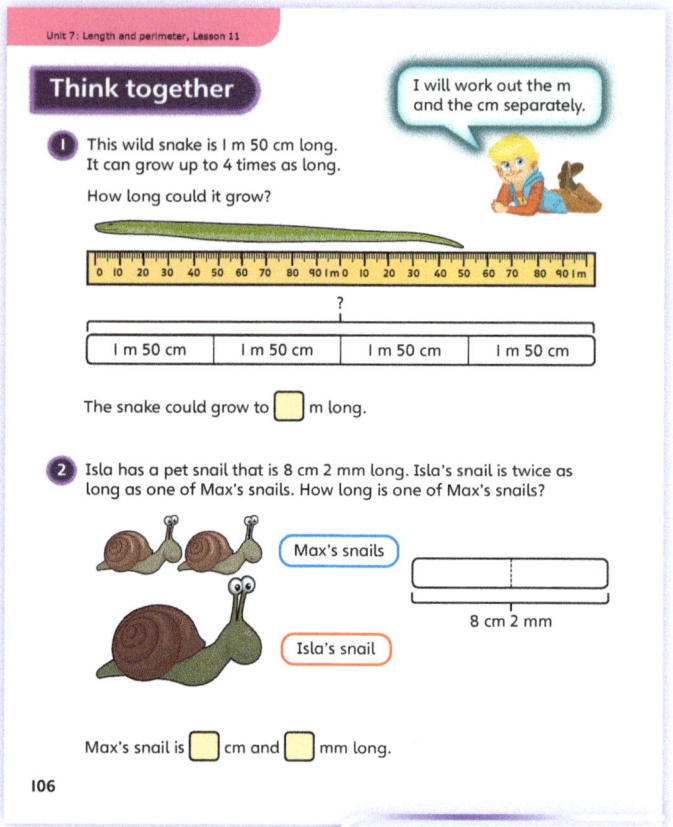

PUPIL TEXTBOOK 3B PAGE 106

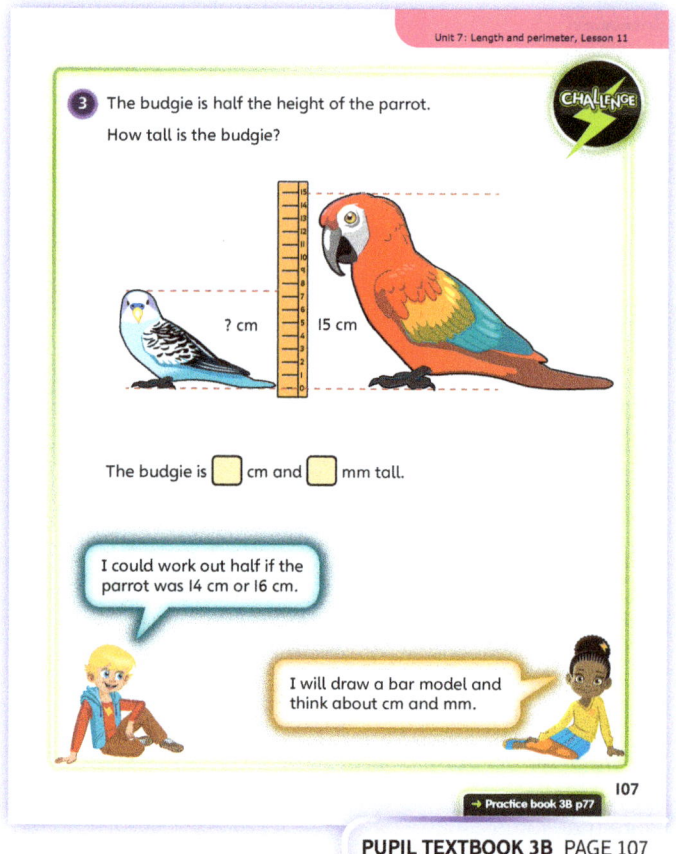

PUPIL TEXTBOOK 3B PAGE 107

Unit 7: Length and perimeter, Lesson 11

Practice

WAYS OF WORKING Independent thinking

IN FOCUS Questions 1, 2 and 3 offer opportunities for children to represent their thinking using bar models. It may be beneficial to remind children that they could use a bar model to support their thinking.

Question 7 could be made more practical by offering children appropriately sized pieces of string. As they cut the string in half, ask them to support their concrete representation using a pictorial representation of a bar model.

STRENGTHEN Ask children to work in pairs to read a problem to each other. When listening to the problem, prompt them to close their eyes and picture it in their heads and then support them to draw it. If necessary, help highlight the key vocabulary with children.

DEEPEN Ask children to compose their own problems, one for each operation. They can then test them out on a partner. Do they try to incorporate some of their learning from the previous lessons?

ASSESSMENT CHECKPOINT By the end of question 7, children should be able to identify the calculation needed to solve a word problem. They should be able to visualise and represent the problem pictorially and then abstractly.

ANSWERS Answers for the **Practice** part of the lesson can be found in the *Power Maths* online subscription.

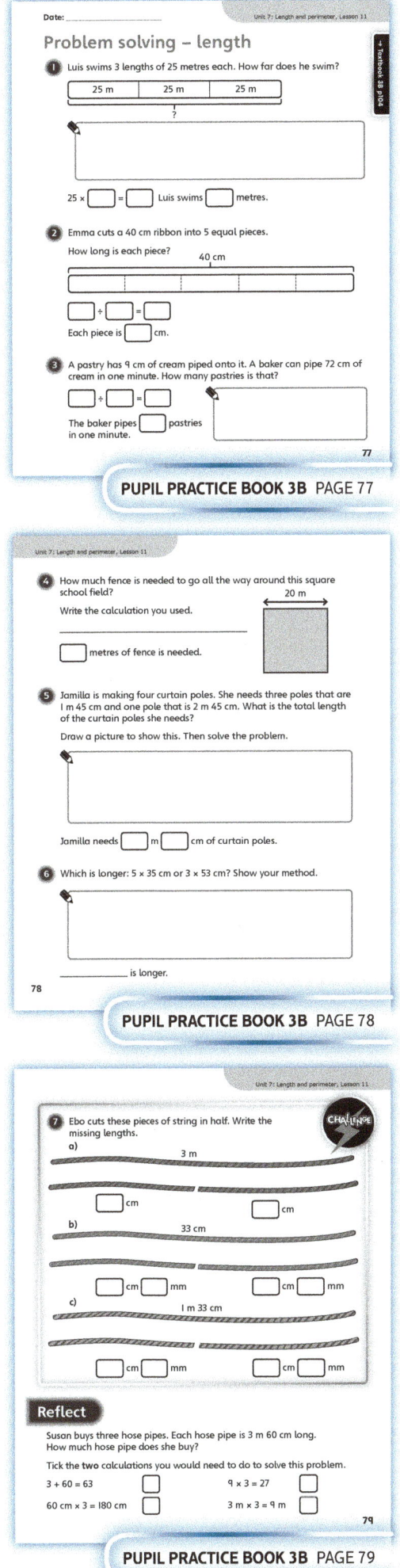

PUPIL PRACTICE BOOK 3B PAGE 77

PUPIL PRACTICE BOOK 3B PAGE 78

PUPIL PRACTICE BOOK 3B PAGE 79

Reflect

WAYS OF WORKING Pair work

IN FOCUS In this activity, children should realise that they can multiply 3 m by 3 and 60 cm by 3 separately, and add their answers. When adding 9 m and 180 cm, they should exchange 100 cm for 1 m, so that they see the total length is 10 m and 80 cm. Begin by giving children time to come up with their own ideas individually. Once they have done this, ask them to share their thoughts with a partner. Discuss why someone might choose one of the incorrect calculations. What mistake (common misconception) might they be making (for example, writing 9 m and 180 cm)? What advice would children give that person to help them correct their mistake?

ASSESSMENT CHECKPOINT At this point in the lesson, children should be able to read and understand a problem and demonstrate their ability to draw out the calculations they need to use to solve it. Children will demonstrate a deeper understanding by identifying the potential misconceptions and how they can be approached and overcome.

ANSWERS Answers for the **Reflect** part of the lesson can be found in the *Power Maths* online subscription.

After the lesson

- Are children secure in identifying the operation needed for each question?
- Are children able to represent the problems using pictures?

Unit 7: Length and perimeter

End of unit check

Don't forget the unit assessment grid in your *Power Maths* online subscription.

WAYS OF WORKING Group work adult led

IN FOCUS

- Question **1** assesses children's ability to convert between different units of measurement and compare measurements.
- Question **2** assesses children's ability to calculate subtraction using measurements. It also assesses their ability to work with number lines and convert between single and mixed units of measure.
- Question **3** assesses children's ability to calculate addition using measurements, but this time without a number line given. It also assesses their ability to convert between single and mixed units of measure.
- Question **4** assesses children's ability to find the perimeter of compound shapes.
- Question **5** assesses children's understanding of the properties of rectangles and how they affect the perimeter. It also assesses children's ability to visualise the perimeter of a shape from a given set of information.
- Question **6** is a SATs-style question where children need to apply their knowledge of multiplication to the context of measurement. This is a multi-step problem that requires children to carry out a series of calculations to find the solution. Some children may struggle to identify the steps involved and might benefit from talking through the problem, drawing a picture to illustrate it, or having concrete apparatus to model it.

ANSWERS AND COMMENTARY

Children who have mastered this unit will be able to confidently measure length in mm, cm and m. They will be able to convert between single units and mixed units and apply this skill when solving measurement problems involving addition, subtraction and finding the perimeter of 2D shapes.

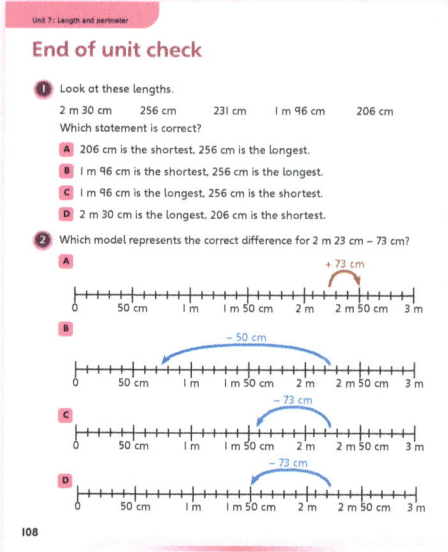

PUPIL TEXTBOOK 3B PAGE 108

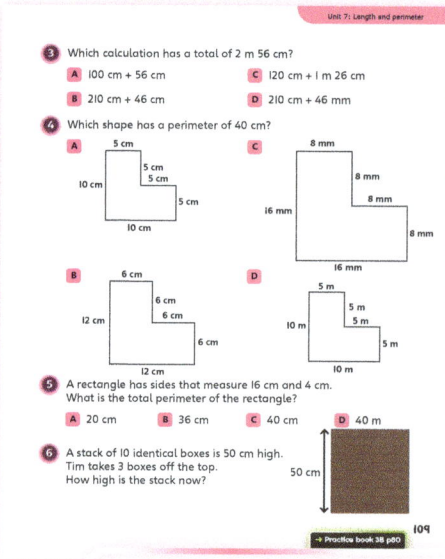

PUPIL TEXTBOOK 3B PAGE 109

Q	A	WRONG ANSWERS AND MISCONCEPTIONS	STRENGTHENING UNDERSTANDING
1	B	A or D suggest that the child has recognised cm as shorter than m but has not converted the mixed measurements.	Ensure that children have access to measuring equipment so that they are able to compare different units of measurement. Base 10 equipment can help to support children in their calculations and when comparing numbers. Support children in drawing the shapes to scale to help them understand the values of the measurements.
2	D	A suggests the child has not recognised the correct operation shown on a number line.	
3	B	D suggests the child has neglected to consider unit of measure. A or C may demonstrate incorrect addition.	
4	A	D suggests the child has neglected to consider units of measure. B or C suggests the child has not grasped the concept of perimeter or has not calculated correctly.	
5	C	A suggests that the child has just added together the two measurements, forgetting that a rectangle has two pairs of equal sides. D suggests the child has confused or ignored the unit of measurement.	
6	35 cm	Ensure children notice this is a 2-step problem. Did they identify the operations needed?	

Unit 7: Length and perimeter

My journal

WAYS OF WORKING Independent thinking

ANSWERS AND COMMENTARY Question 1: The combined height of Reena and Danny is 263 cm 5 mm. Ambika is 134 cm 2 mm tall. The combined height of Richard and Ambika > the combined height of Reena and Danny.

To answer this question, children will need to demonstrate a range of skills. They will need to identify the unit of measurement of the missing values and then convert between different units. They will also need to add and subtract to find the missing values and then use greater than and less than signs to write the expression. Some children may need assistance in one or more of these areas in order to find the answer. Offer number lines and manipulatives and, where needed, support children in converting all the heights to a common unit of measurement.

Question 2: Children may record answers such as: 'No, because there are two new sides that were joined together before, so the perimeter of each rectangle is more than half.'

If children are struggling, offer simpler measurements for the sides of the rectangles. It may help to have rectangles available for the children to manipulate. Ask: *What has happened to the rectangle by cutting it in half? Can you show me what is happening? Which sides are measured for the perimeter of the single rectangle? Which sides are measured for the perimeters after the rectangle has been cut in half? Why?*

Power check

WAYS OF WORKING Independent thinking

ASK
- Can you explain the difference between mm, cm and m?
- What is 'perimeter'?
- How confident are you at adding and subtracting measurements?

Power play

WAYS OF WORKING Independent thinking

IN FOCUS Use this **Power play** to assess children's ability to recognise properties of rectangles and how these properties affect the perimeter of the shape.

ANSWERS AND COMMENTARY There are a number of possibilities. Each answer should have a perimeter of 36 cm and two pairs of equal sides. Possible answers include a square with side lengths of 9 cm.

If children are struggling, they could use cm cubes to make rectangles with total side lengths of 36 cm. Ask: *Can you reuse those cubes to make another rectangle with the same perimeter?* Ensure they measure the perimeter in side lengths, rather than counting the number of cubes used.

After the unit

- How confident are children in finding the perimeter of compound shapes?
- What opportunities can you find to develop this area of mathematical understanding in other areas of the curriculum?

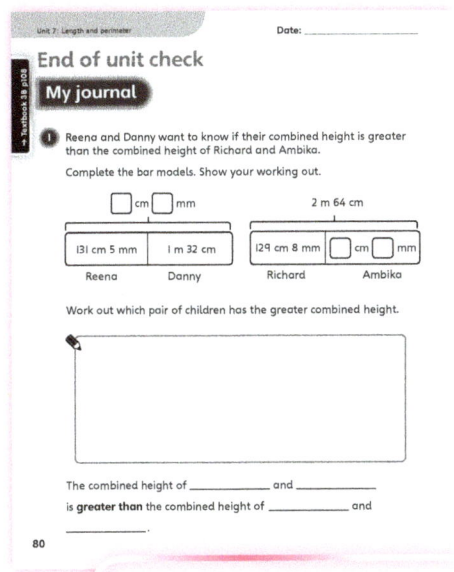

PUPIL PRACTICE BOOK 3B PAGE 80

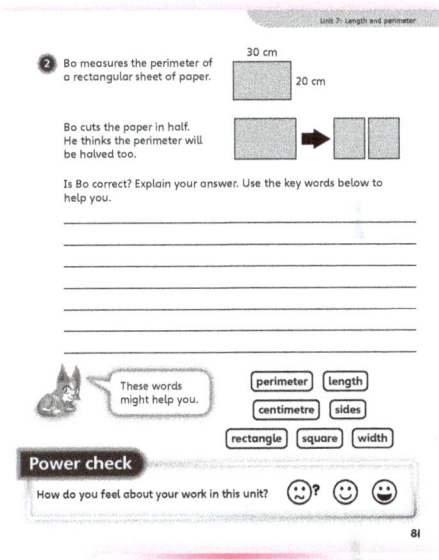

PUPIL PRACTICE BOOK 3B PAGE 81

PUPIL PRACTICE BOOK 3B PAGE 82

Strengthen and **Deepen** activities for this unit can be found in the *Power Maths* online subscription.

Unit 8
Fractions ①

Mastery Expert tip! 'Children need a rich experience of seeing, touching and creating simple fractions in order to gain a more secure understanding of what a fraction actually is. We created a fraction museum on a whiteboard and each lesson we added examples of our learning with lots of colourful resources and drawings. It really helped my class to embed what they learnt, recall what they already knew and ask questions where they were unsure.'

Don't forget to watch the Unit 8 video!

WHY THIS UNIT IS IMPORTANT

In this unit, children will understand the concept of a unit fraction and a non-unit fraction and understand what the numerator and denominator represent. Children will compare and order simple unit fractions and also non-unit fractions where the denominators are equal. In addition to this, children will learn to recognise and show, using diagrams, equivalent fractions with small denominators. They will explore a fraction wall and use it to find equivalent fractions. Children will order fractions on a number line and compare two fractions using bar models and the comparison signs <, > or =. They will learn to add and subtract two or more fractions with the same denominator, answering questions in more than one way and comparing the efficiency of each method.

WHERE THIS UNIT FITS

→ Unit 7: Length and perimeter
→ **Unit 8: Fractions (1)**
→ Unit 9: Mass

Before they start this unit, it is expected that children:
- understand how to say and write simple fractions
- understand the concept of equal and non-equal parts
- understand the whole.

ASSESSING MASTERY

Children who have mastered this unit will be able to draw and explain the concepts of unit and non-unit fractions. They will also be able to compare them using pictorial and abstract methods. Children who have mastered this unit will also be able to describe equivalent fractions and mark fractions with different denominators on a number line. They can compare pairs of unit fractions and fractions with small denominators and explain which is larger. Children can count in fraction steps of a constant size to help make sense of adding and subtracting fractions with the same denominator, and find pairs of fractions that total 1.

COMMON MISCONCEPTIONS	STRENGTHENING UNDERSTANDING	GOING DEEPER
Children may find the concept of numerators and denominators confusing, particularly if the numerator is greater than 1.	Provide opportunities for children to use resources and model fractions, for example, using fraction rods, beads and fraction strips.	Encourage children to make a fraction wall from scratch to deepen their understanding.
Children may find it difficult to recognise fractions as points on a number line.	Revisit counting in fraction intervals by showing children pre-marked number lines with different denominators. Hide some of the fractions and ask children to find them.	Give children number lines labelled from 0 to 1. Ask them to create their own questions regarding the position of different fractions on the number line.
Children who find it difficult to solve problems may not know where to start and so may guess which method to use.	Offer real-life examples of problem solving situations. Encourage children to role-play the problem and sketch something that helps them make sense of it.	Challenge children to find different ways to answer the same question. Often, more able children are less inclined to look for alternative options, but finding the answer is not as important as the way they find the answer.

Unit 8: Fractions 1

UNIT STARTER PAGES

Use these pages to introduce the unit focus to children. You can use the characters to explore different ways of working.

STRUCTURES AND REPRESENTATIONS

Fraction wall: This representation is crucial to allow children to find equivalent fractions. If children become confident using the fraction wall, it will increase their conceptual understanding of fractions. It can be used by itself or with a number line to compare fractions with different denominators.

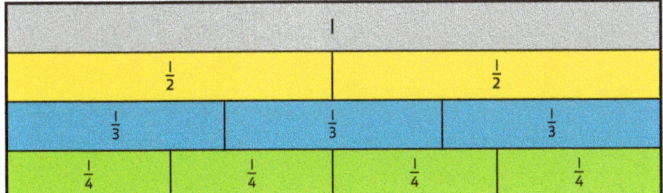

Number line: This model helps children to understand fractions as numbers. Positioning fractions on a number line will require a secure understanding of the role of the numerator and denominator within a fraction.

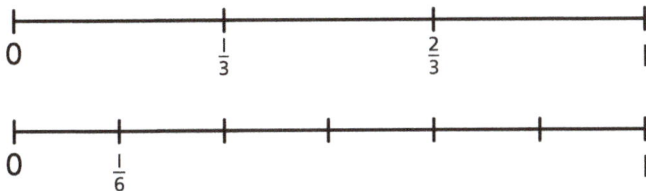

Fraction strip: This is a powerful representation that allows children to organise the information they are given visually, and understand how it should be manipulated in order to find the solution to a problem. It can be used alone, or with a number line to enhance understanding.

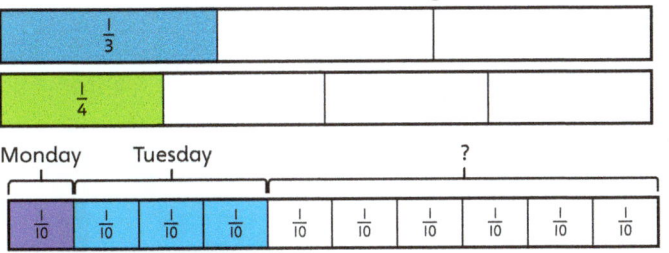

KEY LANGUAGE

There is some key language that children will need to know as part of the learning in this unit:

- part, whole, equal parts, fraction, denominator, numerator, equivalent, equivalent fraction
- partition, split, share, count on, count back, compare, measure, calculate, method
- whole number, add, subtract, difference, multiply, divide, equal to, greater than (>), less than (<), inequality statement

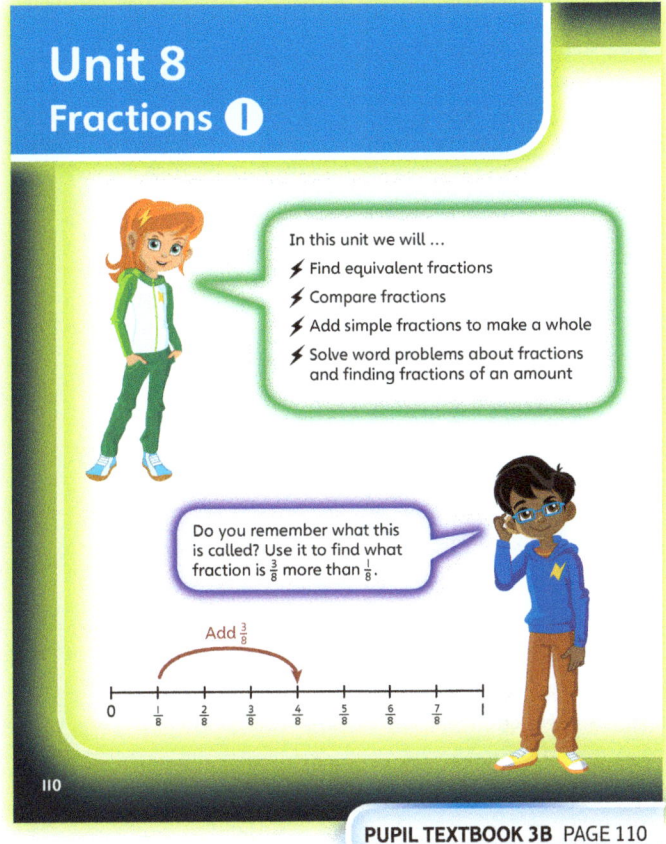

PUPIL TEXTBOOK 3B PAGE 110

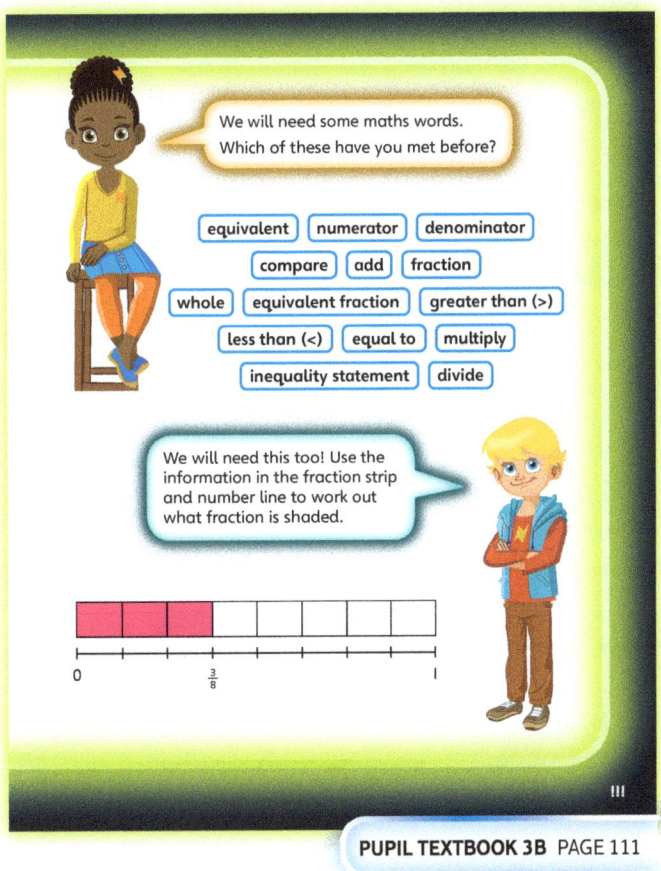

PUPIL TEXTBOOK 3B PAGE 111

Unit 8: Fractions (1), Lesson 1

Understand the denominator of unit fractions

Learning focus

In this lesson, children will learn and understand that the denominator of a unit fraction tells you the number of equal parts the whole is made up of.

Before you teach

- Are children able to read simple fractions such as $\frac{1}{2}$ and $\frac{1}{4}$?
- Do children know how to represent simple fractions such as $\frac{1}{2}$ and $\frac{1}{4}$?

NATIONAL CURRICULUM LINKS

Year 3 Number – fractions

Recognise and use fractions as numbers: unit fractions and non-unit fractions with small denominators.

ASSESSING MASTERY

Children can recognise unit fractions and can explain and represent simple unit fractions using diagrams and objects. Children know that the denominator of a unit fraction tells them the number of equal parts that make up the whole.

COMMON MISCONCEPTIONS

Children may think the numerator of a fraction is the parts shaded and the denominator is the parts that are unshaded. For example, if a shape is divided into four equal parts, with one part shaded, they think it means $\frac{1}{3}$ of the shape is shaded. Ask:
- *What do the shaded parts show? What does the denominator show?*

Children may think that, if a shape is divided into three unequal parts and one part is shaded, then $\frac{1}{3}$ of the shape is shaded. Ask:
- *If $\frac{1}{3}$ of the shape was shaded, would all the parts need to be equal or can they be unequal?*

Be careful not to cause a misconception that a shape can be divided into equal parts, where they look the same.

STRENGTHENING UNDERSTANDING

Ask children to represent $\frac{1}{2}$. Ask them what they notice. Explain that the numerator is the number of shaded parts and that the denominator is the total number of parts that make up the whole. Ask them to represent fractions such as $\frac{1}{3}, \frac{1}{4}$ or $\frac{1}{5}$. Discuss what they notice about all the shapes.

GOING DEEPER

Ask children to draw a five frame on their mini whiteboards. Ask them how they could shade in $\frac{1}{10}$ of this shape. Ask them what do they need to do first. What is the whole divided into?

When the denominator is 10, what should the whole be divided up into? Ask questions similar to these where children need to first divide the whole into more equal parts.

KEY LANGUAGE

In lesson: shaded, fraction, numerators, denominators, unit fraction

Other language to be used by the teacher: same, different

STRUCTURES AND REPRESENTATIONS

Fraction strips

RESOURCES

Optional: fraction shapes, art materials

 In the eTextbook of this lesson, you will find interactive links to a selection of teaching tools.

Quick recap

Ask children to draw a square or rectangle and divide it into two equal parts. Ask them to then draw a square or rectangle and divide it into four equal parts. Discuss methods that children used to divide the shapes.

Unit 8: Fractions (1), Lesson 1

Discover

WAYS OF WORKING Pair work

ASK

- Question 1 a): *Choose one of the shapes. How many parts is it divided into? Do the parts look equal? How many parts are shaded? What fraction of the shape is shaded?*
- Question 1 b): *What do you notice about the numerators of the fraction? What about the denominators?*

IN FOCUS The key focus of questions 1 a) and 1 b) is for children to be able to see representations of unit fractions and write them down. In question 1 a), children should consider how many equal parts each shape is divided into. They should also consider whether the parts are equal and then write down the fraction of the shapes that are shaded. In question 1 b), some children may focus on the shape and colour first, before seeing that in every shape one part is shaded and that there is a 1 as the numerator in each of the fractions. They should notice that the denominators are different because the shapes are divided into different equal parts.

PRACTICAL TIPS Children could make shapes and images using art materials where simple unit fractions are shaded. Flags of the world are great for this, too, when the flag patterns are divided into equal-sized parts.

ANSWERS

Question 1 a): $\frac{1}{2}$ of the rectangle is shaded.
$\frac{1}{5}$ of the pentagon is shaded.
$\frac{1}{4}$ of the circle is shaded.
$\frac{1}{3}$ of the trapezium is shaded.

Question 1 b): The numerators are all equal to 1.
The denominators are all different.

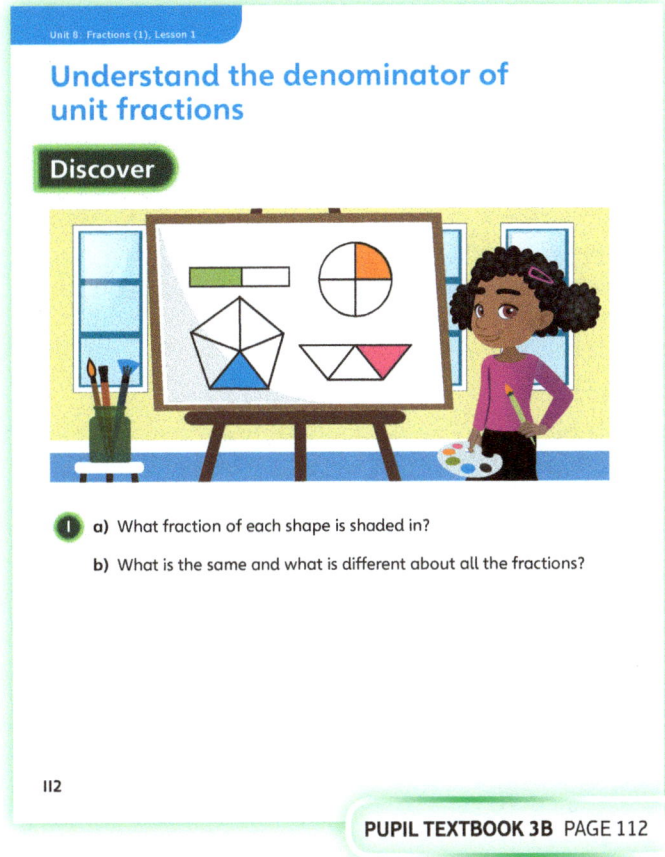

PUPIL TEXTBOOK 3B PAGE 112

Share

WAYS OF WORKING Whole class teacher led

ASK

- Question 1 a): *How many parts is each shape divided into? Do the parts look equal? How many parts are shaded? What fraction of the shape is shaded?*
- Question 1 b): *What is the top number of a fraction called? What do you notice about all the numerators? What is the bottom number called? What do you notice about the denominators?*

IN FOCUS In question 1 a), work through each shape in turn, asking children to consider each shape, the number of equal parts it is divided into, the number of parts that are shaded, and then ask them to write down the fraction. Each time, ensure children are pointing to the shape to help them. In question 1 b), use the language of numerator and denominator to talk about the numbers in the fractions. Children should notice that all the numerators are 1, because one part of each shape is shaded and that all the denominators show the number of equal parts the shape is divided into. Use Sparks's speech to remind children that these shapes show unit fractions, as the numerators are 1.

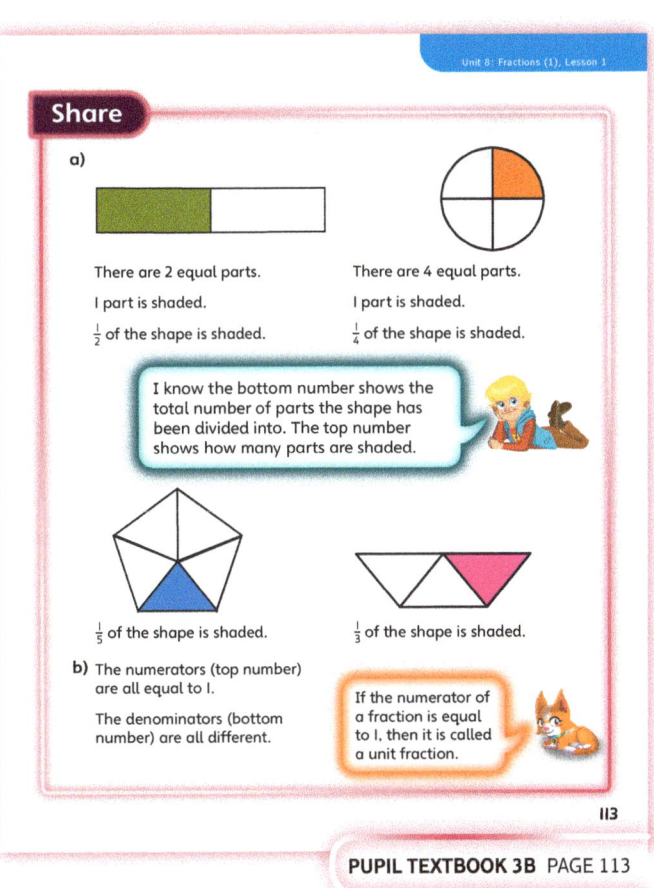

PUPIL TEXTBOOK 3B PAGE 113

Think together

WAYS OF WORKING Whole class teacher led (I do, We do, You do)

ASK

- Question ① a): *How many parts is each shape divided into? What fraction of each shape is shaded?*
- Question ②: *How many parts is the shape divided into? Do the parts look equal in size?*
- Question ③ a): *What does $\frac{1}{4}$ mean? How many equal parts must a shape be made up of if you want to shade $\frac{1}{4}$ of it? Why is the fraction $\frac{1}{4}$ called a unit fraction?*
- Question ③ b): *How many equal parts does the shape need to be divided into to shade $\frac{1}{8}$ of it?*

IN FOCUS Question ① provides further practice and consolidation of writing and representing unit fractions. Children see that each numerator is 1 because one part of each shape is shaded, and that the denominator shows the total number of equal parts. Question ② addresses a key misconception that parts must be equal.

Question ③ a) asks children to shade in $\frac{1}{4}$ of a shape. Ask them to share their shapes and to discuss what is the same and what is different between them.

Question ③ b) asks children to consider how to shade in $\frac{1}{8}$ of a shape that has been split into quarters. It is important to establish that, however children split the quarters, they must each be split in the same way so there are 8 equal parts.

STRENGTHEN Ask children to draw a shape and represent $\frac{1}{2}$. Ask them what they notice. Explain that the numerator is the number of shaded parts and that the denominator is the total number of parts the shape is divided into. Ask children to represent other fractions such as $\frac{1}{3}, \frac{1}{4}, \frac{1}{5}$. Discuss what they notice about all the shapes.

DEEPEN Give children a square or ask them to draw one. How many ways can they shade in $\frac{1}{4}$ of the shape?

ASSESSMENT CHECKPOINT Questions ① and ② will check that children know how to represent a unit fraction and that they know the denominator is the number of equal parts that a shape is divided into.

ANSWERS

Question ① a): $\frac{1}{4}$ of the shape is shaded.
$\frac{1}{8}$ of the shape is shaded.
$\frac{1}{3}$ of the shape is shaded.

Question ① b): If the numerator of a fraction is equal to 1, then it is called a unit fraction.

Question ②: The shape has not been divided into three equal parts, so $\frac{1}{3}$ is not shaded. The bottom section is equal to the two sections on top, so the correct answer would be $\frac{1}{4}$ of the shape is shaded.

Question ③ a): Children should shade 1 square. Any of the four squares could be shaded so the children's shapes may or may not look the same.

Question ③ b): It is possible to shade in $\frac{1}{8}$ of the shape if you divide one of the squares into two equal parts and then shade in one part.

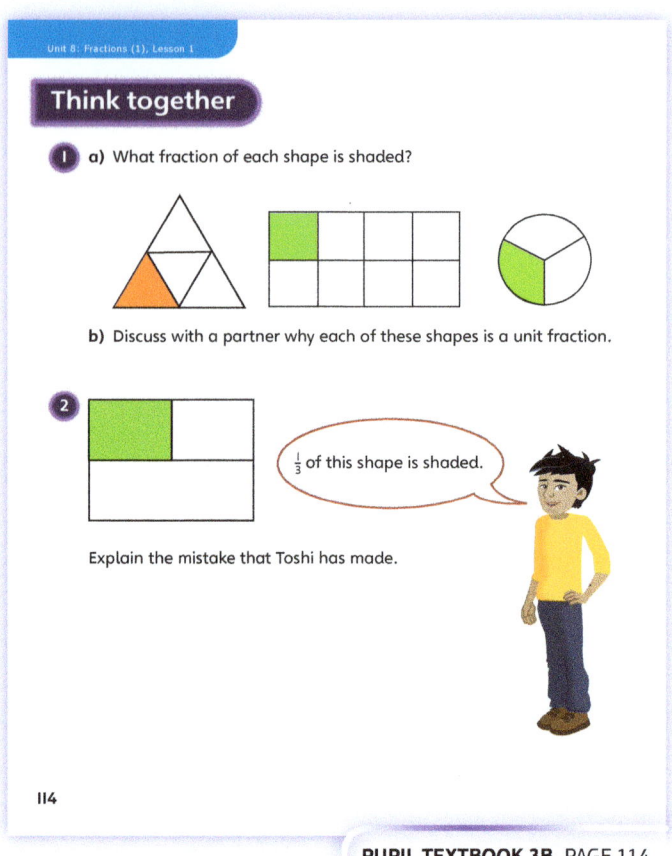

PUPIL TEXTBOOK 3B PAGE 114

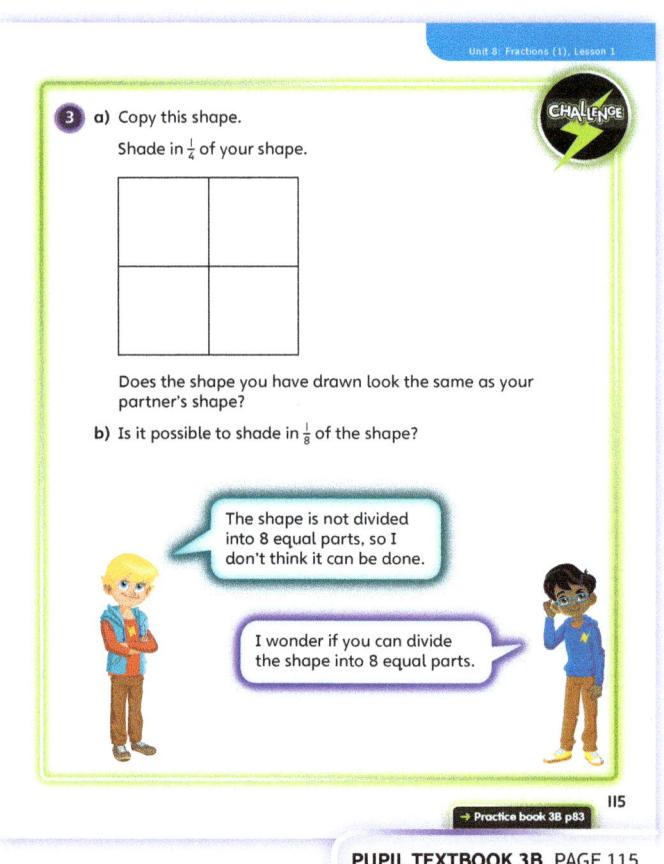

PUPIL TEXTBOOK 3B PAGE 115

Unit 8: Fractions (1), Lesson 1

Practice

WAYS OF WORKING Independent thinking

IN FOCUS Questions ❶ and ❷ ask children to write down what fraction of each shape is shaded. Each fraction is a unit fraction as one equal part of each shape is shaded. Question ❸ asks children to represent given unit fractions by shading in parts of shapes. This builds nicely on from the understanding gained in questions ❶ and ❷. Question ❹ is more abstract and asks children to circle all the unit fractions. They should look for fractions where the numerator is 1. In question ❺, children are asked to shade in $\frac{1}{4}$ of each of three squares, but the parts in all of the squares look different. Ask them what is the same and what is different about the squares.

STRENGTHEN To support understanding in questions ❶ and ❷, ask children how many parts each shape is divided into and whether the parts are equal. Explain that this is the denominator of the fraction. Ask them how many of the parts are shaded, then explain that this is the numerator. Ask them to represent their own unit fractions.

DEEPEN Question ❼ asks children how they could shade in $\frac{1}{10}$ of a shape that has been split into five equal parts. Ask: *What do you need to do first? What is the whole divided into? When the denominator is 10, what should the whole be divided into?*

THINK DIFFERENTLY In question ❻, children demonstrate that $\frac{1}{4}$ of the second shape is not shaded because the shape is not split into equal parts. They may think the same for the first shape, but explain that the parts cover the same space even though they may look different, hence they are equal.

ASSESSMENT CHECKPOINT Questions ❶, ❷ and ❹ are key questions to help children understand unit fractions. Question ❻ checks that children understand that parts must be equal in order for them to show unit fractions.

ANSWERS Answers for the **Practice** part of the lesson can be found in the *Power Maths* online subscription.

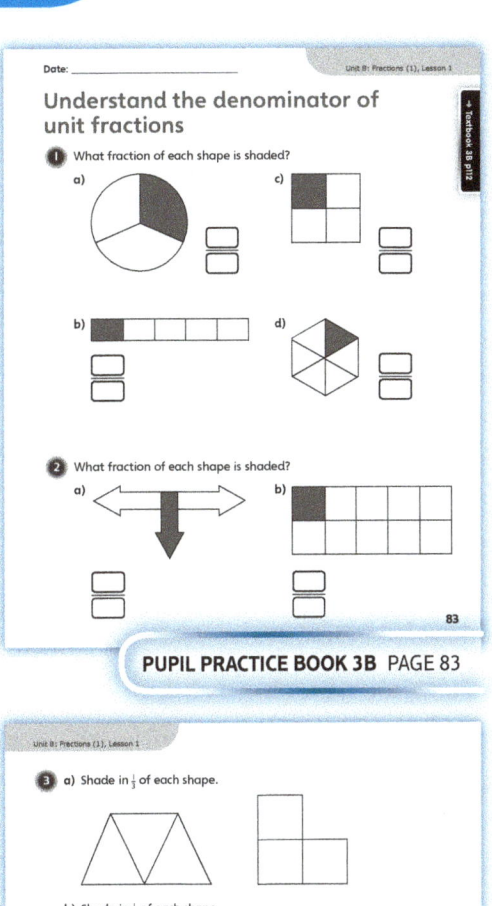

PUPIL PRACTICE BOOK 3B PAGE 83

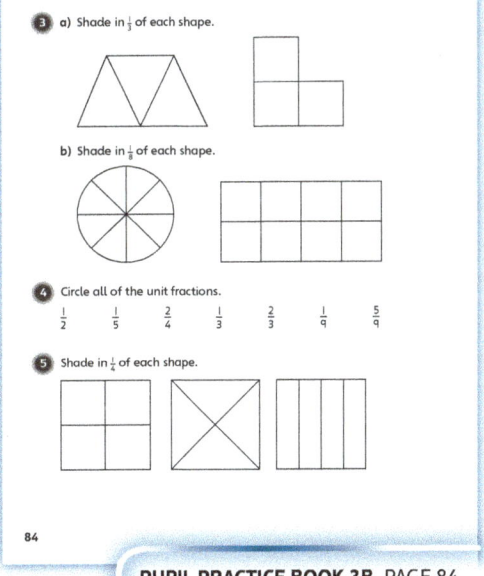

PUPIL PRACTICE BOOK 3B PAGE 84

Reflect

WAYS OF WORKING Independent thinking and then whole class

IN FOCUS Children reflect on their understanding in the lesson and write down or tell a partner two things they have learnt. For example, they may say that in a unit fraction the numerator is always 1. Encourage them to use full sentences and accurate language. They may talk about a question they found particularly easy or challenging.

ASSESSMENT CHECKPOINT Check children know that a unit fraction has a numerator of 1 and they know the denominator tells them the number of equal parts a shape is divided into.

ANSWERS Answers for the **Reflect** part of the lesson can be found in the *Power Maths* online subscription.

After the lesson ⏸

- Do children know that every unit fraction has a numerator equal to 1?
- Do children know that the denominator represents the number of equal parts?
- Are children able to tell you what unit fraction of a shape is shaded?

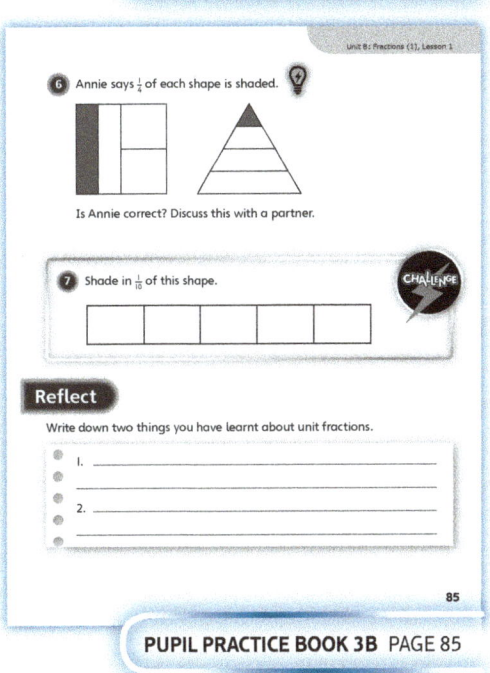

PUPIL PRACTICE BOOK 3B PAGE 85

Unit 8: Fractions (1), Lesson 2

Compare and order unit fractions

Learning focus
In this lesson, children will compare and order unit fractions.

Before you teach
- Can children represent a unit fraction?
- Do children understand what a unit fraction is?
- Do children know and can they use the inequality signs < and >?

NATIONAL CURRICULUM LINKS

Year 3 Number – fractions

Recognise and use fractions as numbers: unit fractions and non-unit fractions with small denominators.

ASSESSING MASTERY

Children can compare and order unit fractions by drawing diagrams and by comparing denominators.

COMMON MISCONCEPTIONS

A very common misconception is that the greater unit fraction is the one with the greater denominator. For example, children think that $\frac{1}{6}$ is greater than $\frac{1}{4}$. Ask:
- Can you draw shapes to show $\frac{1}{6}$ and $\frac{1}{4}$? Which fraction is greatest?

STRENGTHENING UNDERSTANDING

Use diagrams such as fraction strips or bar models to compare unit fractions. For example, draw two shapes that are the same length. Divide the first shape into fifths and the second shape into thirds. Ask children to shade in $\frac{1}{5}$ and $\frac{1}{3}$. They should shade in the first one from the left to help them compare the shapes. Ask them which fraction is smaller and how the diagrams help them see this.

GOING DEEPER

Children should be confidently able to compare and order fractions by considering the denominators. For example, ask them how many unit fractions they can find between $\frac{1}{2}$ and $\frac{1}{10}$.

KEY LANGUAGE

In lesson: compare, order, unit fraction, smaller, greater, smallest, greatest, denominator

Other language to be used by the teacher: less than, greater than, numerator

STRUCTURES AND REPRESENTATIONS

Fraction strips, bar models

RESOURCES

Mandatory: strips of paper that are 12 cm long, rulers

Optional: fraction shapes

 In the eTextbook of this lesson, you will find interactive links to a selection of teaching tools.

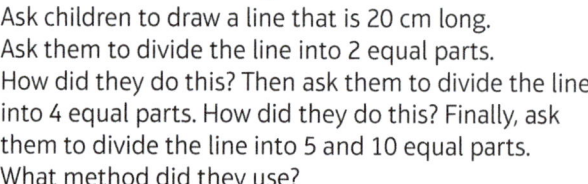

Quick recap

Ask children to draw a line that is 20 cm long. Ask them to divide the line into 2 equal parts. How did they do this? Then ask them to divide the line into 4 equal parts. How did they do this? Finally, ask them to divide the line into 5 and 10 equal parts. What method did they use?

Discover

WAYS OF WORKING Pair work

ASK

- Question 1 a): *How can the pieces of paper be divided into 3 and 4 equal parts? Can the paper be folded or do you need to measure?*
- Question 1 b): *What fraction of each shape is shaded? How can you compare the fractions? Which fraction is greater? Which is smaller? How do you know?*

IN FOCUS The focus of questions 1 a) and b) is for children to use fraction strips to compare unit fractions. For question 1 a), give children small strips of paper, 12 cm long. This will allow them to either divide the fractions by measuring or by folding the paper. You could use 18 cm or 24 cm, too. Encourage children to explore different methods. For question 1 b), they should explore how they can compare the two fractions, possibly by putting one fraction strip above the other. Some children may put them next to each other.

PRACTICAL TIPS Use paper strips to replicate the scenario.

ANSWERS

Question 1 a): Mo can cut his paper into 3 equal parts by measuring the length and dividing by 3.
Emma can cut her paper into 4 equal parts by halving the piece of paper and then halving it again.

Question 1 b): $\frac{1}{3}$ is greater than $\frac{1}{4}$.
Mo has the bigger parts.

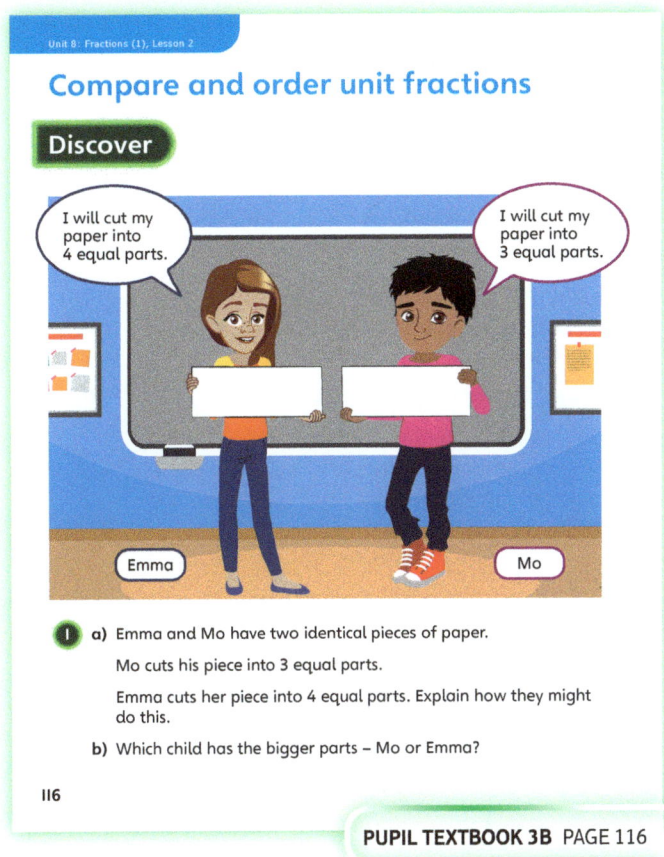

PUPIL TEXTBOOK 3B PAGE 116

Share

WAYS OF WORKING Whole class teacher led

ASK

- Question 1 a): *How long is your sheet of paper? How did you divide the strip of paper into equal parts? What method did you use to divide the strip into 4 parts?*
- Question 1 b): *What fraction of each shape is shaded? How can you compare the fractions? Why should you put them above each other instead of next to each other?*

IN FOCUS In question 1 a), discuss with children the different methods to divide a strip of paper into 3 parts or 4 parts. They may know a method for accurately dividing a piece of paper into 3 equal parts just by folding, but discuss how Mo has done this (by measuring the length of the paper and dividing it into 3 equal parts). Talk about the importance of dividing the strip of paper into equal parts so they can compare them later. Discuss how Emma divided the piece of paper into 4 equal parts by folding, but explain that she could have measured too. In question 1 b), explain how the two unit fractions have been compared by putting the two strips above each other.

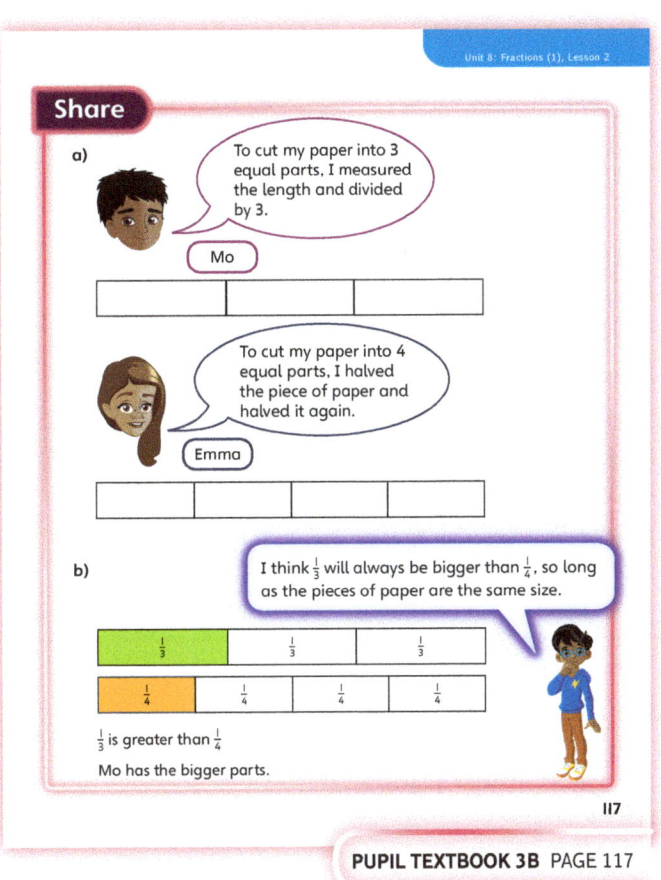

PUPIL TEXTBOOK 3B PAGE 117

Think together

WAYS OF WORKING Whole class teacher led (I do, We do, You do)

ASK

- Question ① a): *How does the diagram show that $\frac{1}{5}$ is smaller than $\frac{1}{3}$?*
- Question ① b): *What can you do to put the fractions in order? Which fraction is the smallest? Why? Which is the greatest?*
- Question ②: *Which fraction do you think is bigger? Why? What diagrams can you draw to help you compare them?*
- Question ③: *What fractions are greater than $\frac{1}{5}$? How do you know without drawing diagrams?*

IN FOCUS Question ① a) provides a similar example to the one that children have just done in the **Discover** and **Share** sections. Discuss why $\frac{1}{5}$ is smaller than $\frac{1}{3}$. In question ① b), children should be able to order the fractions in order of size simply by looking at the fraction strips. This will help to further solidify their understanding of fraction strips and what they show. Question ② provides an opportunity to address the misconception that the fraction $\frac{1}{5}$ is greater than $\frac{1}{4}$ because the denominator of $\frac{1}{5}$ is greater. Ask children to use fraction strips to show that this is correct. Explain that this is a common mistake. In question ③, children are encouraged to use their knowledge of denominators to compare fractions as opposed to using pictorial representations. They should realise, for example, that fractions with a denominator greater than 5 are smaller than $\frac{1}{5}$ and those that have denominators that are smaller than 5 are greater than $\frac{1}{5}$. Encourage children to make generalisations.

STRENGTHEN Use diagrams such as fraction strips or bar models to compare unit fractions. For example, draw two shapes that are the same length. Divide the first shape into a given fraction and then divide the second shape into a given fraction. Ask children to shade in the unit fractions and set the shapes above each other to compare them.

DEEPEN Ask children to make generalisations for comparing unit fractions.

ASSESSMENT CHECKPOINT Children should be able to use question ① to show they can compare unit fractions using diagrams and questions ② and ③ to help them compare fractions using the size of the denominator.

ANSWERS

Question ① a): $\frac{1}{5}$ is smaller than $\frac{1}{3}$.

Question ① b): Smallest to greatest: $\frac{1}{6}, \frac{1}{3}, \frac{1}{2}$.

Question ②: Reena is not correct. Children should show that $\frac{1}{4}$ is greater than $\frac{1}{5}$ by drawing two fraction strips of the same size and dividing one into $\frac{1}{4}$s and the other into $\frac{1}{5}$s.

Question ③: Fractions greater than $\frac{1}{5}$: $\frac{1}{2}$ and $\frac{1}{4}$.
Fractions less than $\frac{1}{5}$: $\frac{1}{6}, \frac{1}{9}, \frac{1}{10}, \frac{1}{15}$.

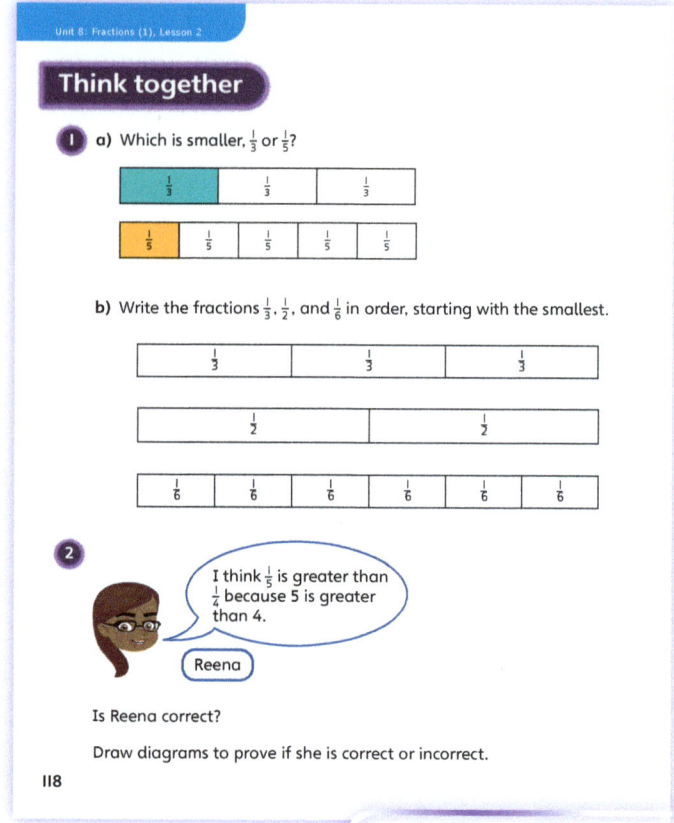

PUPIL TEXTBOOK 3B PAGE 118

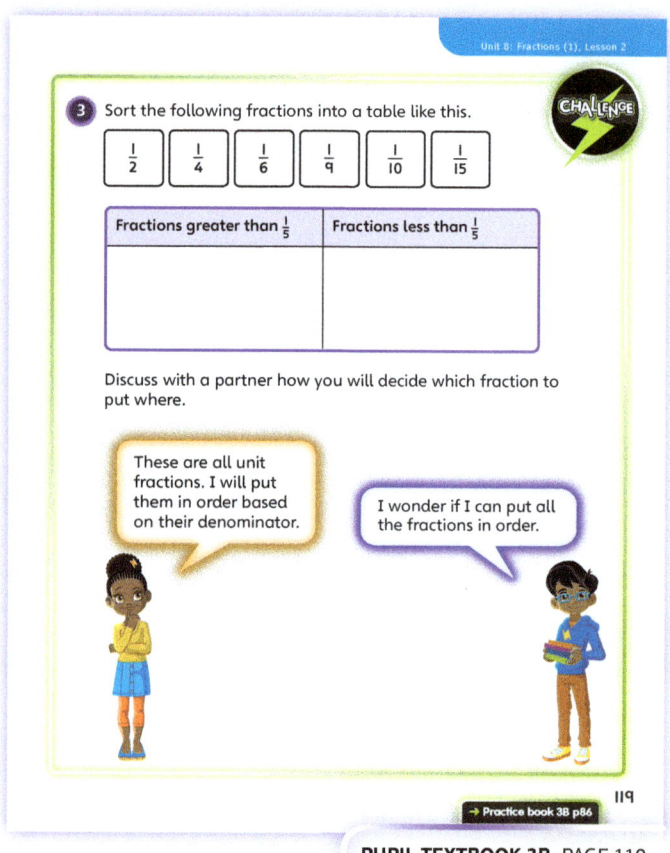

PUPIL TEXTBOOK 3B PAGE 119

Unit 8: Fractions (1), Lesson 2

Practice

WAYS OF WORKING Independent thinking

IN FOCUS In questions ❶ and ❷, children compare fractions using fraction strips that have shaded parts. In question ❸, children use the language 'less than' and 'greater than' to compare unit fractions. They should be encouraged to draw fraction strips to help them if needed. They may use their knowledge of comparing unit fractions by comparing the denominators. In question ❹, children use the < and > signs to compare unit fractions.

In question ❺, children circle all the fractions less than $\frac{1}{6}$. These are all unit fractions, so children should be encouraged to see that, here, they only need to look at denominators. If the denominator is greater than 6, then the fraction is divided into more than 6 parts, and hence the fraction is less than $\frac{1}{6}$.

Throughout these questions, children may use diagrams to help them compare fractions before moving onto comparing fractions by only looking at the denominator. By the end of the lesson, children will be more confident in comparing unit fractions by considering the denominators.

STRENGTHEN Use diagrams such as fraction strips or bar models to compare unit fractions. For example, draw two shapes that are the same length. Divide the first shape into a given fraction and then the second shape into a given fraction. Ask children to shade in the unit fractions and set the shapes above each other to compare them.

DEEPEN Ask children to make generalisations for comparing unit fractions.

ASSESSMENT CHECKPOINT Questions ❶ and ❷ assess whether children know how to compare fractions using diagrams for support. Questions ❸, ❹ and ❺ assess whether children can compare fractions by comparing the value of the denominators.

ANSWERS Answers for the **Practice** part of the lesson can be found in the *Power Maths* online subscription.

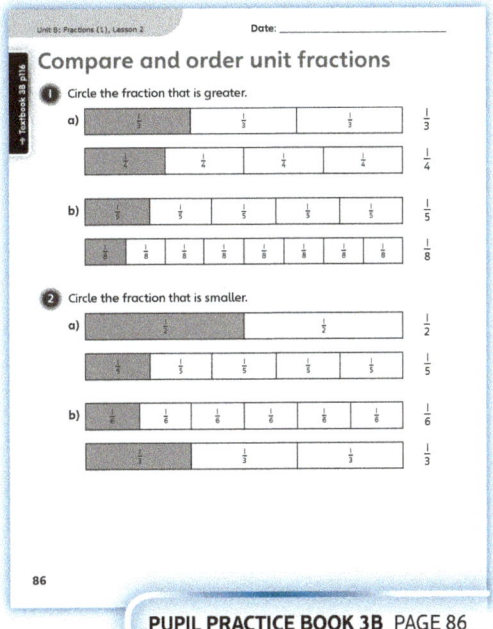

PUPIL PRACTICE BOOK 3B PAGE 86

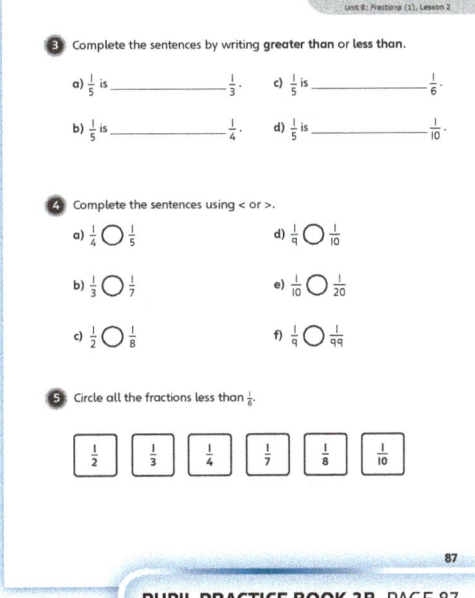

PUPIL PRACTICE BOOK 3B PAGE 87

Reflect

WAYS OF WORKING Independent thinking

IN FOCUS Children compare two unit fractions by drawing diagrams to support them. Look for children drawing the wholes the same length and setting out the fractions underneath one another. Some children may need support with drawing fraction strips and you may consider giving them small strips of paper that they can stick into their book.

ASSESSMENT CHECKPOINT Use this question to check that children have an understanding of comparing two unit fractions.

ANSWERS Answers for the **Reflect** part of the lesson can be found in the *Power Maths* online subscription.

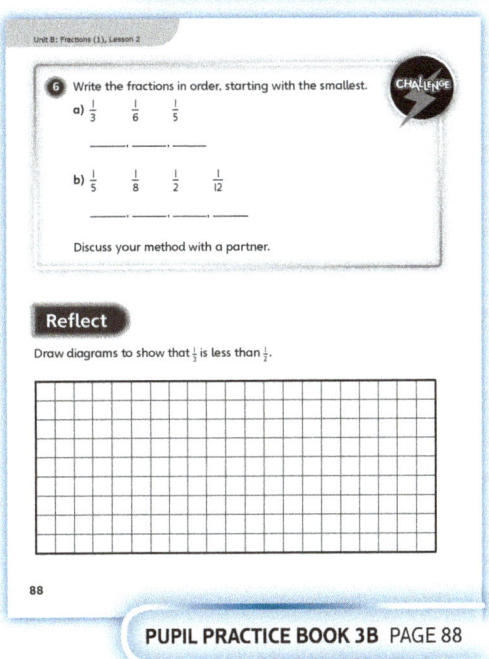

PUPIL PRACTICE BOOK 3B PAGE 88

After the lesson

- Can children use fraction strips to compare two unit fractions?
- Can children compare unit fractions by comparing the denominators?
- Do children have a method to order unit fractions?

Unit 8: Fractions (1), Lesson 3

Understand the numerator of non-unit fractions

Learning focus
In this lesson, children will understand what the numerator of a non-unit fraction represents.

Before you teach
- Can children represent simple unit fractions?
- Do children know that the numerator is the top number in a fraction and the denominator is the bottom number in a fraction?
- Do children know that the denominator can represent the number of equal parts that the whole has been divided into?

NATIONAL CURRICULUM LINKS

Year 3 Number – fractions
Recognise and use fractions as numbers: unit fractions and non-unit fractions with small denominators.

ASSESSING MASTERY
Children can write down what fraction of a whole is shaded or represented.

COMMON MISCONCEPTIONS
Children may think the numerator of a fraction is the parts shaded and the denominator is the parts that are unshaded. Explain that the denominator of a fraction represents the number of equal parts the whole has been divided into. Ask:
- *Can you draw a diagram to show $\frac{1}{5}$? How many equal parts are there? How many equal parts are shaded? How many equal parts are unshaded?*

STRENGTHENING UNDERSTANDING
To help children work out what fraction of a shape is shaded, ask them to first tell you how many equal parts the shape is divided into. You may need to count the parts together. Explain that this is the bottom number of the fraction or the denominator. Ask children how many of these parts are shaded. Explain this is the numerator or the top number of the fraction. Ask them to tell you how many parts are unshaded. Ensure they understand that this is **not** the same as the denominator.

GOING DEEPER
Give children a paper square or rectangle and ask them to shade in a given fraction, for example, $\frac{3}{4}$ or $\frac{5}{8}$. Do not have the rectangle already divided into parts; children have to work out different strategies that they can use.

KEY LANGUAGE

In lesson: fraction, numerator, denominator, unit fraction, non-unit fraction

Other language to be used by the teacher: shaded, unshaded

STRUCTURES AND REPRESENTATIONS
Fraction strips

RESOURCES
Optional: fraction shapes, paper squares or rectangles, double sided counters, coloured pencils

 In the eTextbook of this lesson, you will find interactive links to a selection of teaching tools.

Quick recap

On the board, draw a shape made up of 5 equal parts. Ask children how they can show $\frac{1}{5}$.

Ask them what the numerator represents and what the denominator represents.

Unit 8: Fractions (1), Lesson 3

Discover

WAYS OF WORKING Pair work

ASK

- Question 1 a): *What do you remember about a unit fraction? What is the same about all the kites? Which kite shows a unit fraction?*
- Question 1 b): *How many parts is Isla's kite divided into? Are all the parts equal? How do you know? How many parts are shaded?*

IN FOCUS Up until now, children have only met unit fractions. The focus of question 1 b) is to introduce children to non-unit fractions. Children may represent the kites using fraction shapes. They should apply their learning from previous lessons to help them understand how to find what fraction of each shape is shaded. Children should notice that all the kites are made up of 4 equal parts, but the number of parts shaded is different. Children may want to consider what fraction of Zac's shape is shaded, which could also be written as a unit fraction.

PRACTICAL TIPS Simple wooden fraction shapes could be used to represent the kites.

ANSWERS

Question 1 a): Max's kite shows a unit fraction shaded.

Question 1 b): $\frac{3}{4}$ of Isla's kite is shaded.

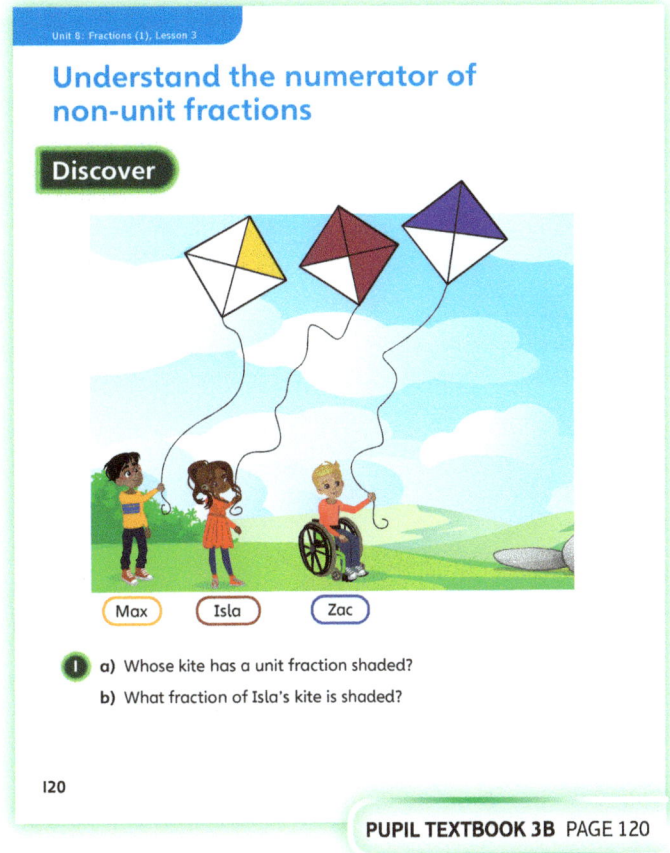

PUPIL TEXTBOOK 3B PAGE 120

Share

WAYS OF WORKING Whole class teacher led

ASK

- Question 1 a): *What do you remember about a unit fraction? What does the numerator have to be? Whose kite has 1 part shaded?*
- Question 1 b): *How many parts is Isla's kite divided into? Are all the parts equal? How do you know? How many parts are shaded?*

IN FOCUS In question 1 a), work through the example, asking children to think back to the previous lesson where they looked at unit fractions. Explain that they are looking for the kite with only 1 part shaded.

Discuss how Max's kite has $\frac{1}{4}$ shaded, which is a unit fraction. You may also want to discuss Zac's shape, which has $\frac{1}{2}$ shaded, even though the shape is split into 4 equal parts. Children may not be ready to see this at this point. In question 1 b), guide children through the example, showing them that the denominator represents the number of equal parts a shape is divided into and the numerator represents the number of shaded parts.

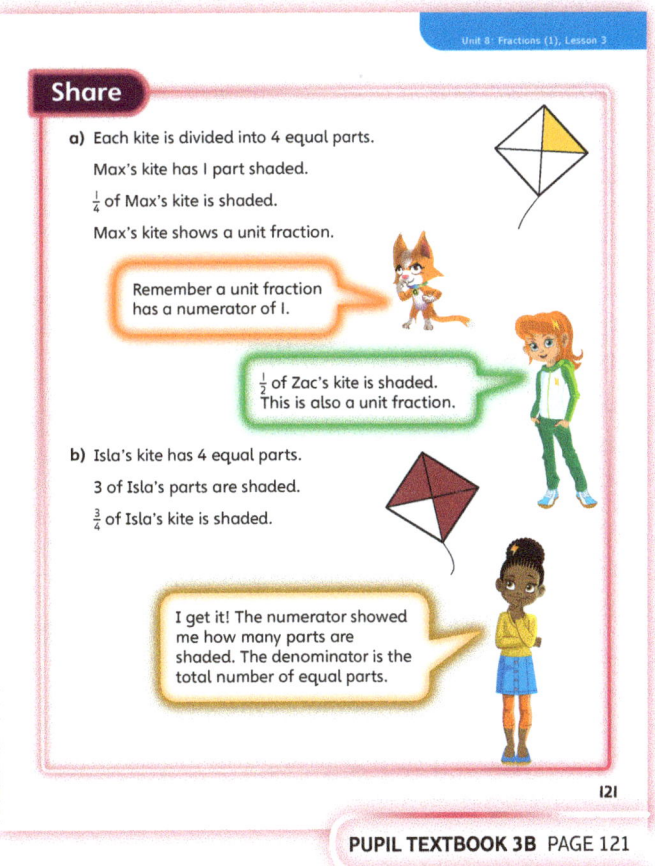

PUPIL TEXTBOOK 3B PAGE 121

Unit 8: Fractions (1), Lesson 3

Think together

WAYS OF WORKING Whole class teacher led (I do, We do, You do)

ASK

- Question ①: *How many parts is each shape divided into? Are all the parts equal in each shape? What fraction of each shape is shaded?*
- Question ②: *How many equal parts is the shape divided into? What fraction of the parts have red stripes? Have blue spots? Are solid purple? What do you notice about your answers? What is the same and what is different? How could you check your answers?*
- Question ③: *Are all the fruit equal in size? Does it matter (when thinking of finding a fraction)?*

IN FOCUS In question ①, children practise what they have been doing in the **Discover** and **Share** sections. They work out how many parts are shaded and how many there are in total to work out the numerator and the denominator of each fraction. In question ②, children are given one diagram and they have to find three fractions from this diagram. You may want to discuss with children what they notice about their answers. Discuss that the three numerators should add up to 6. In question ③, children are presented with pieces of fruit. Usually, when dealing with fractions, we must have equal sized parts that the whole is divided into. Here, however, the individual pieces of fruit are not all the same size. In this context, it does not matter as we think of each piece of fruit as an equal part.

STRENGTHEN To help children work out what fraction of a shape is shaded, ask them to first tell you how many equal parts the shape is divided into. You may need to count the parts together. Explain that this is the bottom number of the fraction or the denominator. Ask: *How many of these parts are shaded?* Explain this is the numerator or the top number of the fraction.

DEEPEN Ask children to represent a non-unit fraction in different ways, for example, how can they represent $\frac{3}{8}$? They could draw a shape and shade it in or they could use double-sided counters.

ASSESSMENT CHECKPOINT These questions will help you see if children can work out accurately what fraction of a shape or of a set of objects matches a given criteria.

ANSWERS

Question ① a): $\frac{3}{5}$ of the shape is shaded.

Question ① b): $\frac{5}{6}$ of the shape is shaded.

Question ① c): $\frac{7}{10}$ of the shape is shaded.

Question ② a): $\frac{2}{6}$ of the shape is red stripes.

Question ② b): $\frac{1}{6}$ of the shape is blue spots.

Question ② c): $\frac{3}{6}$ of the shape is purple.

Question ③ a): $\frac{4}{8}$ of the fruit are oranges.

Question ③ b): $\frac{2}{8}$ of the fruit are red.

Question ③ c): $\frac{7}{8}$ of the fruit are round.

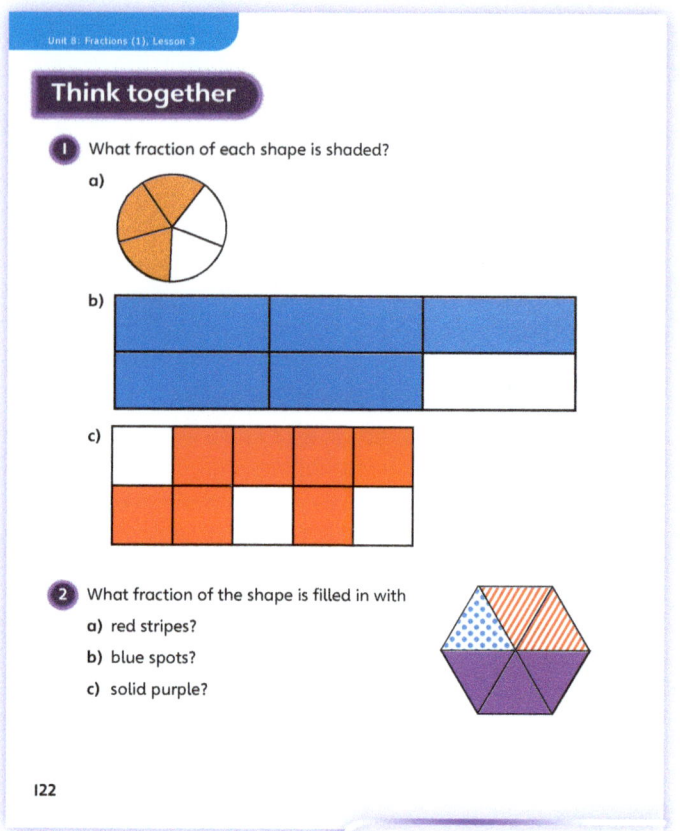

PUPIL TEXTBOOK 3B PAGE 122

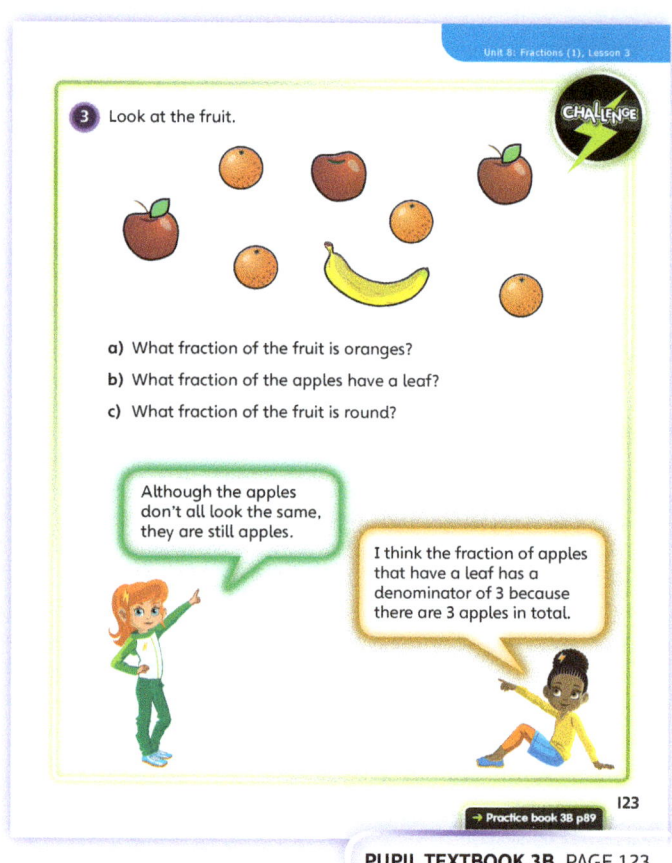

PUPIL TEXTBOOK 3B PAGE 123

Unit 8: Fractions (1), Lesson 3

Practice

WAYS OF WORKING Independent thinking

IN FOCUS Question ❶ asks children to write down what fraction of a circle is shaded, beginning with a unit fraction before moving on to non-unit fractions. Ask them what is the same about each of their fractions and shapes and what is different. This question should help them see that the numerator is the number of parts shaded and the denominator is the total number of equal parts. Questions ❷ and ❸ ask children to work out what fraction of shapes are shaded, all of which are non-unit fractions. Question ❹ asks children to shade in a given fraction of a circle. Ask them to compare their answers: this will show that they can shade in different parts but the answers are the same. Question ❺ reinforces children's understanding that the shape must be divided into equal-sized parts in order to work with fractions.

STRENGTHEN To help children work out what fraction of a shape is shaded, ask them to first tell you how many equal parts the shape is divided into. You may need to count the parts together. Explain that this is the bottom number of the fraction or denominator. Ask children how many of these parts are shaded. Explain this is the numerator or the top number of the fraction.

DEEPEN Give children a paper square or rectangle and ask them to shade in a given fraction, for example, $\frac{3}{4}$ or $\frac{5}{8}$. Do not have the rectangle already divided into parts; children have to work out different strategies that they can use.

ASSESSMENT CHECKPOINT Use these questions to assess if children can work out what fraction of a shape or set of objects is shaded or fits other criteria.

ANSWERS Answers for the **Practice** part of the lesson can be found in the *Power Maths* online subscription.

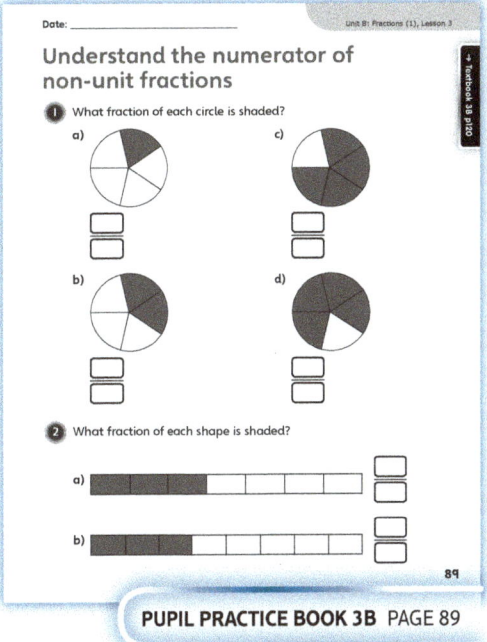

PUPIL PRACTICE BOOK 3B PAGE 89

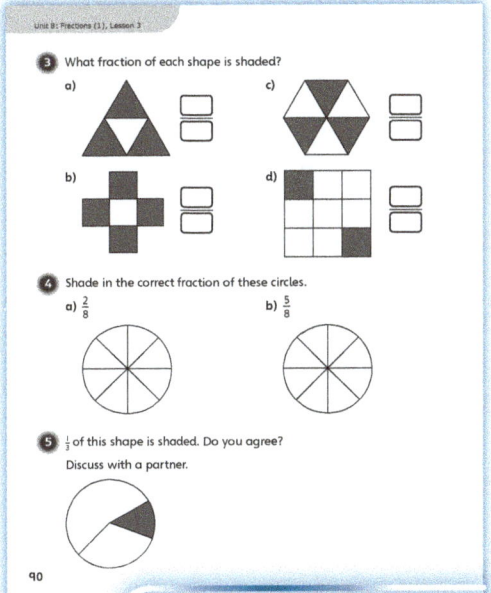

PUPIL PRACTICE BOOK 3B PAGE 90

Reflect

WAYS OF WORKING Whole class

IN FOCUS Children draw two representations of $\frac{2}{3}$. Encourage some children to use objects rather than shapes if needed for support – for example, $\frac{2}{3}$ of the pencils are red. Share answers as a class. Showcase some of the more interesting examples.

ASSESSMENT CHECKPOINT Check that children have made a shape that has 3 equal parts, of which 2 are shaded. If preferred, children can do the same thing using a set of objects.

ANSWERS Answers for the **Reflect** part of the lesson can be found in the *Power Maths* online subscription.

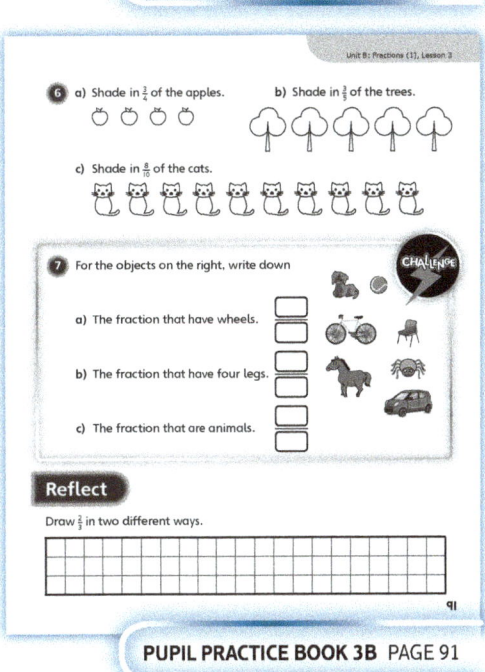

PUPIL PRACTICE BOOK 3B PAGE 91

After the lesson

- Are children able to write down what fraction of a shape is shaded?
- Are children able to write down what fraction of a set of objects meet given criteria, for example, what fraction is round, green, have wheels, etc?
- Are children able to shade in a given fraction of a shape?

Unit 8: Fractions (1), Lesson 4

Understand the whole

Learning focus
In this lesson, children will work out what two fractions they need to add together to make the whole.

Before you teach
- Can children write what fraction of a shape is shaded?
- Do children understand what the numerator and denominator of a fraction tells them?
- Do children understand that a whole can be written as a fraction, depending on how many parts it is split into?

NATIONAL CURRICULUM LINKS

Year 3 Number – fractions
Recognise and use fractions as numbers: unit fractions and non-unit fractions with small denominators.

ASSESSING MASTERY
Children can work out what they need to add on to a given fraction to make a whole. They can also tell if two fractions with the same denominator add up to make the whole.

COMMON MISCONCEPTIONS
As a result of seeing the addition sign, children may believe that both the numerator and denominator are added within the calculation. For example, that $\frac{1}{2} + \frac{1}{2} = \frac{1}{4}$. Ask:
- What does $\frac{1}{2}$ represent? What happens if you have two parts and add one part to the other part? Shade $\frac{1}{2}$ of a shape. Now shade the other half of the shape. What fraction of the shape is shaded?

STRENGTHENING UNDERSTANDING
To support children with questions such as $\frac{2}{7} + \frac{?}{7} = 1$, use fraction circles or fractions strips to help them. Ask them to shade in two of the parts in one colour and explain that this shows the $\frac{2}{7}$. Then ask them to shade in the other parts a different colour. Ask them what fraction of the shape the other colour is. They should say $\frac{5}{7}$.

Explain that these two parts added together make a whole as the whole diagram is shaded.

So, $\frac{2}{7} + \frac{5}{7} = 1$.

GOING DEEPER
Ask children more open questions such as $\frac{?}{7} + \frac{?}{7} = 1$. How many answers can they find? What do they know the numerators add up to?

Extend further by doing $\frac{?}{8} + \frac{?}{8} + \frac{?}{8}$.

KEY LANGUAGE

In lesson: fraction, equal parts, whole, plus, add, numerator, denominator

Other language to be used by the teacher: shaded

STRUCTURES AND REPRESENTATIONS
Fraction circles, fraction strips

RESOURCES
Optional: printed fraction circles, fraction shapes, fraction tiles

 In the eTextbook of this lesson, you will find interactive links to a selection of teaching tools.

Quick recap
On the board, show a fraction wheel split into quarters. Ask a child to come up and shade in a quarter. Ask children to show you what fraction is shaded. Keep shading in one more part and asking, after each part, what fraction is shaded. Discuss when the whole shape is shaded that $\frac{4}{4}$ is equal to 1.

Unit 8: Fractions (1), Lesson 4

Discover

WAYS OF WORKING Pair work

ASK

- Question 1 a): *What fraction of each circle is shaded?*
- Question 1 b): *What does it mean to add them together? Why do you think they make a whole?*

IN FOCUS Question 1 a) practises what children have learnt in previous lessons. They may need reminding what the numerator and denominator represent. In question 1 b), children understand that when they add two fractions like this they can put the fractions together. They may need to use their own copy of the fraction circles and shade in the two fractions in different colours to make sure they are the whole.

PRACTICAL TIPS For this question, you might want to give children printed fraction circles that are divided into tenths. They can then replicate the scenario.

ANSWERS

Question 1 a): $\frac{3}{10}$ of Jamie's shape is shaded.
$\frac{7}{10}$ of Max's shape is shaded.

Question 1 b): $\frac{3}{10} + \frac{7}{10} = 1$ whole

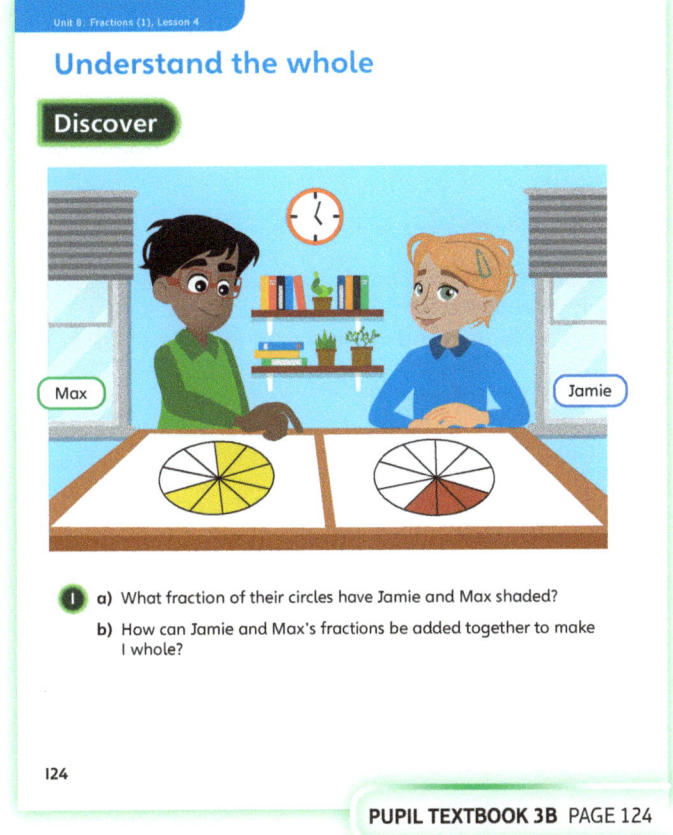

PUPIL TEXTBOOK 3B PAGE 124

Share

WAYS OF WORKING Whole class teacher led

ASK

- Question 1 a): *How many equal parts is each shape divided into? How many parts are shaded? What fraction of each shape is shaded?*
- Question 1 b): *Can you see where Max's fraction is? Can you see where Jamie's fraction is? Can you see why they add together to make 1? What number sentence can you write down?*

IN FOCUS In question 1 a), work through each fraction, considering how many parts the shape is divided into and how many parts are shaded to show that $\frac{3}{10}$ and $\frac{7}{10}$ are shaded. In question 1 b), explain to children that when they add the two fractions together, they can put them together in one circle. Ensure that they can see where Jamie and Max's fractions are. Use the circle to explain why the fraction sentence $\frac{3}{10} + \frac{7}{10} = 1$ is correct.

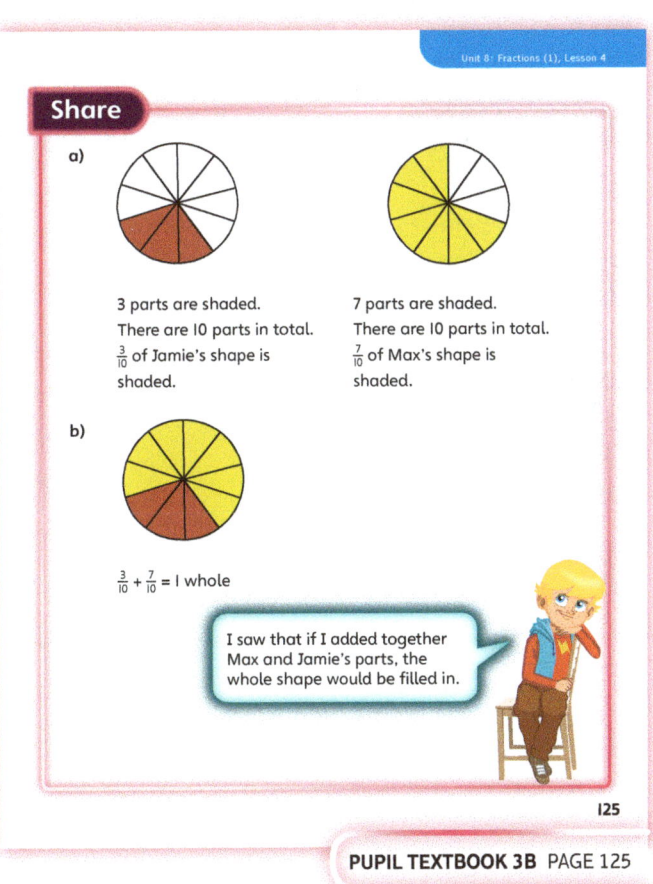

PUPIL TEXTBOOK 3B PAGE 125

163

Think together

WAYS OF WORKING Whole class teacher led (I do, We do, You do)

ASK

- Question ① a): *What fraction of the shape is red? What fraction of the shape is yellow? What fraction of the whole is shaded? How do these help you work out the missing number?*
- Question ②: *What might help you answer this question?*

IN FOCUS Question ① practises what children have just been doing in the **Discover** and **Share** sections. They work through a series of examples. By the end of the question, children should start to see that the numerators of the fractions should add together to make the whole. Question ② becomes more abstract where children are asked to find different answers to the fractions that make the whole. If children are struggling with this concept, provide them with a fraction circle split into six parts. In question ③, children are given information about an iceberg and they have to use their knowledge to realise that the iceberg is the whole. They then use the fact that $\frac{1}{8}$ is showing to work out the remainder.

STRENGTHEN If children struggle with question ②, ask them to draw or use a fraction circle that is split into six parts. Ask them to shade in one part. Then ask them how many more parts they need to shade in to make the whole. Use these two numbers to help children complete the sentence.

DEEPEN Ask children more open questions like question ②, such as $\frac{?}{8} + \frac{?}{8} = 1$. How many answers can they find? What do they know that the numerators add up to? Extend this further by doing $\frac{?}{8} + \frac{?}{8} + \frac{?}{8}$.

ASSESSMENT CHECKPOINT Use question ① to see if children can work out the missing fractions to make a whole. Use question ② to assess whether children can find pairs of fractions with the same denominator that add up to make a whole. Children should understand that the sum of the two numerators must add up to the denominator, so here they are looking for pairs of numbers that add to make 6.

ANSWERS

Question ① a): $\frac{3}{5} + \frac{2}{5} = 1$

Question ① b): $\frac{5}{9} + \frac{4}{9} = 1$

Question ① c): $\frac{3}{7} + \frac{4}{7} = 1$

Question ②: Children should find: $\frac{1}{6} + \frac{5}{6} = 1$, $\frac{2}{6} + \frac{4}{6} = 1$ and $\frac{3}{6} + \frac{3}{6} = 1$.

Question ③: $\frac{7}{8}$ of the iceberg is under the water.

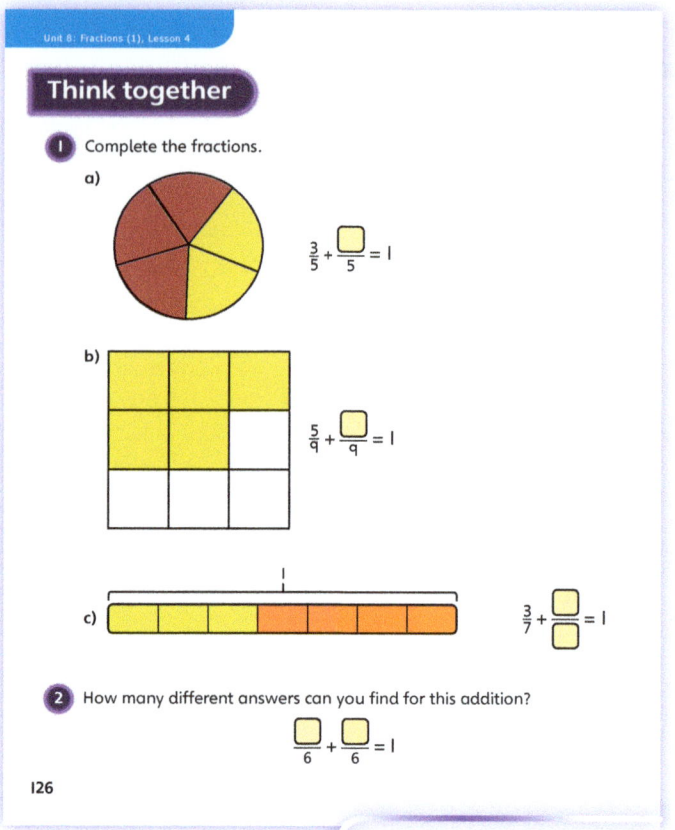

PUPIL TEXTBOOK 3B PAGE 126

PUPIL TEXTBOOK 3B PAGE 127

Unit 8: Fractions (1), Lesson 4

Practice

WAYS OF WORKING Independent thinking

IN FOCUS In question ❶, children use the fraction circles and fraction strips to work out the missing numerators. In question ❷, children are presented with abstract calculations to complete. By this time, children should realise that the numerators need to sum to the value of the denominator. Question ❸ provides some context-based examples where children need to use their knowledge to work out the missing fraction to make the whole. In question ❹, children have to find multiple solutions. They can use the fraction strip to help them. Question ❺ addresses a key misconception that children sometimes have when adding fractions. From the work earlier in the exercise, they should realise that the fraction at the end should be a whole.

STRENGTHEN Children who are demonstrating that they do not understand the concept or are struggling to get beyond a particular misconception may find it helpful to work with concrete manipulatives, such as fraction tiles.

DEEPEN Question ❹ provides an opportunity for children to find multiple solutions to the same question. Ensure they understand that this question is not about sharing equally (dividing) but is about finding complementary fractions. Present children with further similarly open questions, varying the number of parts in the whole. Challenge children to find all of the different ways the calculation could be completed.

ASSESSMENT CHECKPOINT Children should be able to use resources to explain the calculations that they have completed. Equally, they should be able to draw different representations to show the solutions to the problems.

ANSWERS Answers for the **Practice** part of the lesson can be found in the *Power Maths* online subscription.

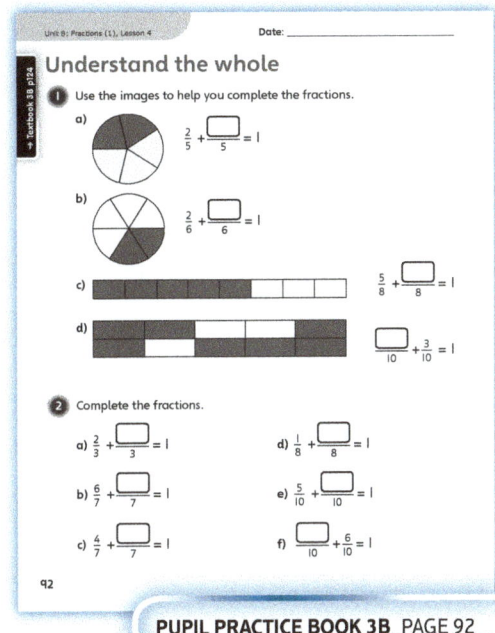

PUPIL PRACTICE BOOK 3B PAGE 92

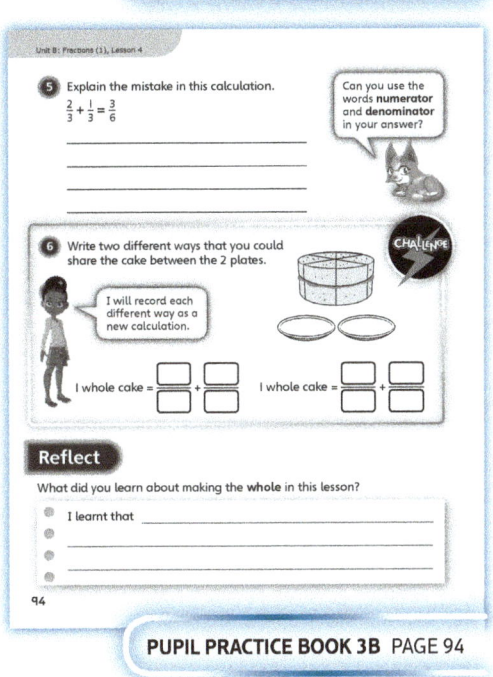

PUPIL PRACTICE BOOK 3B PAGE 93

Reflect

WAYS OF WORKING Independent thinking

IN FOCUS Encourage children to use accurate vocabulary that they have been exposed to during the lesson to fully explain what they have learnt in this lesson and show their understanding.

ASSESSMENT CHECKPOINT The strength of children's written responses should allow assessments to be made about whether they have a secure understanding of the concept or whether they are relying on a shortcut to find the correct solution to the problems presented to them.

ANSWERS Answers for the **Reflect** part of the lesson can be found in the *Power Maths* online subscription.

PUPIL PRACTICE BOOK 3B PAGE 94

After the lesson

- Are children confident finding a fraction complementary to another fraction to make a whole?
- Were children able to use a variety of different representations to show their solutions?

Unit 8: Fractions (1), Lesson 5

Compare and order non-unit fractions

Learning focus
In this lesson, children will compare and order non-unit fractions where the denominators are equal.

Before you teach
- Can children represent a fraction on a fraction strip?
- Can children use the inequality signs < and >?

NATIONAL CURRICULUM LINKS

Year 3 Number – fractions

Compare and order unit fractions, and fractions with the same denominators.

ASSESSING MASTERY

Children can compare and order two or more non-unit fractions where the denominators of the fractions are equal. Children should be able to use fractions strips to help them understand but then compare the two numerators of the fraction.

COMMON MISCONCEPTIONS

Some children may be unsure of the meaning of numerator and denominator, particularly if the numerator is greater than 1. Ask:
- *What does the number of equal parts represent? Is this the numerator or the denominator? What does the number of shaded parts represent? Is this the numerator or the denominator?*

When using resources, they may find it difficult to identify the fraction strips they need to use. Ask:
- *Show me a fraction strip that represents $\frac{3}{5}$. How many parts will the bar be split into? How many parts are shaded?*

Some children may also struggle to see why $\frac{5}{8} > \frac{4}{8}$ using fraction strips. Ask:
- *Why is it important for the two fraction strips to be the same length and same width? Does it help to line up the equal parts?*

STRENGTHENING UNDERSTANDING

In order to support children comparing two non-unit fractions, provide them with fraction strips divided into the correct number of parts. Ask them what they notice about the fraction strips. Now ask them to shade in the two fractions. Ask them to set them directly above each other to compare them. Ask: *Which of the two fraction strips has the greater/smaller amount shaded? How do you know?*

GOING DEEPER

Some children may want to begin to compare very simple examples where the denominators are not equal. For example, they may compare $\frac{1}{2}$ and $\frac{3}{4}$ using their knowledge of fractions or children could compare $\frac{2}{5}$ and $\frac{1}{2}$ using fraction strips. All comparison at this stage should be by shading in fraction strips.

KEY LANGUAGE

In lesson: compare, greater, smaller, greatest, smallest, numerator, denominator, equal, order

Other language to be used by the teacher: fraction strip

STRUCTURES AND REPRESENTATIONS

Fraction strips

RESOURCES

Optional: printed fraction strips

 In the eTextbook of this lesson, you will find interactive links to a selection of teaching tools.

Quick recap
Ask children to show an example of a non-unit fraction by shading in or drawing their own fraction strip. Ask them to then challenge the class on their fraction. Use ten of the fractions produced and see if the class can get all ten.

Unit 8: Fractions (1), Lesson 5

Discover

WAYS OF WORKING Pair work

ASK
- Question 1 a): *What do you think Zac is doing? What are Alex and Richard waving? Why might this be the case? How many parts are there in each scarf? Are the parts equal?*
- Question 1 b): *How can you compare the two fractions?*

IN FOCUS Questions 1 a) and b) introduce children to the concept of comparing non-unit fractions. There are two scarves that each represent one whole. Children may draw them out or use printed fraction strips to work out what fraction of each scarf is shaded. In order to compare the two fractions in question 1 b), they should realise the best way to compare them is to put the two fraction strips above each other. Some children may start to simply look at the numerators as the denominators are the same.

PRACTICAL TIPS Use cut out fraction strips to represent the scarves.

ANSWERS

Question 1 a): $\frac{6}{8}$ of Richard's scarf is shaded.
$\frac{4}{8}$ of Alex's scarf is shaded.

Question 1 b): $\frac{6}{8}$ is greater than $\frac{4}{8}$. Richard's scarf has a greater fraction of shaded parts.

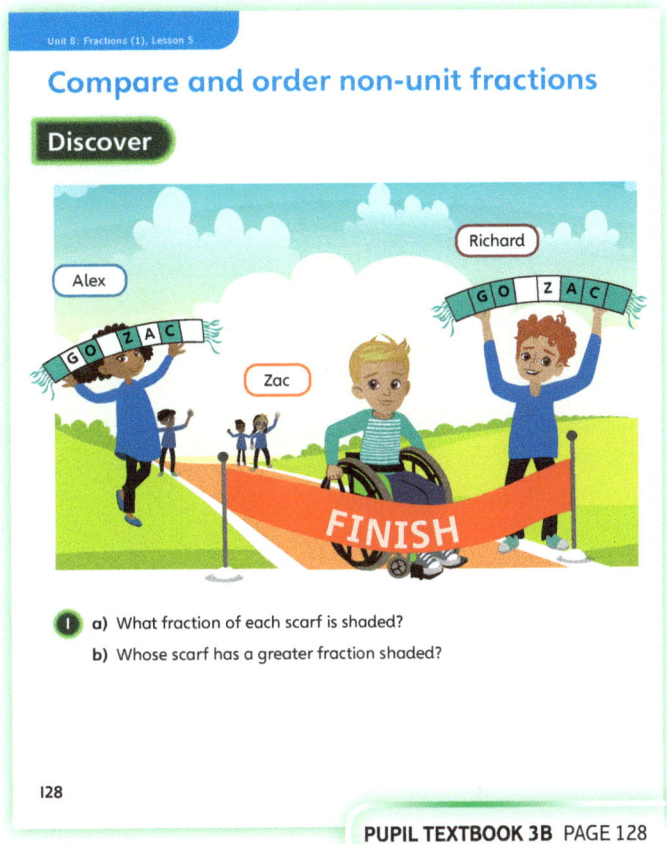

PUPIL TEXTBOOK 3B PAGE 128

Share

WAYS OF WORKING Whole class teacher led

ASK
- Question 1 a): *How many parts does each scarf have? Are the parts equal? What fraction of each scarf is shaded? How do you know?*
- Question 1 b): *How can we compare the two fraction strips? What do you notice about the wholes? What do you notice about the fraction strips? Can we compare the fractions without having to draw the fraction strips?*

IN FOCUS Question 1 a) draws on teaching from earlier in this unit by asking children to work out what fraction of the whole is shaded. Some children may say the parts do not look the same, but this is more to do with the perspective of the scarf. If Alex's scarf was flat, the scarves would be made up of equal parts. Ask children to draw on their white boards or shade in a fraction strip to show both scarves. This prepares them for question 1 b), where children compare which fraction is greater by lining up fraction strips, one above the other. They should be able to see from this that $\frac{6}{8}$ is greater than $\frac{4}{8}$ as a greater amount of the fraction strip is shaded.

Explain that we shaded in the bars from the left as this makes the comparison easier. Discuss at this stage what they notice about the numerators and that, if the denominators are the same, then you can just compare the numerators.

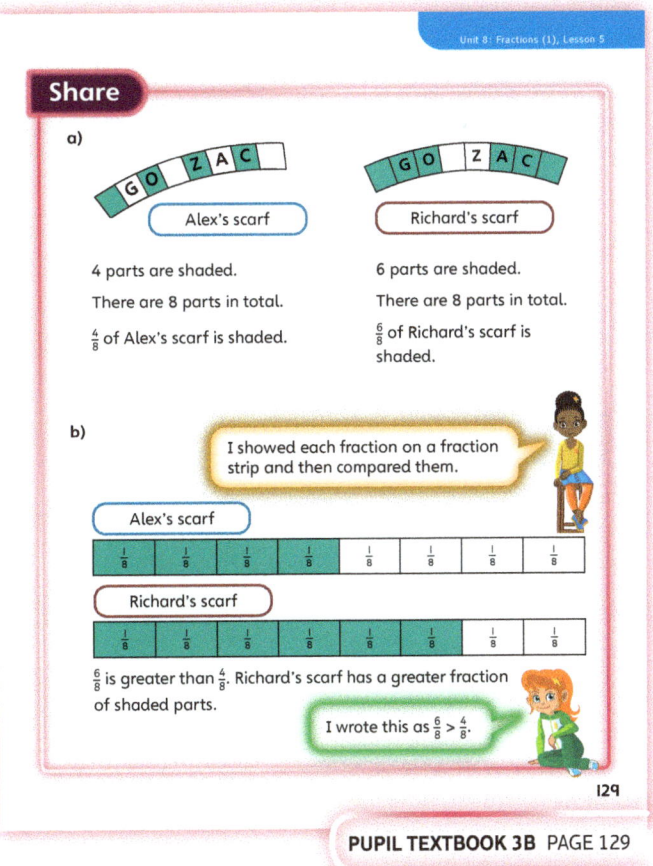

PUPIL TEXTBOOK 3B PAGE 129

Think together

WAYS OF WORKING Whole class teacher led (I do, We do, You do)

ASK

- Question ❶: *What fractions are shown? Which fraction is greater? Which is smaller? How do you know?*
- Question ❷: *How can you compare the fractions? Can you do this without drawing fraction strips? Is it necessary to check?*
- Question ❸: *What must you know about the missing numerator? How do you know this? What could the missing numbers be?*

IN FOCUS Question ❶ shows two fraction strips divided into sevenths to compare two fractions. Children should use the same method that they used in the **Discover** and **Share** sections. You may want to stress that the wholes are the same size and it is easier to compare if the fraction strips are above each other, as in this example. In question ❷, children do not have fraction strips and are asked to compare fractions that have the same denominator. Check that children are confident that they can just compare the numerators of the fractions, as the denominators are the same. For question ❷ a), you may want to still show this as fraction strips, however children need to move towards a more abstract understanding. Question ❸ develops this understanding further by asking children to find possible missing numerators. In question ❸ a), for example, if children compare the numerators, then the missing number must be a number less than 7.

STRENGTHEN For question ❷, in order to support children comparing two non-unit fractions, provide them with fraction strips divided into the correct number of parts. Ask them what they notice about the strips. Then ask them to shade in the two fractions. They should set them directly above each other to make the comparison easier. Ask: *Which of the two strips has the greater/smaller amount shaded? How do you know?*

DEEPEN Ask children to develop their own rules for comparing fractions with the same denominator. Ask them how they would teach someone this concept.

ASSESSMENT CHECKPOINT Children should be able to confidently answer questions ❷ and ❸ a) to show an understanding of comparing fractions with the same denominator. Although children can continue to draw diagrams to support them, the goal is for children to be able to do it without this support.

ANSWERS

Question ❶: $\frac{3}{7}$ is smaller than $\frac{5}{7}$.

Question ❷ a): $\frac{7}{10} > \frac{3}{10}$

Question ❷ b): $\frac{4}{5} > \frac{2}{5}$

Question ❷ c): $\frac{1}{6} < \frac{4}{6}$

Question ❸ a): The missing number could be 8 or 9: $\frac{7}{9} > \frac{8}{9}$ and $\frac{9}{9}$.

Question ❸ b): The missing number could be 3, 4, 5, 6 or 7: $\frac{2}{10} < \frac{3}{10}, \frac{4}{10}, \frac{5}{10}, \frac{6}{10}, \frac{7}{10} < \frac{8}{10}$.

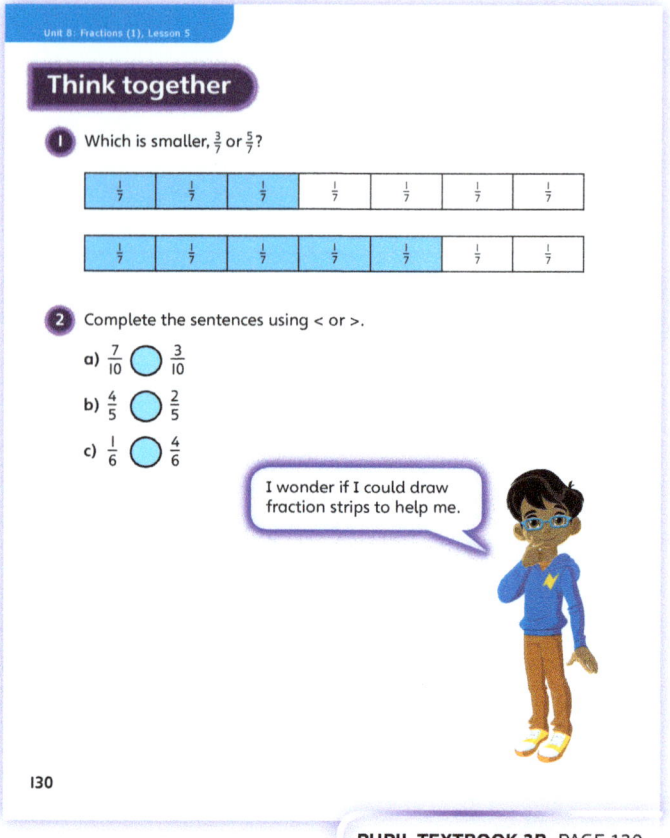

PUPIL TEXTBOOK 3B PAGE 130

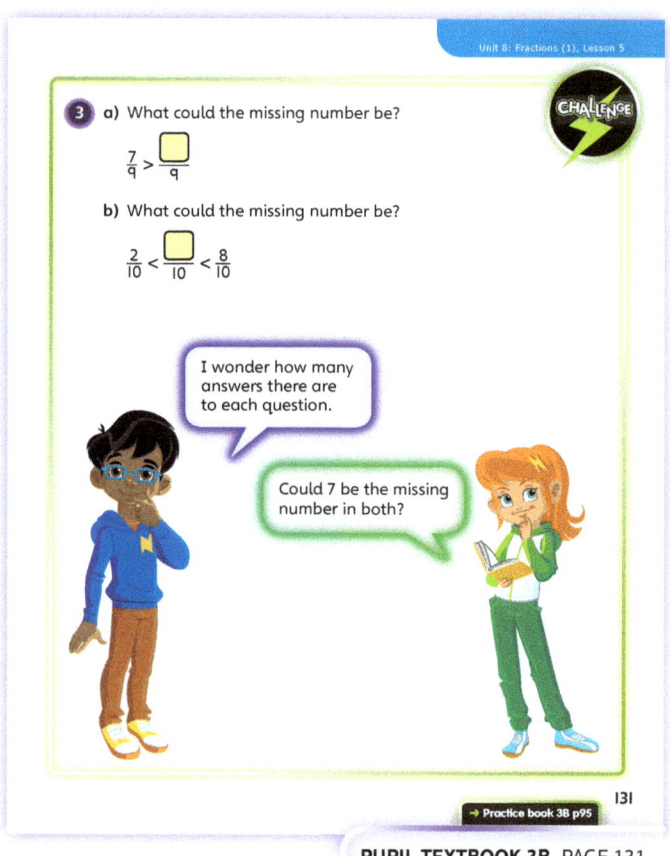

PUPIL TEXTBOOK 3B PAGE 131

Unit 8: Fractions (1), Lesson 5

Practice

WAYS OF WORKING Independent thinking

IN FOCUS Questions **1** and **2** ask children to compare two non-unit fractions with the same denominator using fraction strips to support them. In question **3**, the pictorial support is removed and children need to compare non-unit fractions by looking at the size of the numerators. They complete the sentences using less than and greater than statements. In question **4**, children follow a similar process, but this time they are using inequality signs. In question **5**, children order a set of fractions. They should quickly realise that, again, because the denominators are the same, they can just look at the numerators.

STRENGTHEN In order to support children comparing two non-unit fractions, provide them with fraction strips divided into the correct number of parts. Ask them what they notice about the strips. Then ask them to shade in the two fractions. They should set them directly above each other to compare them. Ask: *Which of the two fraction strips has the greater amount shaded? Which has the smaller amount shaded? How do you know?*

DEEPEN Question **7** provides an example of a question where children could go deeper by comparing fractions where the denominators are different but the numerators are the same. This builds on the work they did when comparing unit fractions earlier in the unit. They should notice that, this time, they need to compare the denominators; the smaller the denominator, the greater the fraction. It is the opposite to when the denominators are equal. Provide children with a few more examples for them to form generalisations.

THINK DIFFERENTLY Question **6** requires children to recognise that, as the denominators are the same, they only need to look at the numerators. They should use reasoning to be able to provide three possible answers for each question.

ASSESSMENT CHECKPOINT Look for children who are confident solving problems from questions **1** to **5**. These show whether children can use pictorial and more abstract methods to compare non-unit fractions.

ANSWERS Answers for the **Practice** part of the lesson can be found in the *Power Maths* online subscription.

Reflect

WAYS OF WORKING Pair work

IN FOCUS Although children may have grasped the key methods and concepts in this lesson, the reflect question may need a little explaining. This question is trying to get children to explain that they know that the circle must be greater than the triangle. Talk through the example to make sure children understand what they are being asked to do.

ASSESSMENT CHECKPOINT Check that children know how to compare two non-unit fractions with the same denominator by comparing the numerators.

ANSWERS Answers for the **Reflect** part of the lesson can be found in the *Power Maths* online subscription.

After the lesson

- Are children able to compare two non-unit fractions with the same denominator by using fraction strips and by comparing numerators?
- Are children able to order a similar set of fractions using the same methods?

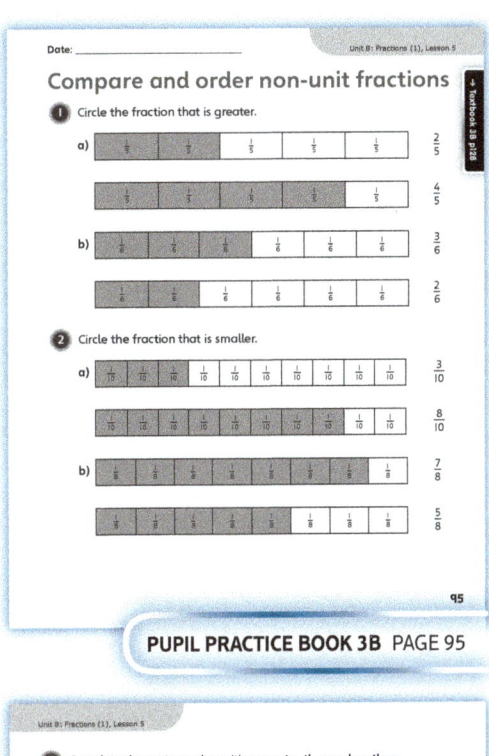

PUPIL PRACTICE BOOK 3B PAGE 95

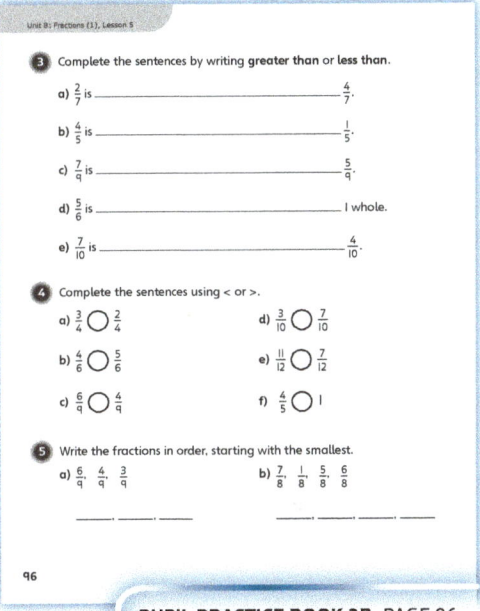

PUPIL PRACTICE BOOK 3B PAGE 96

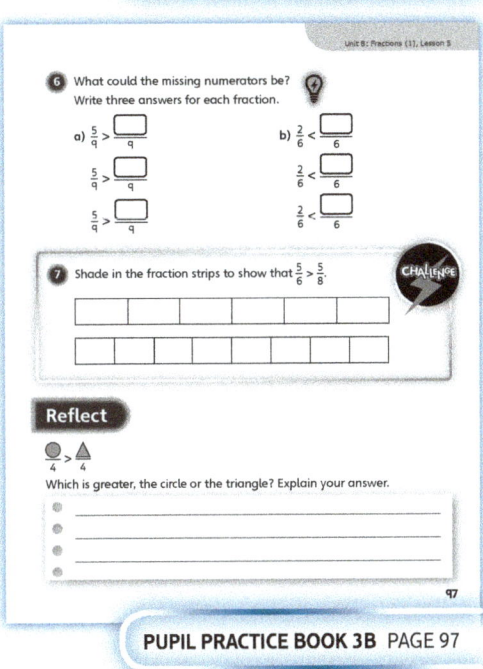

PUPIL PRACTICE BOOK 3B PAGE 97

Unit 8: Fractions (1), Lesson 6

Divisions on a number line

Learning focus
In this lesson, children will use number lines between 0 and 1 and understand what they increase by each time.

Before you teach
- Are children able to draw a number line from 0 to 10 going up in 1s and a number line from 0 to 100 going up in 10s?
- Do children understand the concept of a fraction?
- Do children know how to use a fraction strip to represent a fraction?

NATIONAL CURRICULUM LINKS

Year 3 Number – fractions

Compare and order unit fractions, and fractions with the same denominators.

ASSESSING MASTERY

Children can work out what fraction a number line increases by each time, by considering the number of intervals on a number line.

COMMON MISCONCEPTIONS

The most common misconception is that children think that if there are five marks between 0 and 1 then the number line increases by $\frac{1}{5}$ each time. Ask:
- *Can you count up in fifths from 0 to 1? Does this match what the number line shows?*

Ask children to draw a fraction strip above the number line and to divide the fraction strip into intervals, using the marks on the number line as the places where the parts of the fraction strip should be. Children will see that there are now six intervals. Explain that this means the number line goes up in sixths, not fifths.

STRENGTHENING UNDERSTANDING

To support children with this difficult concept, use a fraction strip to work out the number of intervals on the number line between 0 and 1. Ask: *How many intervals are there? What does this mean the number line increases by each time?*

Count aloud, pointing and marking the fractions on the number line as you count to check they are correct.

GOING DEEPER

Ask children to mark $\frac{3}{8}$ on a blank number line from 0 to 1. Ask them to consider how they will divide the number line. How many intervals will they need? How will they make sure that they divide the line into equal intervals?

KEY LANGUAGE

In lesson: fraction strip, intervals, increase, fraction

Other language to be used by the teacher: divide

STRUCTURES AND REPRESENTATIONS

Fraction strips, number lines

RESOURCES

Mandatory: mini whiteboards

Optional: wipeable number lines, fraction strips divided into various equal parts

 In the eTextbook of this lesson, you will find interactive links to a selection of teaching tools.

Quick recap
As a class, count in tenths from 0 to 1. Ask children to count aloud in other fractions from 0 to 1. What do they notice when they get to 1?

Discover

WAYS OF WORKING Pair work

ASK

- Question 1 a): *What does Aki's number line start and end with? How can we use a fraction strip to help us work out what his number line increases by each time?*
- Question 1 b): *Is this the same as Aki's number line? What is the same and what is different? Why is it important that the marks are equally spaced?*

IN FOCUS This task asks children to consider two number lines between 0 and 1. In question 1 a), encourage children to draw the number line on their mini-whiteboards, with a fraction strip above it. The fraction strip will help them see how many intervals there are on the number line and, therefore, will help them work out what the number line increases by. In question 1 b), children compare and contrast the two number lines. They should notice that Jamie's number line does not go up in quarters because the marks are not equally spaced. If they draw a fraction strip above the number line, they will know from their earlier work in this unit that they have not got equal parts.

PRACTICAL TIPS Recreate the situation in your classroom with the number lines set up on the board.

ANSWERS

Question 1 a): Aki's number line goes up in $\frac{1}{3}$s.

Question 1 b): Jamie's number line does not go up in $\frac{1}{4}$s.

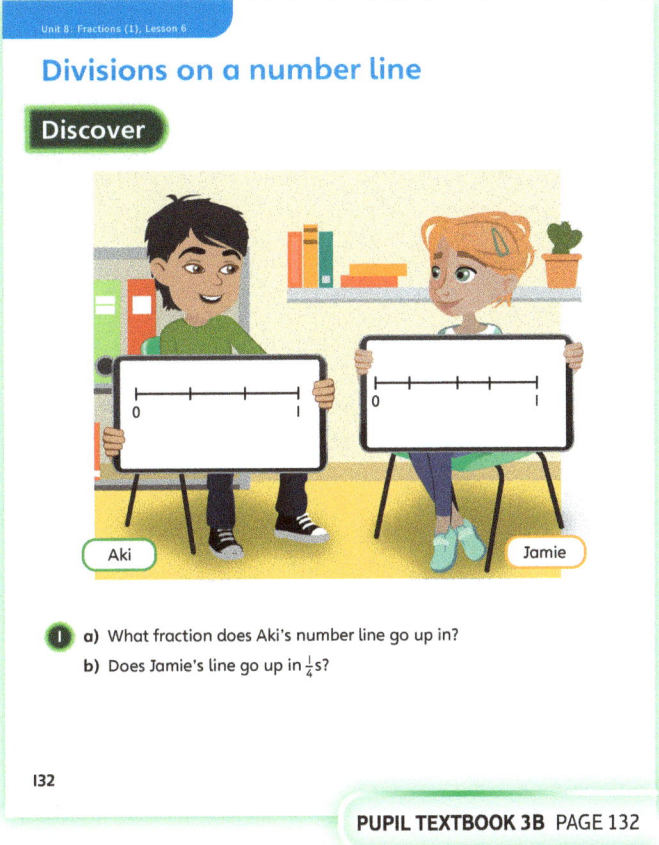

PUPIL TEXTBOOK 3B PAGE 132

Share

WAYS OF WORKING Whole class teacher led

ASK

- Question 1 a): *What does the number line start and end with? Can you see how it is made up of three intervals? What is each interval worth? What does the number line go up in?*
- Question 1 b): *How many intervals does the number line have? Are the intervals equal? Why does this number line not go up in quarters?*

IN FOCUS In question 1 a), talk children through the method of finding out what a number line goes up in by drawing a fraction strip above and working out the number of intervals. Question 1 b) provides an example where the intervals are not equal and so the fraction does not go up in quarters. Go through the steps carefully with children. Use the number lines showing the jumps to check the answer.

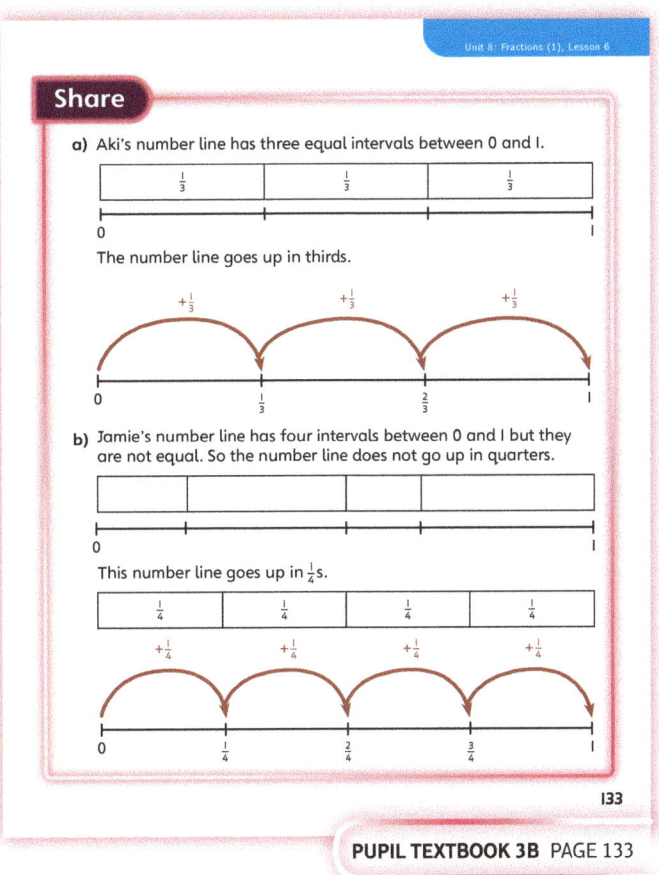

PUPIL TEXTBOOK 3B PAGE 133

Think together

WAYS OF WORKING Whole class teacher led (I do, We do, You do)

ASK
- Question ❶: *How many intervals are there on the number line? Are the intervals equal? What does the number line go up in? How can you check?*
- Question ❷: *Is there a way you can check each one? Is there a way you can work out which number line goes up in sixths by counting the intervals?*
- Question ❸: *Why do you think Alex is incorrect? How does placing a marker in the middle of the number line help Luis?*

IN FOCUS In question ❶, children follow the steps that they used to work out what the number lines in the **Discover** section increased by each time. Encourage children to check their answers by counting up. In question ❷, different number lines are given. See if children can work them out without drawing fraction strips. They should be able to see that one number line goes up in 6s to get to 30. The other two number lines are made up of one that is correct and then one that addresses a key misconception. Discuss that, because a number line has six marks between 0 and 1, it does not mean it goes up in sixths. Check again that it is not correct by counting up in sixths together. Question ❸ discusses a method that children might use to divide a number line into quarters.

STRENGTHEN To support children with this difficult concept, use a fraction strip to work out the number of intervals on the number line between 0 and 1. Ask children how many intervals there are. Then ask them what the number line increases by each time. Count aloud, pointing and marking the fractions on the number line as you count to check they are correct.

DEEPEN Ask children if they can discover a connection between the number of intervals between 0 and 1 and the fraction amount that the lines increase by each time.

ASSESSMENT CHECKPOINT Use questions ❶ and ❷ to check children's understanding of the concept and whether they can work out what a number lines increases by each time.

ANSWERS

Question ❶: The number line goes up in $\frac{1}{5}$s.

Question ❷: B and C go up in $\frac{1}{6}$s.

Question ❸ a): Alex has put her first marker too near the start of the line.
Luis has put his mark half-way along the line or at $\frac{2}{4}$.

Question ❸ b): Luis has a better method for marking a line in quarters. Now he can split each half in half to make quarters.

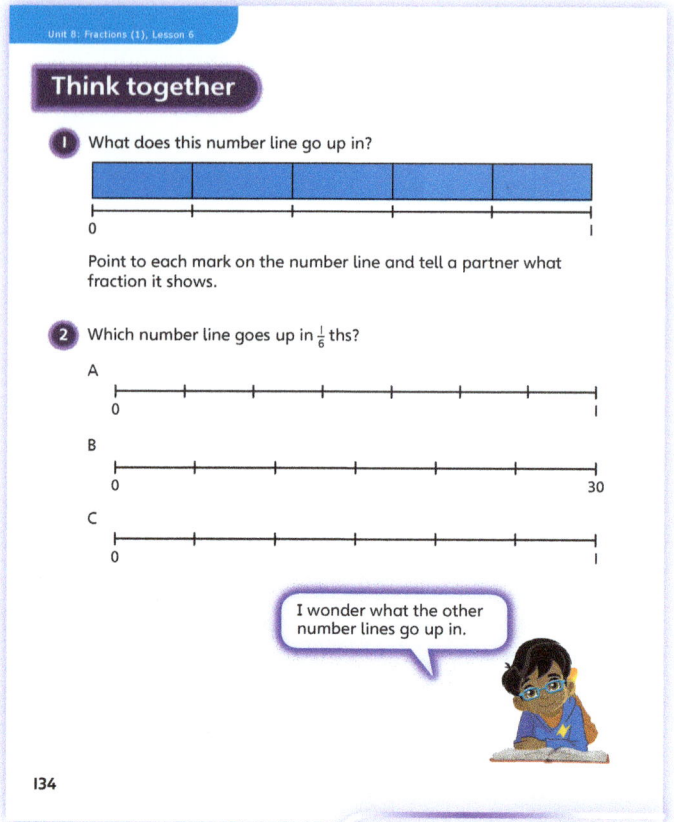

PUPIL TEXTBOOK 3B PAGE 134

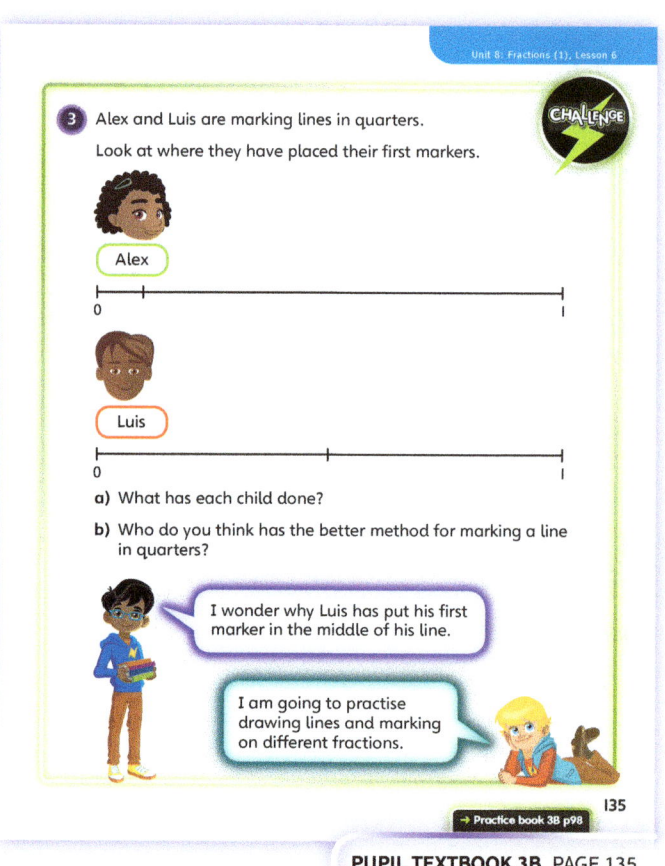

PUPIL TEXTBOOK 3B PAGE 135

Unit 8: Fractions (1), Lesson 6

Practice

WAYS OF WORKING Independent thinking

IN FOCUS Question ❶ asks children to find what the number lines increase by each time. Fraction strips showing the intervals are given for support. In question ❷, fraction strips are given but not the size of the strip, so children will need to work this out. This leads nicely on from question ❶. Remind children that the number of equal parts will help them work out the fraction. Question ❸ provides a real-life context where children may see scales and they have to use the techniques in this lesson to work out the scales.

STRENGTHEN To support children with this difficult concept, use a fraction strip to work out the number of intervals on the number line between 0 and 1. Ask them how many intervals there are. Then ask them what this means the number line increases by each time. Count aloud, pointing to and marking the fractions on the number line as you count to check they are correct.

DEEPEN Use question ❺ to check if children have effective strategies of dividing a number line. Some children may use strategies such as measuring, others may use interval division.

THINK DIFFERENTLY Question ❹ addresses the misconception of when a number line goes up in quarters and when it does not. Children should compare the two number lines and recognise that with the second number line, although it has 3 interval markers, these are not evenly spread across the number line.

ASSESSMENT CHECKPOINT Questions ❶ and ❷ will help you to see if children can work out by what fraction a number line between 0 and 1 goes up in. Some children may see that the denominator of the fraction is 1 more than the number of marks between 0 and 1.

ANSWERS Answers for the **Practice** part of the lesson can be found in the *Power Maths* online subscription.

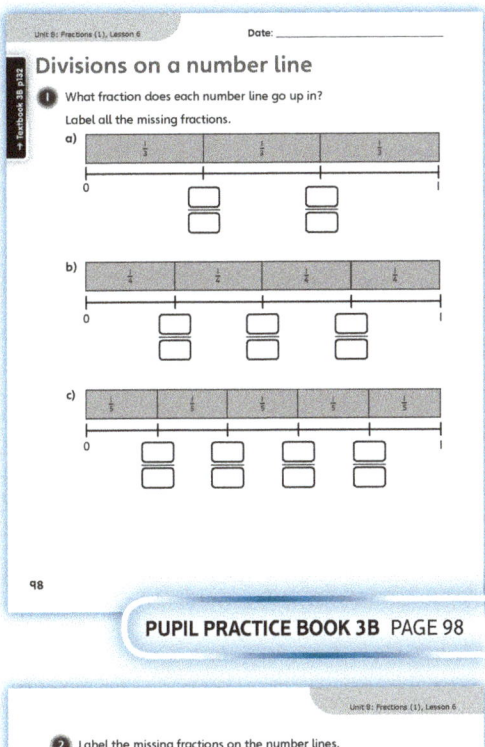

PUPIL PRACTICE BOOK 3B PAGE 98

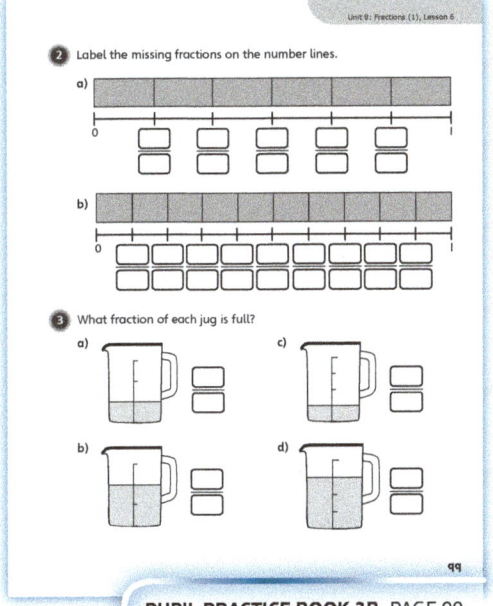

PUPIL PRACTICE BOOK 3B PAGE 99

Reflect

WAYS OF WORKING Pair work

IN FOCUS Children reflect on the methods that they have used in this lesson. Some children will need to draw a fraction strip above the number line to help them and then work out how many intervals there are. Other children should, by the end of the lesson, be able to work it out from the number of intervals on a number line.

ASSESSMENT CHECKPOINT Check that children have a method for working out what a number line increases by each time.

ANSWERS Answers for the **Reflect** part of the lesson can be found in the *Power Maths* online subscription.

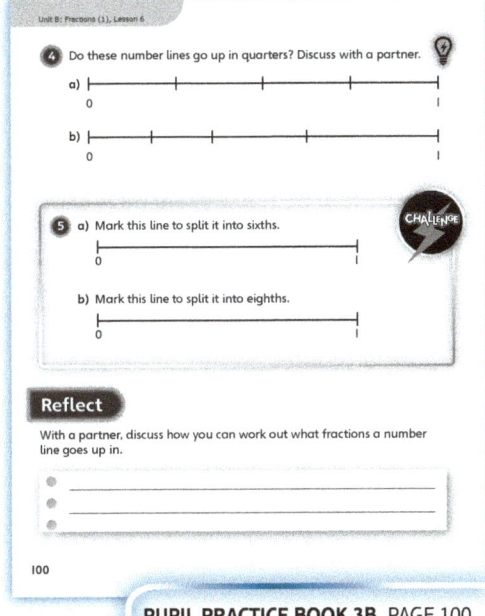

PUPIL PRACTICE BOOK 3B PAGE 100

After the lesson

- Can children work out what interval a number line increases by each time?

Unit 8: Fractions (1), Lesson 7

Count in fractions on a number line

Learning focus
In this lesson, children will place fractions on a number line, remaining within the whole. They will recognise that the denominator represents the number of parts the number line must be partitioned into.

Before you teach
- Do children remember that fractions are equal to one whole when the numerator and denominator are the same?

NATIONAL CURRICULUM LINKS

Year 3 Number – fractions
Compare and order unit fractions, and fractions with the same denominators.

ASSESSING MASTERY

Children can see fractions as a number as well as a quantity of an amount. They can accurately place fractions on a number line from 0 to 1 and understand the role of the denominator and numerator within this process.

COMMON MISCONCEPTIONS

Having worked with tenths earlier on in this unit, children may initially want to continue working with this denominator. Ask:
- *How many equal parts has the whole been split into? What denominator does this mean the number has?*

STRENGTHENING UNDERSTANDING

Children may find number lines pre-marked with various denominators useful to see where different fractions are placed between 0 and 1. To understand how far from 0 a fraction is placed, children can fold strips of paper into the required number of parts. This will physically demonstrate the position of a fraction in relation to 0.

GOING DEEPER

Based on the location of different fractions on a number line, children can begin to compare the size of fractions and record their comparisons using the signs < and >. During these comparisons, children may recognise that fractions with different denominators are equivalent to each other (but this is not the focus of the lesson).

KEY LANGUAGE

In lesson: part, whole, interval, denominator, numerator, equal, represent, previous, greater than, less than, calculation

Other language to be used by the teacher: partition

STRUCTURES AND REPRESENTATIONS

Number line, bar model

RESOURCES

Mandatory: fraction tiles, strips of paper
Optional: blank number lines

 In the eTextbook of this lesson, you will find interactive links to a selection of teaching tools.

Quick recap

On the board, draw a fraction circle split into 8 parts. Shade in 1 part and ask children what fraction is shaded. Keep shading in a part, one at a time, and asking children what fraction of the circle is shaded. Stop when you get to 1 whole.

Unit 8: Fractions (1), Lesson 7

Discover

WAYS OF WORKING Pair work

ASK

- Question 1 a) and b): *Can fractions be numbers?*
- Question 1 a): *Could the arrows be fractions? What could the denominators be?*
- Question 1 a): *How will the size of the numerator relate to the position of the fraction on a number line?*
- Question 1 b): *What will the denominator be for all the numbers that the arrows are pointing to?*

IN FOCUS In questions 1 a) and 1 b), it is important for children to understand that a fraction is a number as well as a division of an amount. Children have mainly encountered a fraction that is representing part of a quantity ($\frac{1}{2}$ a cake, for example). When this is shown on a fraction strip, the fraction is written in the middle of the division. However, when the fraction is a number, it is labelled on the number line at the end of the part.

PRACTICAL TIPS Using strips of paper will help children understand how to split a whole into the required number of parts. Ask children to fold the paper. Once the paper is unfolded, the positions of the creases along the whole of the strip will illustrate the locations of fractions along the number line.

ANSWERS

Question 1 a): The number line goes up in $\frac{1}{7}$s.

Question 1 b): The arrows are pointing to $\frac{1}{7}, \frac{2}{7}, \frac{4}{7}, \frac{5}{7}$ and $\frac{6}{7}$.

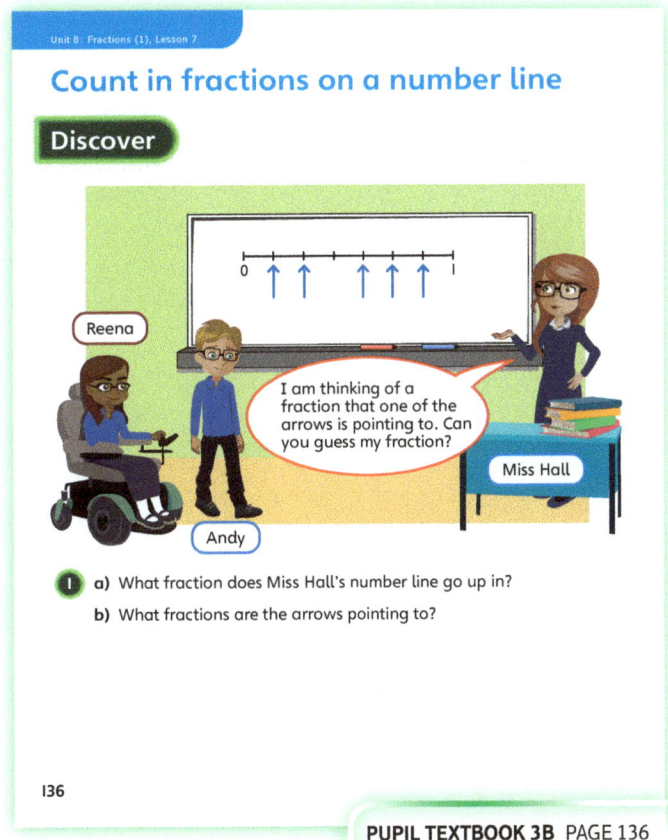

PUPIL TEXTBOOK 3B PAGE 136

Share

WAYS OF WORKING Whole class teacher led

ASK

- Question 1 a): *Where on the number line, in relation to each part, is each seventh shown?*
- Question 1 a): *Why is this the location of each seventh?*
- Question 1 b): *Which numbers have an arrow pointing to them?*

IN FOCUS The images used here to represent the problems provide a familiar visual link between previous learning and this lesson. They visually show where each seventh is marked on the number line and that this is at the end of each part. This is also consistent if a bar model, fraction tiles or folded strips of paper are used as physical models.

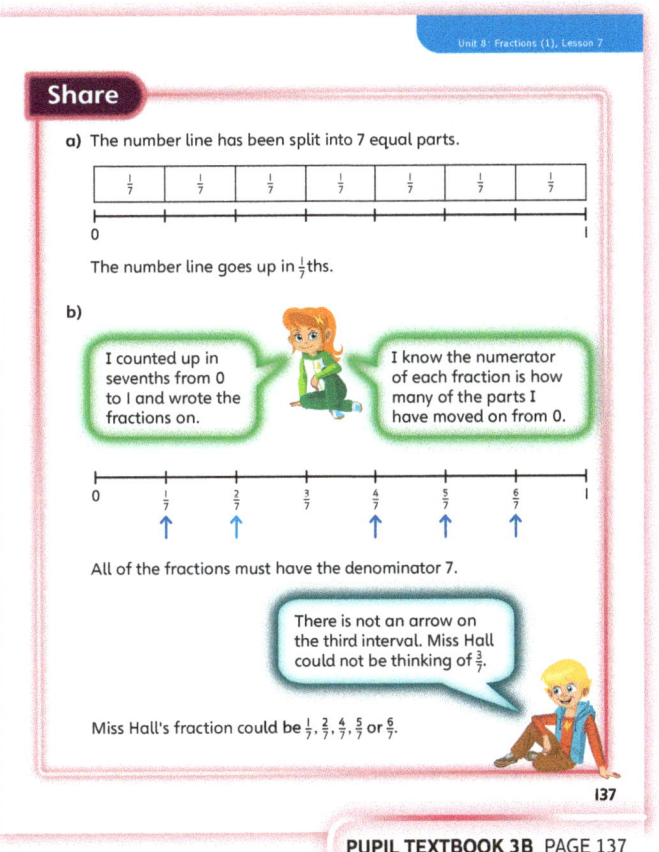

PUPIL TEXTBOOK 3B PAGE 137

Think together

WAYS OF WORKING Whole class teacher led (I do, We do, You do)

ASK
- Question ①: *What do the number lines increase by each time?*
- Question ②: *How many intervals are there on the number lines? What do the number lines go up in?*

IN FOCUS In question ①, count aloud as a class going up in fifths and eighths. As you count, mark the fractions on the number line. In question ②, children may think the number of marks tells them the fraction that the number line increases by. Explain that it is the number of *intervals* that matters and not the number of marks. To check if they are correct, children can count aloud from 0 and see if they get to 1. In question ③, children use the fraction strips to help them work out where fractions should be placed on a number line.

STRENGTHEN Children can use fraction tiles or strips of paper to ensure the length of the whole is the same length as 0 to 1 on the number line. Children can then place the necessary number of tiles on the number line, or see how many folds along the line they should mark the fraction.

DEEPEN Children could be encouraged to compare the fractions that they have placed on a number line throughout the lesson, to begin to understand the relative size of each number. They could use the signs > and < to record these comparisons.

ASSESSMENT CHECKPOINT Children should be able to identify the number of intervals that are on a number line in order to ascertain the denominator of a fraction. They should then be able to work out how many intervals along the number line their number or fraction occurs and label their fraction accurately.

ANSWERS

Question ① a): Children should count: $\frac{1}{5}, \frac{2}{5}, \frac{3}{5}, \frac{4}{5}, \frac{5}{5}$.

Question ① b): Children should count: $\frac{1}{8}, \frac{2}{8}, \frac{3}{8}, \frac{4}{8}, \frac{5}{8}, \frac{6}{8}, \frac{7}{8}, \frac{8}{8}$ and $\frac{8}{8}, \frac{7}{8}, \frac{6}{8}, \frac{5}{8}, \frac{4}{8}, \frac{3}{8}, \frac{2}{8}, \frac{1}{8}$.

Question ② a): The arrows are pointing to $\frac{1}{6}$ and $\frac{5}{6}$.

Question ② b): The arrows are pointing to $\frac{3}{10}$ and $\frac{9}{10}$.

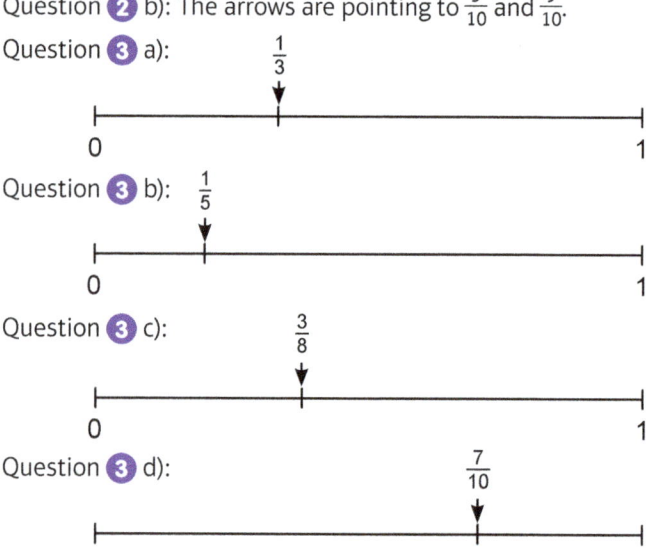

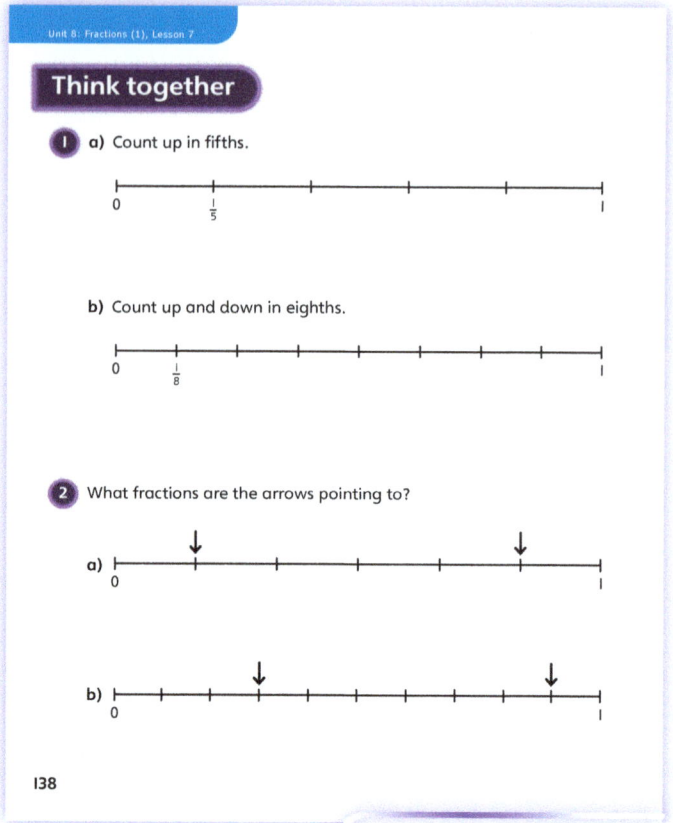

PUPIL TEXTBOOK 3B PAGE 138

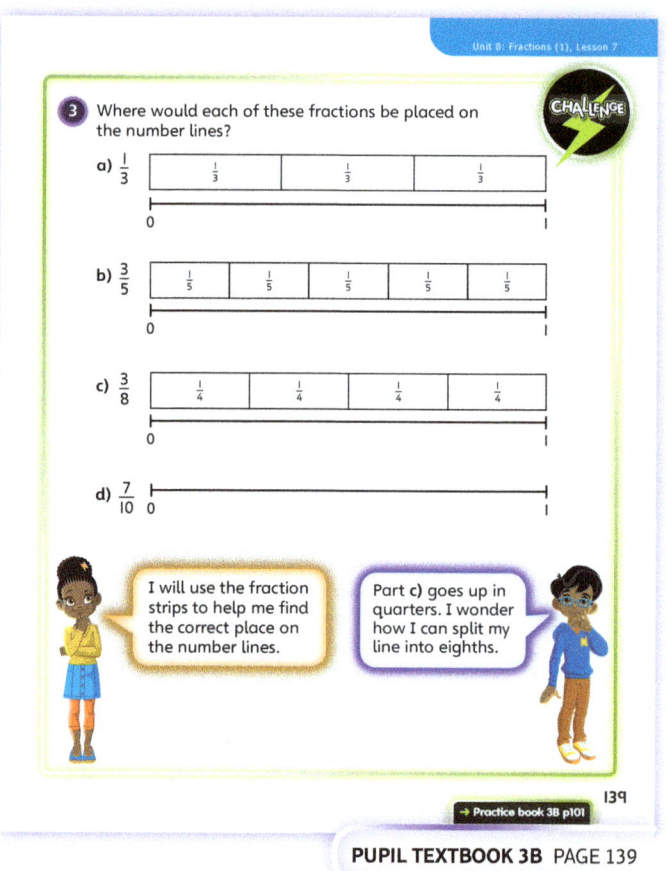

PUPIL TEXTBOOK 3B PAGE 139

Unit 8: Fractions (1), Lesson 7

Practice

WAYS OF WORKING Independent thinking

IN FOCUS Question ❶ starts by asking children to continue the number lines and fill in the missing fractions. Question ❷ asks children to draw arrows at the relevant parts on the number line. In question ❸, children need to work out what the number lines are going up in and then work out what the arrows are pointing to. Children may think the number of marks tells them the fraction it goes up by. Explain that it is the number of intervals that matter and not the number of marks. Use a method similar to that used in the **Share** section to help children's understanding. Discuss how they can check if they are correct.

STRENGTHEN Children should continue to use fraction tiles to help them visualise the contexts they are given in more depth. Encourage children to tackle the problems one fraction at a time, rather than with different fraction tiles on show at the same time.

DEEPEN Give children blank number lines and the opportunity to explore how to accurately position fractions on the number line without the aid of fraction tiles. This will encourage them to begin to think about how to split a whole into a number of parts. This may include measuring the length of the whole and dividing it accordingly.

ASSESSMENT CHECKPOINT Children should be able to explain how they positioned each fraction on the number line and in which order they used the information given to them. They should also be able to demonstrate, with resources, the steps they took to answer a nominated question.

ANSWERS Answers for the **Practice** part of the lesson can be found in the *Power Maths* online subscription.

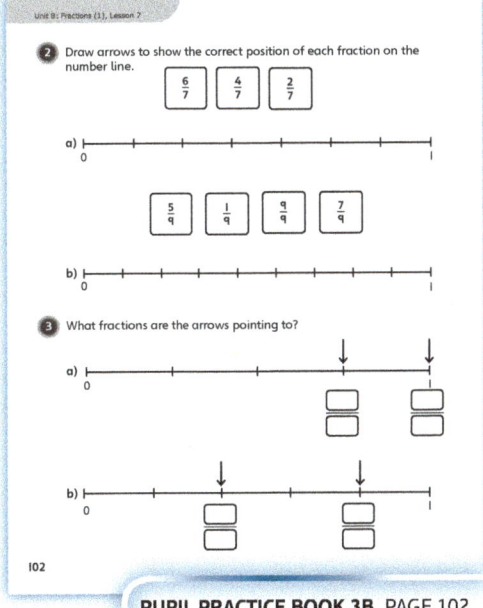

PUPIL PRACTICE BOOK 3B PAGE 101

PUPIL PRACTICE BOOK 3B PAGE 102

Reflect

WAYS OF WORKING Pair work

IN FOCUS Children need to draw their own number line that goes up in sixths. Children may put on six marks as opposed to five marks and they should double-check their answers.

ASSESSMENT CHECKPOINT The clarity of children's responses will allow assessments to be made on how secure children are with their understanding of how to accurately place a fraction on a number line.

ANSWERS Answers for the **Reflect** part of the lesson can be found in the *Power Maths* online subscription.

PUPIL PRACTICE BOOK 3B PAGE 103

After the lesson

- Can children continue a number line, counting up in single fractions?
- Can children work out the fractions that given arrows are pointing to on number lines that have marks?

177

Unit 8: Fractions (1), Lesson 8

Equivalent fractions as bar models

Learning focus
In this lesson, children will learn to recognise equivalent fractions with small denominators. They will use diagrams to represent equivalent fractions.

Before you teach
- Do children know what the numerator and denominator represent?
- Are they familiar with resources such as fraction strips?
- What opportunities will you be providing for children to gain hands-on experience with fractions?

NATIONAL CURRICULUM LINKS
Year 3 Number – fractions
Recognise and show, using diagrams, equivalent fractions with small denominators.

ASSESSING MASTERY
Children can explain and demonstrate when two fractions are equivalent.

COMMON MISCONCEPTIONS
Some children may be unsure of the meaning of numerator and denominator, particularly if the denominator is greater than 1. When using resources, they may find it difficult to identify the fraction strips they need to use by referring to the denominator. Ask:
- *Show me a fraction strip that represents $\frac{3}{5}$. How many parts will the bar be split into? How many parts are shaded?*

Some children may find the concept of an equivalent fraction tricky, as they may have found a rule that works for some fractions but not for others. For example, to find an equivalent fraction, all you do is double the top and the bottom ($\frac{1}{2} = \frac{2}{4} = \frac{4}{8}$), but that rule cannot be used to complete the fractions $\frac{1}{2} = \frac{3}{?} = \frac{6}{?}$. Ask:
- *Tell me a fraction that is equivalent to $\frac{1}{2}$. How do you know? Are there others?*

STRENGTHENING UNDERSTANDING
Rather than simply learning a rule in order to solve a calculation, children need to know why the rule works as it does. Children should be given the opportunity to use fraction tiles, coloured rods and paper strips to show fractions in different ways. The more practical experience they have, the quicker they will understand the concept of equivalent fractions: that equivalent fractions are the same size, but can be split into different equal parts.

GOING DEEPER
Provide children with a fraction wall and ask them to create their own equivalent questions to show the depth of their understanding. How many equivalent fractions can they find?

KEY LANGUAGE
In lesson: whole, fraction, numerator, denominator, **equivalent fractions**, equal, equal parts

Other language to be used by the teacher: shared, multiply, divide

STRUCTURES AND REPRESENTATIONS
Bar model, fraction wall

RESOURCES
Mandatory: fraction strips, bar model

Optional: paper strips, coloured rods, printed fraction walls, scissors

 In the eTextbook of this lesson, you will find interactive links to a selection of teaching tools.

Quick recap
Draw some shapes on the board, each with some parts shaded. Ask children to write down what fraction of the shape is shaded. Discuss their answers.

Unit 8: Fractions (1), Lesson 8

Discover

WAYS OF WORKING Pair work

ASK

- Question 1 a): *What does the picture show? How far has Lee run?*
- Question 1 b): *What markings can you see on the tracks above and below the track Lee is running on? What do you think these markings show?*

IN FOCUS Question 1 a) brings attention to the fact that a fraction can be the same size, but split into a different number of equal parts. The picture should encourage children to see that $\frac{1}{2}$ and $\frac{2}{4}$ are equal.

PRACTICAL TIPS Give children the opportunity to use a variety of different resources to create $\frac{1}{2}$, for example coloured rods and paper fraction strips. Children should use these resources to establish that $\frac{1}{2}$ can be presented in different ways.

ANSWERS

Question 1 a): Both Lee and Mr Lopez are correct.

Question 1 b): $\frac{1}{2} = \frac{2}{4} = \frac{4}{8} = \frac{8}{16}$

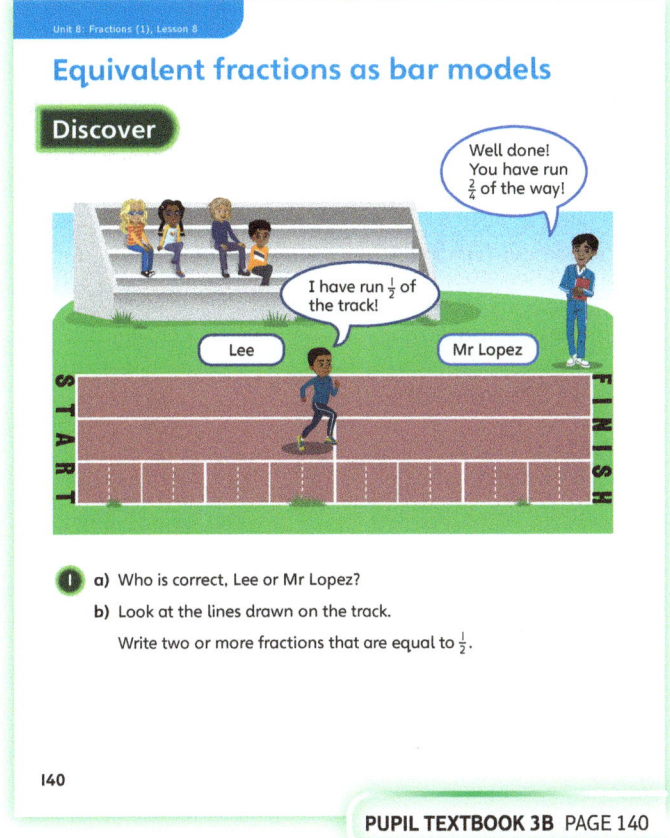

PUPIL TEXTBOOK 3B PAGE 140

Share

WAYS OF WORKING Whole class teacher led

ASK

- Question 1 a): *How many parts is the first bar split into? How many of the parts are shaded? What fraction does the first model represent? What fraction does the second model represent? What do you notice about the size of the shaded part in each bar? How many $\frac{1}{4}$s do you think may be equivalent to $\frac{1}{2}$?*

IN FOCUS This may be a good opportunity to clarify any misconceptions children may have. Ensure they are confident that the first fraction represents $\frac{1}{2}$ and the second fraction represents $\frac{2}{4}$. Children should notice that both lengths are equal, and establish that $\frac{1}{2} = \frac{2}{4}$.

Discuss the relationship between the numerator and denominator. Ask: *What do you notice about the numerator and denominator of each fraction? What is the relationship between 1 and 2? 2 and 4? Are there other fractions that are equal to $\frac{1}{2}$? What is the relationship between the numerator and the denominator of these fractions?*

Provide children with two paper strips. Ask them to fold the first strip in half and the second strip into quarters. They then use scissors to cut the first strip into two parts, and the second strip into four parts. Ask children to align the two quarters below the half. What do they notice? How can they record their findings? Record that $\frac{1}{2} = \frac{2}{4}$, explaining that the = sign means 'equivalent' or 'the same as'.

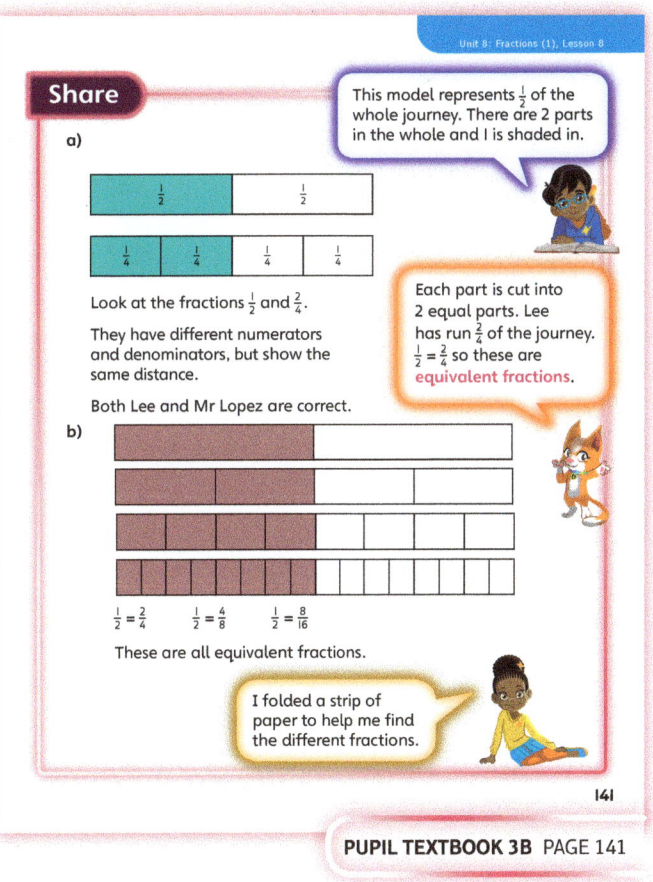

PUPIL TEXTBOOK 3B PAGE 141

Think together

WAYS OF WORKING Whole class teacher led (I do, We do, You do)

ASK

- Question ❶: *How many parts are there? How many of them are coloured? If the paper is folded again, how many parts will there be? What do all the parts have in common? What happened to the original coloured parts? How many parts are coloured now?*
- Question ❷: *What fraction of the paper is coloured? How can the information be recorded?*

IN FOCUS In questions ❶ and ❷, children may notice that $\frac{1}{6}$ is half of $\frac{1}{3}$, and $\frac{1}{10}$ is half of $\frac{1}{5}$. The questions highlight the fact that two $\frac{1}{6}$s are the same as $\frac{1}{3}$, and two $\frac{1}{10}$s are the same as $\frac{1}{5}$.

Question ❸ asks children to use their knowledge of fractions to find the missing fractions on the fraction wall. They should then look for other fractions equivalent to $\frac{1}{2}$, $\frac{1}{3}$, $\frac{1}{4}$, $\frac{1}{6}$ and $\frac{1}{12}$.

STRENGTHEN When children are working on questions ❶ and ❷, strengthen their understanding by giving them paper strips similar to the ones used in the questions. Encourage discussion on how the paper should be folded so that all the parts are equal. This may be an opportunity to clarify that in a fraction all the fraction parts are of equal size. Encourage children to use the correct mathematical vocabulary to explain what they notice during this activity.

DEEPEN Deepen children's understanding by asking them to build their own fraction wall. Ask: *What fractions will they choose? How many of each unit fraction will they need to make a whole?* Encourage them to explain their reasoning clearly. Ask children to write as many fractions equivalent to $\frac{1}{4}$ as they can. They should try to be systematic in their approach.

ASSESSMENT CHECKPOINT At this point in the lesson, children should understand what equivalent fractions are, and recognise and show equivalent fractions by using paper fraction strips.

ANSWERS

Question ❶ a): $\frac{2}{6}$ or $\frac{1}{3}$ of the strip is shaded.
Question ❶ b): $\frac{1}{3} = \frac{2}{6}$
Question ❷ a): $\frac{2}{10}$ or $\frac{1}{5}$ of the strip is shaded.
Question ❷ b): $\frac{2}{10} = \frac{1}{5}$
Question ❸: $\frac{1}{3}, \frac{1}{4}, \frac{1}{6}, \frac{1}{12}$

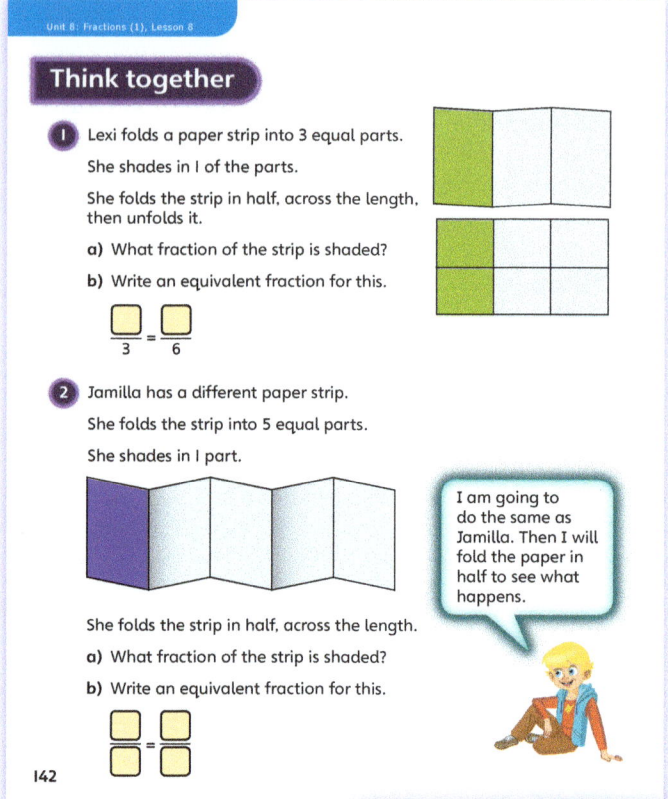

PUPIL TEXTBOOK 3B PAGE 142

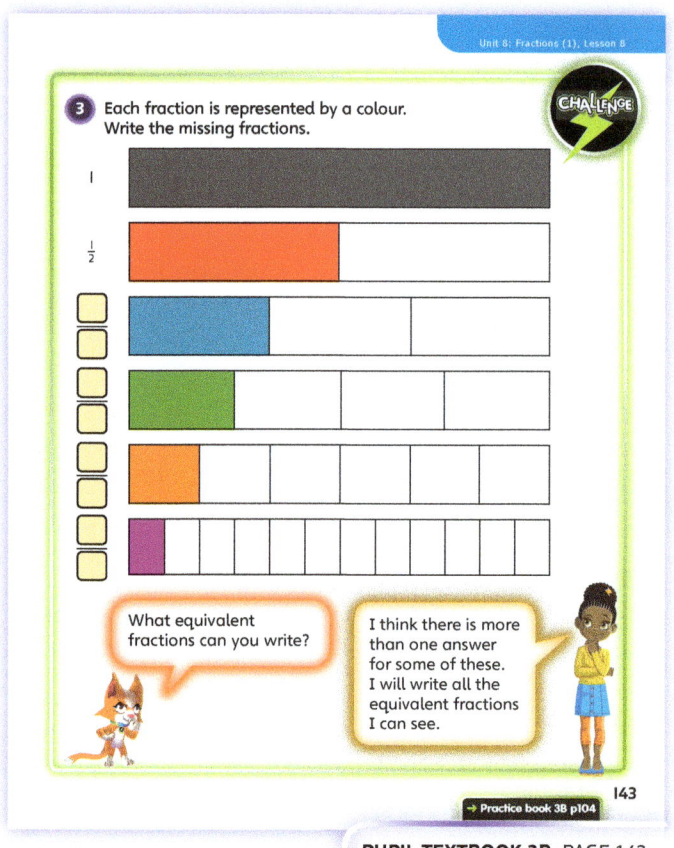

PUPIL TEXTBOOK 3B PAGE 143

Unit 8: Fractions (1), Lesson 8

Practice

WAYS OF WORKING Independent thinking

IN FOCUS When working independently on the questions in this section, children should continue to secure their understanding of identifying equivalent fractions using diagrams. They use bar models to find the missing numerator or denominator from the equivalent fractions given.

STRENGTHEN Strengthen understanding by giving children paper fraction strips so they can see a concrete representation of the question. In question ❶, ask: *What is the same and what is different? Think of the example used in the lesson: what is the question asking? What does 'equivalent fraction' mean?*

DEEPEN Provide children with similar equivalent fraction walls to the one in question ❹, but for different equivalent fractions. The more children have the opportunity to show a fraction in different ways, the deeper their understanding will be.

Prompt children to explore the concept presented in question ❺ in more depth. When discussing fractions, they need to pay attention not only to the number of parts shaded, but also to the size of each part. Ask: *Is each part equal? What is the numerator/denominator? What fraction is being represented?*

ASSESSMENT CHECKPOINT Children should be confident in identifying and showing equivalent fractions using diagrams. They can explain when two fractions are equal using the correct mathematical language. Successful answers and discussion around question ❸ should offer an indication of the depth of their understanding.

ANSWERS Answers for the **Practice** part of the lesson can be found in the *Power Maths* online subscription.

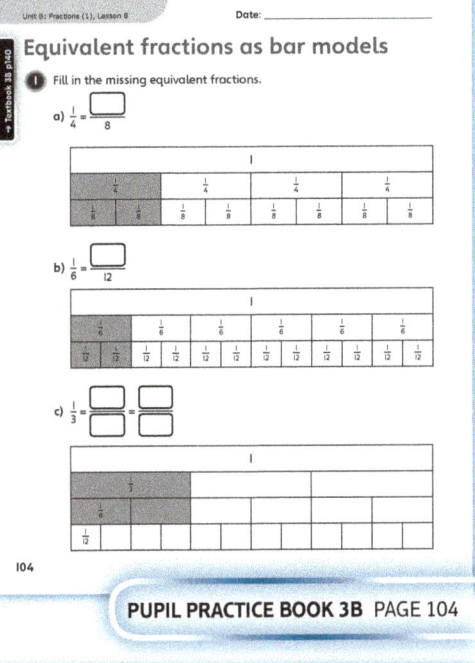
PUPIL PRACTICE BOOK 3B PAGE 104

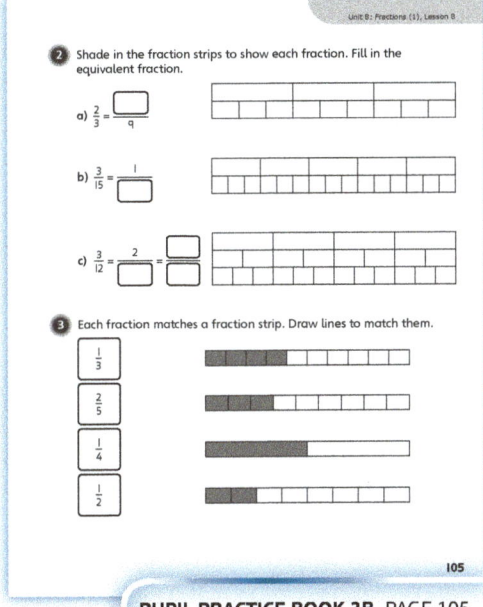
PUPIL PRACTICE BOOK 3B PAGE 105

Reflect

WAYS OF WORKING Pair work

IN FOCUS Begin this activity by giving children time to discuss with a partner how they can make equivalent fractions by folding a piece of paper. Once children have discussed their methods, give them time to write their explanations. Can they draw a diagram to support their reasoning?

ASSESSMENT CHECKPOINT Look for clarity in children's explanations. They should refer to the fact that equivalent fractions are the same size, but can be split into a different number of equal parts. Some children may want to model the activity with a piece of paper to cement their understanding.

ANSWERS Answers for the **Reflect** part of the lesson can be found in the *Power Maths* online subscription.

After the lesson

- Do children know what equivalent fractions are?
- Can children explain how to make equivalent fractions?
- Which resources did children find most useful?

181

Unit 8: Fractions (1), Lesson 9

Equivalent fractions on a number line

Learning focus
In this lesson, children will learn to recognise and show equivalent fractions with small denominators, predominantly represented on number lines.

Before you teach
- Do children know how to place unit fractions on a number line?
- What other experiences can you offer children so they understand how to recognise equivalent fractions?

NATIONAL CURRICULUM LINKS

Year 3 Number – fractions

Recognise and show, using diagrams, equivalent fractions with small denominators.

ASSESSING MASTERY

Children can show equivalent fractions on a number line. They can confidently describe how to use number lines to find equivalent fractions. Children use their knowledge of equivalence to mark fractions with different denominators on a number line. For example, on a number line labelled from 0 to 1, they can mark $\frac{1}{6}, \frac{1}{3}, \frac{1}{2}$.

COMMON MISCONCEPTIONS

Children may not have complete mastery of fractions, for example, they may find it difficult to place fractions on a number line. Some children may count the marks on a number line, rather than the intervals, and may be unsure of how to count in different fractional amounts. Ask:
- How many equal parts has the number line been split into? What denominator will the fractions in this number line have?

Some children may believe that to find equivalent fractions, you 'do the same to the top and the bottom', such as adding or subtracting the same number. For instance, some children may believe that $\frac{7}{8} = \frac{8}{9}$, since 7 + 1 = 8 and 8 + 1 = 9. Ask:
- Can you think of a fraction that is equivalent to $\frac{1}{4}$? What did you do to find this equivalent fraction? What did you do to the numerator and the denominator?

STRENGTHENING UNDERSTANDING

Allow time for all children to be confident when using a variety of resources. Revisit counting in fraction intervals, by showing children pre-marked number lines with different denominators. Hide some of the fractions and ask children to find them. If children generalise incorrectly, rather than telling them what is wrong, provide time for them to explain their findings. Ask: Show me $\frac{7}{8}$. Show me $\frac{8}{9}$. Are they equal? How do you know?

GOING DEEPER

Provide children with pairs of fractions that are equal and some that are not equal. Ask them to group them into 'equivalent fractions' and 'non-equivalent fractions'. Ask: How can you prove your answers? What resources will you use? By choosing their own resources and explaining their findings, children will gain a deeper understanding.

KEY LANGUAGE

In lesson: fraction, equivalent, equal, number line

Other language to be used by the teacher: whole, compare, intervals, shared, equal parts, numerator, denominator, multiply (×), divide (÷)

STRUCTURES AND REPRESENTATIONS

Fraction wall, number line, bar model

RESOURCES

Mandatory: fraction wall, number line, fraction strips, bar model

Optional: paper strips, fraction tiles, fraction rods, string

 In the eTextbook of this lesson, you will find interactive links to a selection of teaching tools.

Quick recap
Draw a number line on the board with 5 marks. Label the start 0 and the end 1. Ask children to come up one at a time to write fractions that they know onto the number line. Repeat for other number lines with a different number of markings.

Unit 8: Fractions (1), Lesson 9

Discover

WAYS OF WORKING Pair work

ASK

- Question 1 a): *What does the picture show? How many parts does the bracelet have? What fraction of the bracelet is blue and thick? What fraction of the bracelet is yellow and thin? Which fraction strip on the wall is the same length as the blue, thicker part of the bracelet?*
- Question 1 b): *Is there another fraction strip that is the same length? How can you show your results?*

IN FOCUS Question 1 b) allows children to find pairs of equivalent fractions. The picture should encourage children to notice the fraction strips that are of equal length. Ask: *What does 'equivalent fraction' mean? What are you looking for on the fraction wall?*

PRACTICAL TIPS Give children the opportunity to recreate the question using concrete resources, such as fraction tiles or fraction rods, to build the fraction strips. They can use string or a paper strip to create a number line. They can measure the distance each bar represents and find the equivalent fractions. Moving from concrete to pictorial representations will deepen their understanding of equivalent fractions.

ANSWERS

Question 1 a): $\frac{3}{4}$ is equivalent to $\frac{6}{8}$.

Question 1 b): The other equivalent fractions on the fraction wall are: $\frac{1}{4} = \frac{2}{8}$ and $\frac{1}{2} = \frac{2}{4} = \frac{4}{8}$.

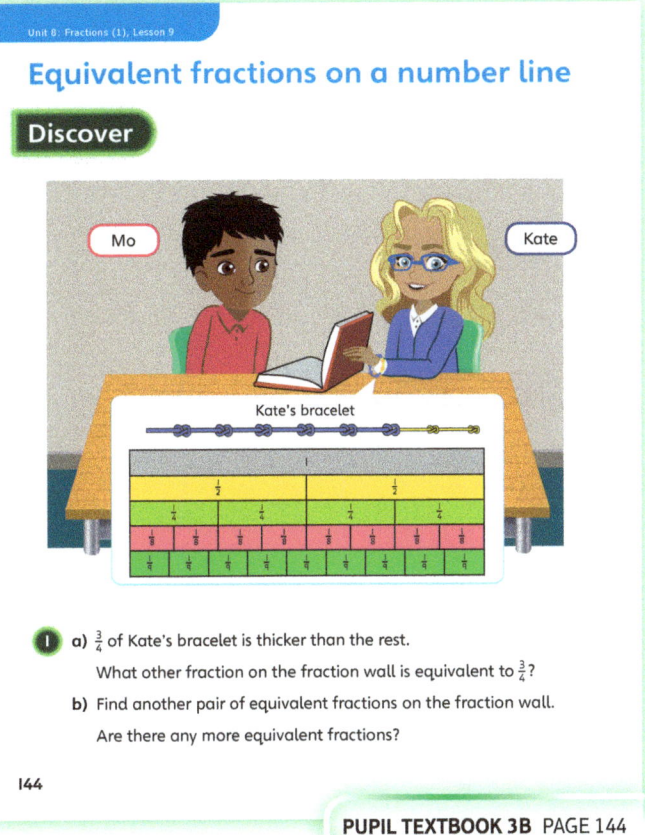

PUPIL TEXTBOOK 3B PAGE 144

Share

WAYS OF WORKING Whole class teacher led

ASK

- Question 1 a): *How many parts are there in total on the bar that shows $\frac{3}{4}$? What does the arrow show? What fraction do the top number line and fraction strip represent? What fraction do the second number line and fraction strip represent? Tell me a fraction that is equivalent to $\frac{1}{4}$. How do you know? Are there others?*
- Question 1 b): *What are the coloured arrows pointing to? Can you see any other fractions that are equivalent?*

IN FOCUS Use this opportunity to recap the potential misconceptions listed in the common misconceptions section. Allow children to exercise their understanding of using a number line to show equivalent fractions.

In question 1 b), discuss the different equivalent fractions that children may have found. Ask: *How do you know these fractions are equal? How did you use the number line and the fraction wall to find the equivalent fractions? Choose one pair of equivalent fractions. What do these fractions have in common? How do they differ?*

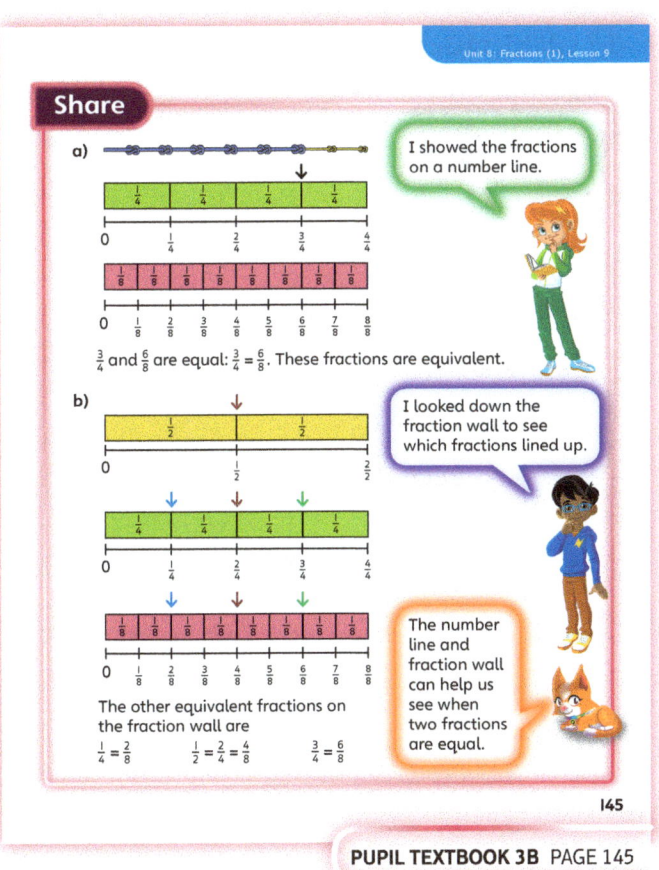

PUPIL TEXTBOOK 3B PAGE 145

Think together

WAYS OF WORKING Whole class teacher led (I do, We do, You do)

ASK

- Question ❶: *How many equal parts are there in each of the number lines?*
- Question ❶: *How will you decide which of the number lines you can use to help you?*

IN FOCUS In question ❶, children are required to use the number lines to find the missing numerators and denominators. Listen to children's reasoning for their answers, and encourage discussion of how the number lines can be used to identify equivalent fractions.

Question ❷ builds on the knowledge that children have gathered so far in the lesson and requires children to show how they can use number lines to find fractions equivalent to $\frac{2}{3}$. Children need to draw and label the number lines independently and use their understanding of equivalence to solve the question.

STRENGTHEN In question ❸, children are reminded that when you add fractions you do not add both the numerators and the denominators. Encourage children to explain why Zac is wrong, using diagrams and the correct mathematical vocabulary.

DEEPEN Deepen children's understanding by asking them to compare $\frac{3}{4}$, $\frac{7}{8}$ and $\frac{11}{12}$. Ask: *Which fraction is the biggest? Which fraction is the smallest? Can you place all these fractions on the same number line labelled from 0 to 1?*

ASSESSMENT CHECKPOINT At this point in the lesson, children should know how to order fractions on a number line. They understand how to use number lines to find equivalent fractions. For an indication of their understanding, observe their work as they attempt to solve question ❸.

ANSWERS

Question ❶ a): $\frac{1}{4} = \frac{2}{8}$

Question ❶ b): $\frac{1}{3} = \frac{2}{6}$

Question ❶ c): $\frac{2}{3} = \frac{4}{6}$

Question ❶ d): $\frac{3}{4} = \frac{6}{8}$

Question ❶ e): $\frac{2}{4} = \frac{4}{8}$ or $\frac{3}{6}$

Question ❷ : $\frac{2}{3} = \frac{4}{6} = \frac{6}{9}$

Question ❸ a): Zac is wrong because $\frac{3}{4}$ is smaller than $\frac{7}{8}$ and $\frac{11}{12}$.

Question ❸ b): $\frac{3}{4} = \frac{6}{8} = \frac{9}{12}$

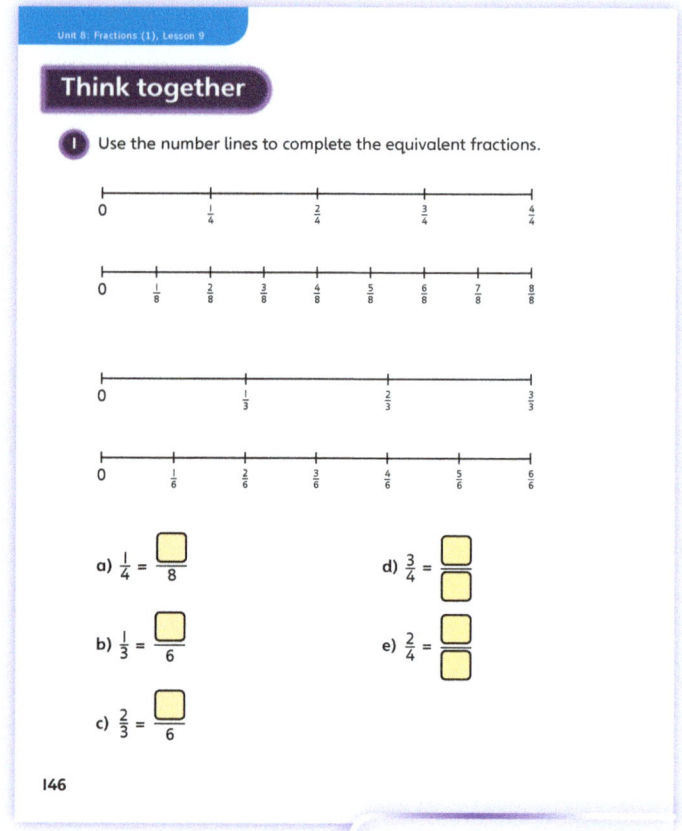

PUPIL TEXTBOOK 3B PAGE 146

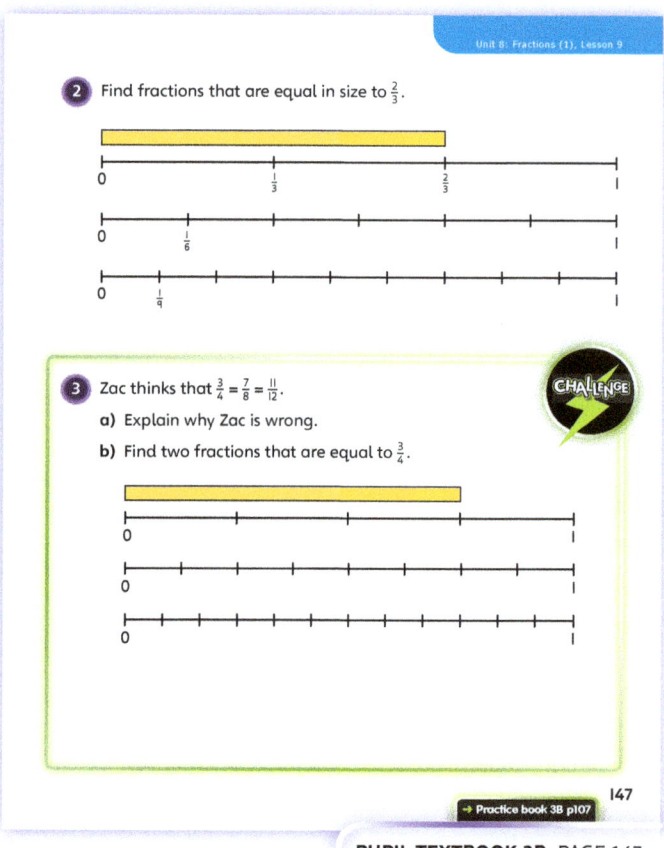

PUPIL TEXTBOOK 3B PAGE 147

Unit 8: Fractions (1), Lesson 9

Practice

WAYS OF WORKING Independent thinking

IN FOCUS These questions will help children to secure their understanding of identifying equivalent fractions using diagrams. They use number lines to find the missing numerator and/or denominator from the given equivalent fractions.

STRENGTHEN If children are struggling, use paper fraction strips, so that they can see a concrete representation of the question. In question ❶, ask: *What is $\frac{1}{2}$ equal to (think of the example in the lesson)? How can you use the number lines to find equivalent fractions? What are you looking for? What does 'equivalent fraction' mean?*

DEEPEN Deepen children's understanding in question ❹ by asking them: *How big is the interval from $\frac{1}{4}$ to $\frac{1}{2}$? How big is the interval from $\frac{1}{8}$ to $\frac{1}{2}$?* Ask them to demonstrate how they know they are correct.

Children could also explore the concept presented in question ❻ in more depth. When discussing the fractions in the question, children need to think about how different parts of the whole are joined together to make one. Ask: *What does a whole look like in a diagram? How can it be represented as a fraction? What other equivalent fractions can you find?*

THINK DIFFERENTLY Question ❺ encourages children to think differently by identifying the fractions that are not equivalent.

ASSESSMENT CHECKPOINT Children should be confident in identifying and showing equivalent fractions using diagrams. They can explain how number lines can be used to find equivalent fractions, using the correct mathematical language.

ANSWERS Answers for the **Practice** part of the lesson can be found in the *Power Maths* online subscription.

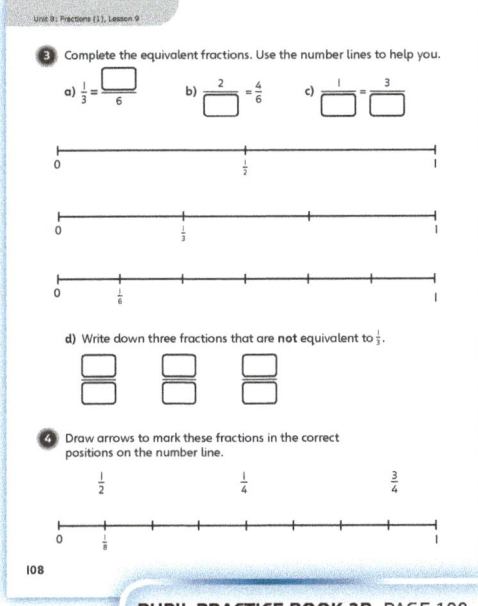

PUPIL PRACTICE BOOK 3B PAGE 107

PUPIL PRACTICE BOOK 3B PAGE 108

Reflect

WAYS OF WORKING Pair work

IN FOCUS Start by giving children time to discuss with a partner how to use number lines to find equivalent fractions. Once children have discussed their methods, give them time to write their explanations. Challenge them to make a missing number question for their partner.

ASSESSMENT CHECKPOINT Look for clarity in children's explanations. They should make reference to the fact that equivalent fractions are the same size, but can be split into a different number of equal parts.

ANSWERS Answers for the **Reflect** part of the lesson can be found in the *Power Maths* online subscription.

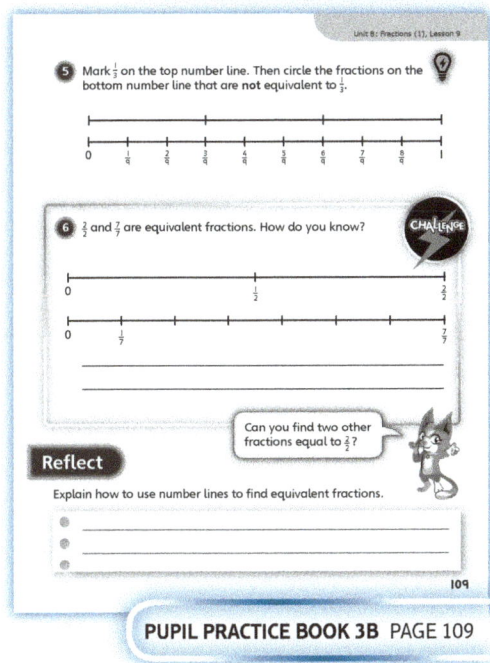

PUPIL PRACTICE BOOK 3B PAGE 109

After the lesson

- Can children show equivalent fractions using number lines and use number lines to help them find equivalent fractions?
- Were children able to use diagrams to support their findings?
- Were children able to explain their reasoning?

185

Unit 8: Fractions (1), Lesson 10

Equivalent fractions

Learning focus

In this lesson, children will continue to develop their ability to find equivalent fractions using proportional reasoning. Diagrams are used to build children's understanding of pattern and numerical reasoning.

Before you teach

- Are children confident in showing equivalent fractions?
- Are children confident using number lines and fraction strips?
- What resources will you provide beyond the number lines and fraction strips so that the lesson is as hands-on as possible?

NATIONAL CURRICULUM LINKS

Year 3 Number – fractions

Recognise and show, using diagrams, equivalent fractions with small denominators.

ASSESSING MASTERY

Children can use proportional reasoning to understand equivalent fractions and pairs of fractions, through the relationship between the numerator and denominator of each fraction.

COMMON MISCONCEPTIONS

Some children will use methods that do not apply to all scenarios. For example, if you double the fraction to find families of equivalent fractions, this will work for $\frac{1}{3} = \frac{2}{6} = \frac{4}{12}$, but not for finding $\frac{5}{15}$ or $\frac{6}{18}$. Children need to understand the meaning of the digits in a fraction, rather than relying on rules they cannot explain. Ask:

- *What are the numerator and denominator in $\frac{2}{4}$ showing us? What is the relationship between them? Can you think of an equivalent fraction that uses different numbers, but shows the same thing?*

STRENGTHENING UNDERSTANDING

Show children concrete representations of $\frac{1}{3}$, such as $\frac{1}{3}$ of a chocolate bar. Split each third into three equal parts. Children establish that $\frac{1}{3} = \frac{3}{9}$. Move from concrete to pictorial representations, by asking children to find $\frac{1}{3}$ on a fraction wall or number line.

GOING DEEPER

Ask children to look at the relationship between the numerator and denominator of $\frac{1}{3}$. 1 is a third of 3, or 3 is three times greater than 1. Now ask children to look at $\frac{3}{9}$. Ask: *What relationship is there between the numerator and denominator of $\frac{3}{9}$?* Give children opportunities to develop their proportional reasoning when considering the relationship between the numerator and denominator in a fraction.

KEY LANGUAGE

In lesson: numerator, denominator, equivalent fraction, pattern, multiply by, divide by

Other language to be used by the teacher: proportional reasoning, missing fraction, relationship, unit fraction, whole

STRUCTURES AND REPRESENTATIONS

Fraction strip model, number line, fraction wall

RESOURCES

Mandatory: fraction cards, number lines, fraction strips, paper strips

Optional: number cards

 In the eTextbook of this lesson, you will find interactive links to a selection of teaching tools.

Quick recap

Display a fraction wall on the board. Ask children to use the fraction wall to write down as many equivalent fractions to $\frac{1}{4}$ as they can see.

Repeat for equivalent fractions for $\frac{1}{5}$ and $\frac{1}{3}$. Discuss methods that children use.

186

Discover

WAYS OF WORKING Pair work

ASK

- Question 1 a): *What are equivalent fractions? What fraction does Reena need to find an equivalent to? What is the relationship between the numerator and denominator in $\frac{1}{2}$?*
- Question 1 b): *What fractions can you make with the numbers shown?*

IN FOCUS Use this task to encourage children to use the visual diagrams they have seen in previous lessons to find other fractions that are equal to $\frac{1}{2}$. This is an opportunity to recap and briefly assess children's current understanding of the numerator and denominator in a fraction. In question 1 a), encourage children to realise that they can use any of the four available number cards, but only some combinations will make an equivalent for $\frac{1}{2}$. In question 1 b), they have the numbers 1 to 6 available. Ask: *Can you think of a pair of equivalent fractions, or suggest a fraction to start with?*

PRACTICAL TIPS Make sure each pair or group of children has 5 paper strips folded into halves, thirds, quarters, fifths and sixths. Remind children that they are looking to make $\frac{1}{2}$. Children could test for themselves how many $\frac{1}{3}$s, $\frac{1}{4}$s, $\frac{1}{5}$s or $\frac{1}{6}$s they need to make $\frac{1}{2}$ (if any).

ANSWERS

Question 1 a): Reena can complete her puzzle with: $\frac{1}{2} = \frac{3}{6}$.

Question 1 b): Danny could complete his puzzle with: $\frac{1}{2} = \frac{3}{6}$, $\frac{2}{3} = \frac{4}{6}$ or $\frac{1}{3} = \frac{2}{6}$.

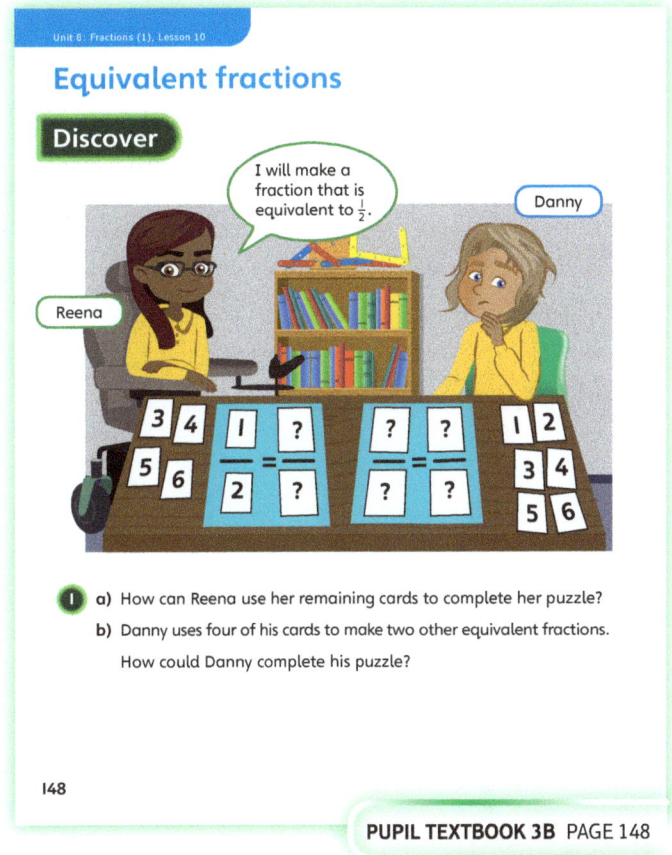

PUPIL TEXTBOOK 3B PAGE 148

Share

WAYS OF WORKING Whole class teacher led

ASK

- Question 1 a): *Tell me a fraction that is equivalent to $\frac{1}{2}$. What cards does Reena have? Can you make the equivalent of $\frac{1}{2}$ a different way, using the remaining cards? How do you know without checking the diagram?*
- Question 1 b): *Can you make another fraction? Can you find an equivalent fraction to that? Are there any other pairs of equivalent fractions you could make?*

IN FOCUS Children can show their understanding of equivalent fractions in these activities. Ask: *Can you explain why these fractions are equal? Are there other fractions equal to them?* Watch for, and discuss, any potential misconceptions children may have.

Question 1 b) is a good opportunity to reinforce the relationship that may exist between the numerator and denominator. For example, children could be looking for fractions where the numerator is 2 times or 3 times less than the denominator (or where the denominator is 2 times or 3 times greater).

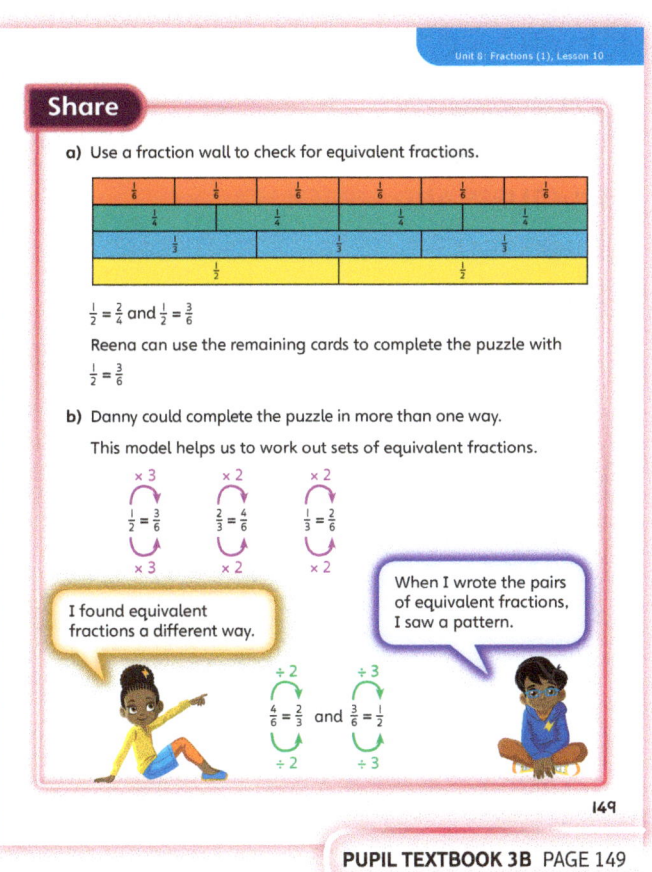

PUPIL TEXTBOOK 3B PAGE 149

Think together

WAYS OF WORKING Whole class teacher led (I do, We do, You do)

ASK
- Question ① a): *What are equivalent fractions? How can you use number lines to find equivalent fractions?*
- Question ②: *What does each numerator show? What does each denominator show? How can you use this information to find the missing numerators and denominators?*

IN FOCUS Questions ① and ② will help scaffold the children's use of diagrams. Ask children to match the number lines with the fractions they are making.

STRENGTHEN In question ①, ask children to describe the pattern in each pair of fractions they make. Ask: *How many times bigger is the denominator than the numerator?* In question ②, look for the pattern that exists between both fractions. Ask: *How many times bigger is one denominator than the other? Will you multiply or divide to find the missing numbers?*

DEEPEN Challenge children in question ③ by asking them to think about different ways in which the question can be answered. Not only will they practise finding equivalent fractions, but they will also have an opportunity to consider which method they prefer and why. To deepen their understanding even further, ask them to draw diagrams to support their answers.

ASSESSMENT CHECKPOINT At this point in the lesson, children should be more confident in finding equivalent fractions using proportional reasoning. Visual diagrams are used to build their understanding of pattern and numerical reasoning.

Children are becoming increasingly fluent in keeping the same relationship between numerator and denominator in all equivalent fractions.

ANSWERS

Question ① a): $\frac{1}{2} = \frac{5}{10}$

Question ① b): $\frac{1}{5} = \frac{2}{10}$

Question ② a): $\frac{2}{5} = \frac{4}{10}$

Question ② b): $\frac{3}{10} = \frac{6}{20}$

Question ② c): $\frac{8}{10} = \frac{4}{5}$

Question ② d): $\frac{6}{8} = \frac{3}{4}$

Question ③ a): $\frac{1}{5} = \frac{4}{20}$

Question ③ b): $\frac{8}{20} = \frac{4}{10}$

Question ③ c): $\frac{8}{16} = \frac{1}{2}$

Question ③ d): $\frac{6}{9} = \frac{2}{3}$

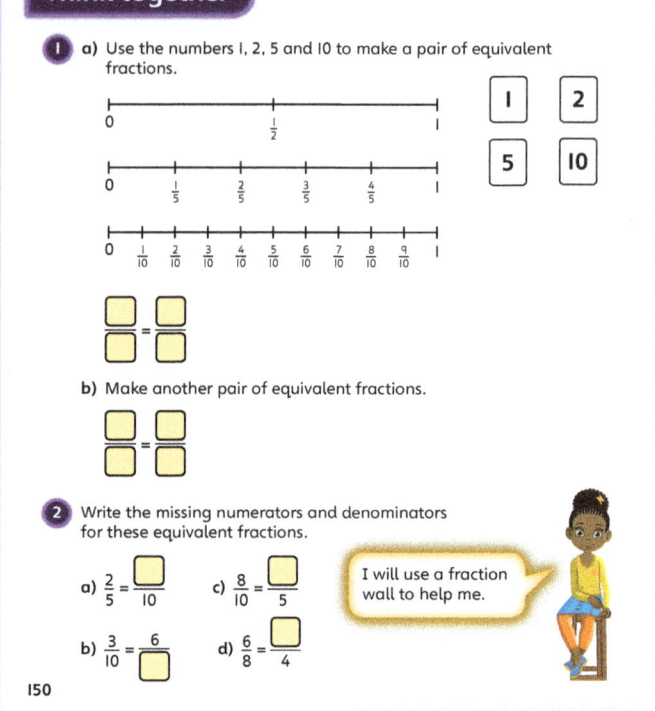

PUPIL TEXTBOOK 3B PAGE 150

PUPIL TEXTBOOK 3B PAGE 151

Unit 8: Fractions (1), Lesson 10

Practice

WAYS OF WORKING Independent thinking

IN FOCUS Question ❶ scaffolds children's understanding of using diagrams to find equivalent fractions. Shading representations of the fractions is very beneficial in giving children an opportunity to develop their fluency in finding missing numerators.

STRENGTHEN If children need support with question ❷, it may be helpful to use number lines and fraction strips to make each of the fractions presented in the question. Ask: *How do you know you are right?* Once they have a clear visual representation in their heads, help them develop their proportional reasoning. Ask: *What can you divide 12 by to get 3? What can you multiply 2 by to get 8?*

DEEPEN Use question ❺ to deepen children's explanation and reasoning skills. Ask: *What part of the lesson did Emma not understand? How will you explain the mistake? What should the answer have been? Can you draw a diagram to show your answer?*

THINK DIFFERENTLY Question ❹ offers children an opportunity to problem solve and think independently. Children have to work systematically to find the answer. Ask: *Can you find another answer? Can you explain your answer in two ways?* This is a good opportunity to clarify any misconceptions that children may have. Ask: *What clues did you use to find the answer? Can you use a number line to support your answer?*

ASSESSMENT CHECKPOINT At this point in the lesson, children should be able to confidently use their understanding of pattern and numerical reasoning to find equivalent fractions.

ANSWERS Answers for the **Practice** part of the lesson can be found in the *Power Maths* online subscription.

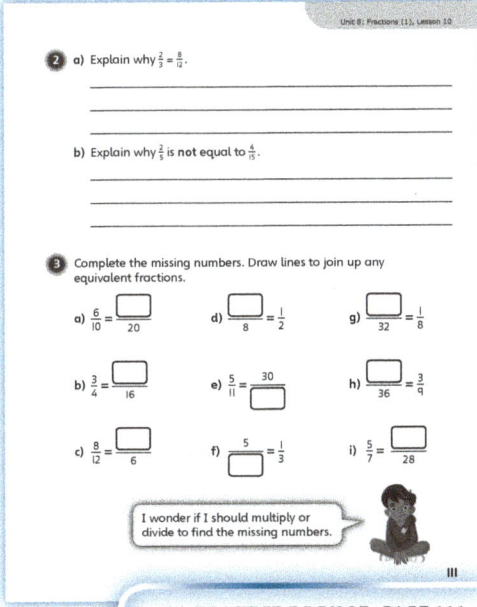

Reflect

WAYS OF WORKING Independent thinking

IN FOCUS This question offers a good opportunity to observe children's reasoning. Pay particular attention to their thinking about the relationship between the numerator and denominator of a single fraction, or the proportional relationship between the numerators and denominators in a pair of fractions.

ASSESSMENT CHECKPOINT Look for children who are able to clearly explain their reasoning. They can use a concrete representation or picture to justify their answer, rather than reverting to a preferred rule or shortcut to try to find the answer.

ANSWERS Answers for the **Reflect** part of the lesson can be found in the *Power Maths* online subscription.

After the lesson

- Have children recognised the links between the concepts explored in the past three lessons?
- How will you reinforce these links?

Unit 8: Fractions (1)

End of unit check

Don't forget the unit assessment grid in your *Power Maths* online subscription.

WAYS OF WORKING Group work adult led

IN FOCUS

- Question ① assesses children's ability to recognise equivalent fractions to $\frac{2}{3}$. Ask: *How can you use the number line to check your answer?*
- Question ② assesses children's ability to find the missing numerator or denominator of equivalent fractions. Ask: *What representations can you use to help you with this question?*
- Question ③ assesses children's ability to compare unit fractions. Ask: *Can you explain how you found the answer? How do you compare unit fractions?*
- Question ④ assesses children's ability to add and subtract fractions to make 1. Ask: *For each of these calculations, how can 1 be represented as a fraction? Why?*
- Question ⑤ assesses children's ability to place a unit fraction on a number line, from 0 to 1.
- Question ⑥ assesses children's ability to use a bar model to compare fractions.
- Question ⑦ is a SATS-style question and assesses children's ability to order non-unit fractions based on their size. Ask: *Which is the smallest fraction? How do you know?*

ANSWERS AND COMMENTARY

Children will demonstrate mastery by finding equivalent fractions and comparing fractions. They can use bar models and number lines to support their answers. Children can add and subtract fractions with confidence and can solve fraction problems.

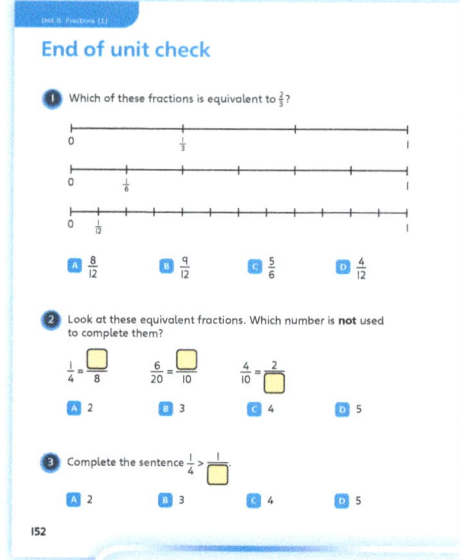

PUPIL TEXTBOOK 3B PAGE 152

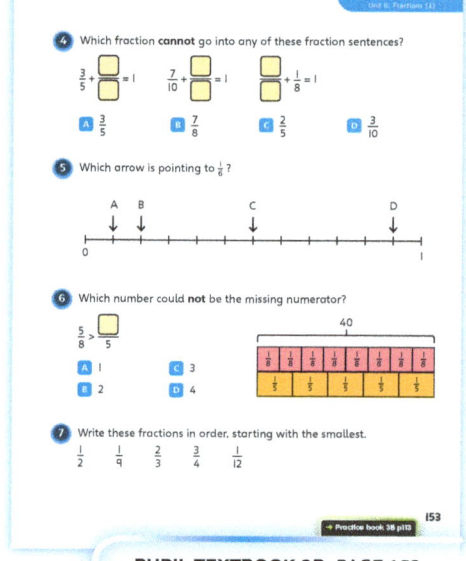

PUPIL TEXTBOOK 3B PAGE 153

Q	A	WRONG ANSWERS AND MISCONCEPTIONS	STRENGTHENING UNDERSTANDING
1	A	Choosing B or D may indicate a lack of understanding about using number lines.	To help children gain fluency in their understanding of fractions: • Make sure they have access to a fraction wall and fraction rods throughout the day. • Ensure all the representations of number lines are labelled clearly. • Ensure children have access to unit fractions made of different numbers of equal parts.
2	C	Choosing A, B or D may indicate a lack of understanding about identifying equivalent fractions.	
3	D	Choosing A or B may indicate a lack of understanding about how to compare unit fractions. Choosing C may indicate that children misread the question.	
4	A	Choosing B, C or D may indicate that children lack experience in adding and subtracting fractions to make one whole.	
5	B	Choosing any of the other arrows suggests that children are not confident at recognising what markers represent on a number line from 0 to 1.	
6	D	Choosing A, B or C indicates that children do not know what the > sign means or they are not able to compare fractions using a bar model.	
7	$\frac{1}{12}, \frac{1}{9}, \frac{1}{2}, \frac{2}{3}, \frac{3}{4}$	Check children have looked at both the numerator and denominator.	

Unit 8: Fractions (1)

My journal

WAYS OF WORKING Independent thinking

ANSWERS AND COMMENTARY

Children may record answers such as those shown below.

When comparing the circle to the square:
- They are unit fractions and the first fraction is smaller than the second.
- The more parts a unit is divided into, the smaller the size of each part.
- Looking at a fraction wall, the bigger the denominator, the smaller the size of the bar. Some children may prove this using real examples and show that, for example, $\frac{1}{3} < \frac{1}{2}$ or $\frac{1}{10} < \frac{1}{8}$.

When comparing the triangle with the pentagon:
- The denominators are the same so the greater the numerator, the greater the fraction.
- If I look at a fraction strip split into five equal parts, the more parts I have, the bigger the fraction is.
- Some children may prove this by using real examples and show that, for example, $\frac{4}{5} > \frac{2}{5}$.

If children are finding it difficult to give reasons for their answers, ask: *What does the numerator show? What does the denominator show? Show me the fractions on a number line. Which one is bigger?*

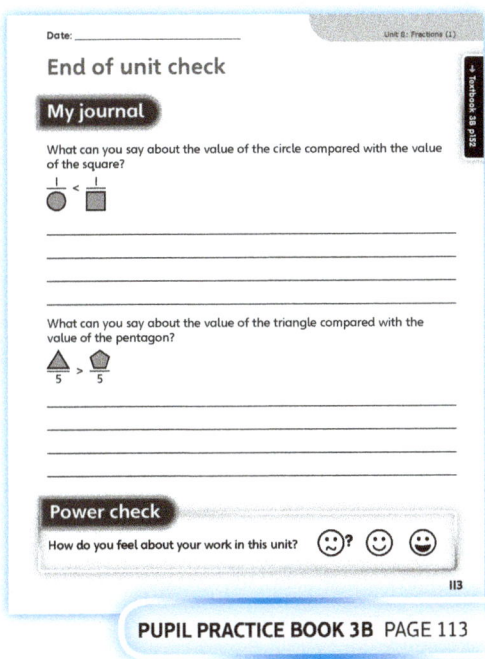

PUPIL PRACTICE BOOK 3B PAGE 113

Power check

WAYS OF WORKING Independent thinking

ASK
- What did you know about numerators and denominators before you started this unit? What new things have you learnt?
- How confident do you feel about comparing fractions?
- What do you feel you could improve on in this unit? What do you need more help or practice with?

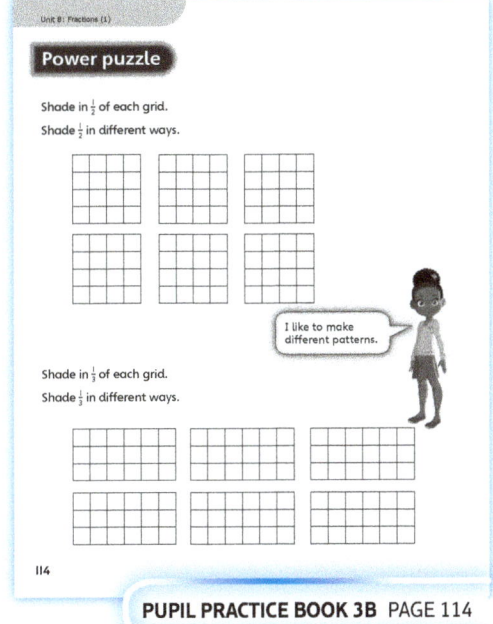

PUPIL PRACTICE BOOK 3B PAGE 114

Power puzzle

WAYS OF WORKING Pair work

IN FOCUS This puzzle will assess children's ability to show equivalent fractions. Children can use a bar model or number line to support their answers and show their reasoning clearly.

ANSWERS AND COMMENTARY Some children may not be sure how to find more than one equivalent fraction. Provide more opportunities to practise counting on and back on a number line using different fraction steps and regular exposure to fraction walls.

After the unit ⏸
- Is your classroom a 'fraction-rich' environment? How many opportunities are there for children to engage with fractions around the school? Are the fraction representatives visible and available for all children to use?

Strengthen and **Deepen** activities for this unit can be found in the *Power Maths* online subscription.

Unit 9
Mass

Mastery Expert tip! 'Throughout this unit, make sure you do lots of practical weighing and measuring. This hands-on learning is vital to achieve mastery!'

Don't forget to watch the Unit 9 video!

WHY THIS UNIT IS IMPORTANT

This unit is important because it strengthens children's knowledge of mass: an important area of learning which has many real-life applications.

First, children will learn how to measure and read a scale, focusing upon unmarked intervals. Next, different masses will be compared and ordered. Following this, children will learn to add and subtract different amounts using a range of strategies. Finally, they will apply their knowledge to real-life problems – an important skill for children to learn in order to work towards mastery.

WHERE THIS UNIT FITS

→ Unit 8 – Fractions (1)
→ **Unit 9 – Mass**
→ Unit 10 – Capacity

This unit involves the application of skills such as addition and subtraction in a measurement context. Children will have covered these strategies in Key Stage 1, and in previous Year 3 units, but will require support when applying them. Measurements were covered in Year 2 Unit 9, in which mass, capacity and temperature were the focus.

Before they start this unit, it is expected that children:
- can use scales to compare, estimate and measure the mass of an object
- are able to measure mass in grams and also in kilograms
- can count in hundreds to 1,000 to link grams to kilograms.

ASSESSING MASTERY

Children who have mastered this unit will be able to read scales accurately, including when there are missing intervals. Children should also be able to apply their understanding to solve problems involving mass. Finally, ideas and methods will be explained effectively, using the correct mathematical vocabulary and representations.

COMMON MISCONCEPTIONS	STRENGTHENING UNDERSTANDING	GOING DEEPER
Children may confuse kilograms and grams and not add them up separately, or convert once they meet the 1,000 g barrier.	Use place value counters to represent different amounts.	Solve problems involving kilograms and grams, in which children must convert the amounts. Challenge children to find midpoints between two intervals. Provide children with some multi-step word problems. Can they explain their steps and solutions?
Children may work out missing intervals incorrectly or think it is not possible to find a value between masses such as 1 kg and 2 kg.	Practise weighing objects on a range of scales.	
Children may solve problems incorrectly, applying the wrong calculations through misinterpretation.	Ask children to represent calculations in a bar model or use the column method.	

Unit 9: Mass

UNIT STARTER PAGES

Use these pages to introduce the unit focus to children. Talk through the key learning points that the characters mention, and the key vocabulary.

STRUCTURES AND REPRESENTATIONS

Number line: The number line is effective when looking at scales and finding missing intervals. Children can count on and back too.

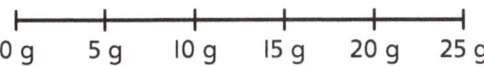

Bar model: The bar model helps children gain a visual understanding of word problems involving measurements.

1 kg 500 g	
1 kg 300 g	200 g

Part-whole models: The part-whole model allows children to convert between units of measure effectively.

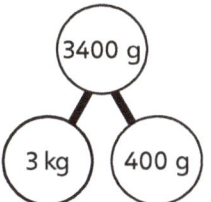

KEY LANGUAGE

There is some key language that children will need to know as part of the learning in this unit:

- mass, measure, grams (g), kilograms (kg)
- interval, scale
- midpoint
- convert
- order
- estimate

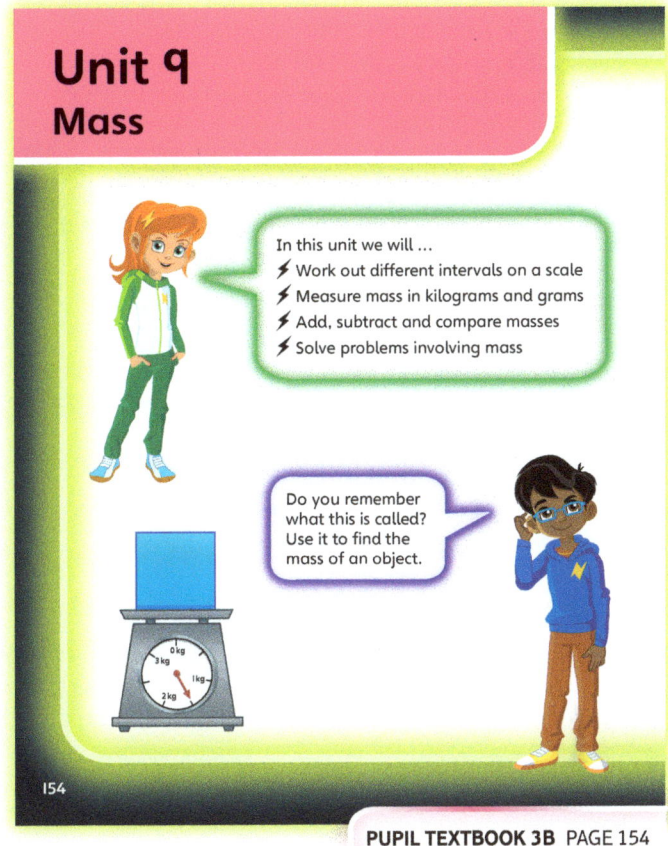

PUPIL TEXTBOOK 3B PAGE 154

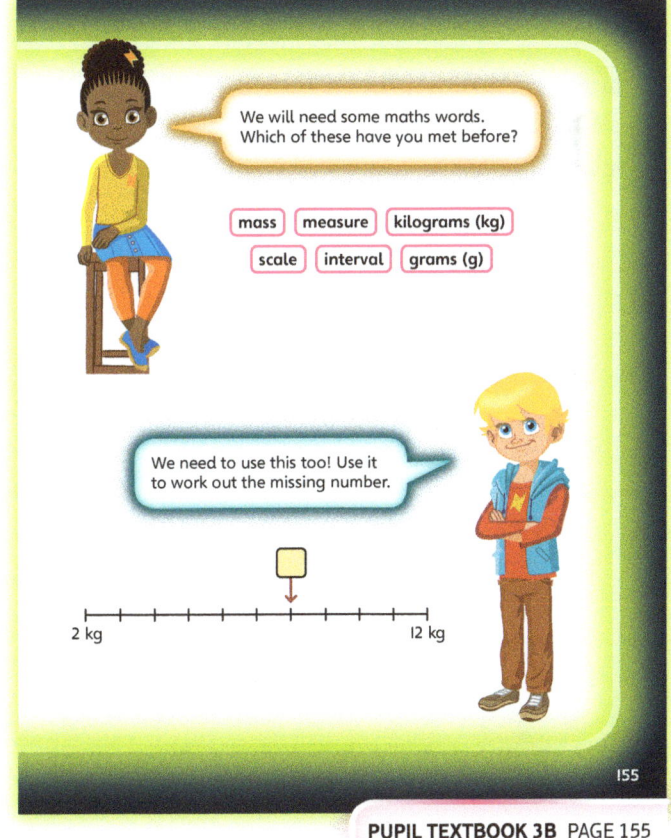

PUPIL TEXTBOOK 3B PAGE 155

Unit 9: Mass, Lesson 1

Use scales

Learning focus

In this lesson, children will recap work on number lines and explore their connection to scales on different measuring devices, such as weighing scales or jugs.

Before you teach

- Can children work out how many intervals there are on a number line?
- Can children count in 10s and 100s?
- Do children know different units of measurement for mass?

NATIONAL CURRICULUM LINKS

Year 3 Measurement

Measure, compare, add and subtract: lengths (m/cm/mm); mass (kg/g); volume/capacity (l/ml).

ASSESSING MASTERY

Children can find what a number line and different scales increase by, by looking at the end points of the number line and considering the number of intervals.

COMMON MISCONCEPTIONS

If they are not confident, it is common for children to guess what a number line goes up in, as opposed to accurately working it out. They often make assumptions that are not correct. Show them two different number lines; the first one going up in 10s from 100 to 200, but only the end points are marked, and the second one from 0 to 1,000, going up in 100s, but again with just the end points marked. Ask:
- *What is the same and what is different about the number lines? How can you work out and then check what a number lines goes up in?*

STRENGTHENING UNDERSTANDING

To support children as they work out what the intervals on a number line increase by, draw a bar model above the number line so it is clear how many sections or intervals the line is split into. Then ask them to work out the difference between the start and end points. Ask them how they can use division to work out what the number line increases by each time. It is important to encourage children to check if they are correct. Count aloud with children from the start point to the end point to check the count is correct.

GOING DEEPER

Ask children to find examples of scales in real life and ask them to work out what they go up in.

KEY LANGUAGE

In lesson: *intervals*, end point, start point, divide, number line, *scale*

Other language to be used by the teacher: increase, decrease

STRUCTURES AND REPRESENTATIONS

Number line, bar model

RESOURCES

Optional: weighing scales, printed number lines

 In the eTextbook of this lesson, you will find interactive links to a selection of teaching tools.

Quick recap

Ask children to draw a number line that goes up from 0 to 100 in 10s. Ask them to show all their marks. Then ask them to change the number line so that it goes up in 5s. What could they do?

194

Unit 9: Mass, Lesson 1

Discover

WAYS OF WORKING Pair work

ASK

- Question 1 a): *Do you remember what these lines are called? Where does the top number line start? Where does it end? What does it go up in? How can you work it out? How can you check?*
- Question 1 b): *How many intervals are there on the bottom number line? How can you work out what it goes up in?*

IN FOCUS In questions 1 a) and 1 b), children are reminded about number lines. Children may approach question 1 a) by guessing that the number line goes up in 1s or 10s, and they should be encouraged to explain how they know. They might count aloud to check, or work it out by dividing the difference between the end and start points by 10. For question 1 b), it may be more of a challenge to guess what the intervals are, and typically children will first check 1, 10 and 100, which do not work. Ask them to consider the midpoint first and then the points between the midpoint and the start and end points.

PRACTICAL TIPS Demonstrate this in the classroom by drawing the number lines on the board.

ANSWERS

Question 1 a): Count up in 10s from 0 to 100 to check if the number line goes up in 10s.

Question 1 b): The number line goes up in 25s.

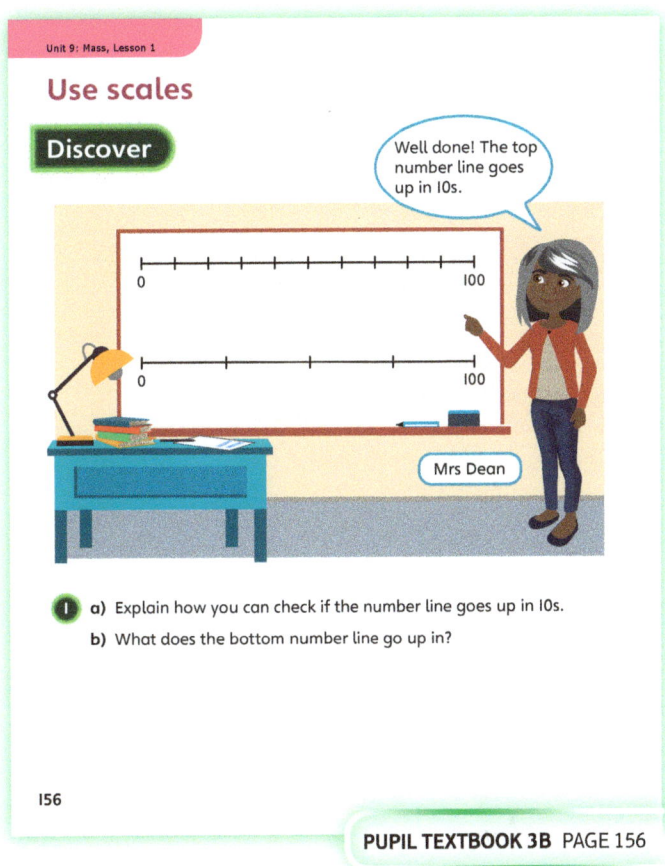

PUPIL TEXTBOOK 3B PAGE 156

Share

WAYS OF WORKING Whole class teacher led

ASK

- Question 1 a): *How did you work out what the number line went up in? How can you check?*
- Question 1 b): *What number should be the middle of the number line? How do you know? How can you use this to work out what the number line goes up in?*

IN FOCUS In question 1 a), discuss methods that children may have used to work out that this number line goes up in 10s. Some children may just know it because it is a familiar model and they have seen it before. Some children may have taken a sensible guess, while others may have worked it out by considering the start and end points and the number of intervals. Use Astrid's strategy to help you check the start and end points and that the number line goes up in 10s. Explain to children that this is always a good thing to do on scales to check their working out is correct.

In question 1 b), you may want to consider the interval method that is shown using a bar model. Some children may tell you points they know. For example, most children will say that the midpoint is 50. This will help them work out that the number line goes up in 25s.

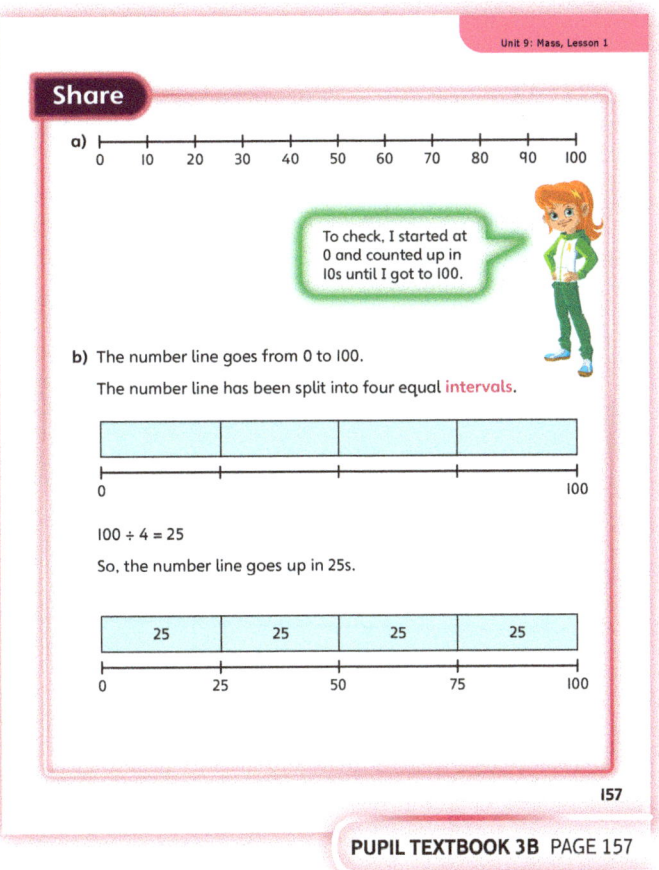

PUPIL TEXTBOOK 3B PAGE 157

Think together

WAYS OF WORKING Whole class teacher led (I do, We do, You do)

ASK

- Question ❶: *How many equal intervals is the number line made up of? How does the bar model help you? How can you work out what the number line goes up in?*
- Question ❷: *What is the difference between these number lines and the ones you have seen so far? How can you work out what these number lines go up in?*
- Question ❸: *How could number lines help you work out what the scales go up in? How can you check you are correct?*

IN FOCUS Question ❶ provides children with another number line between 0 and 100, this time with five intervals. Children may use different approaches to work out that the number line increases by 20 each time. The bar model showing the intervals will help them. Throughout these questions, ask children to count out loud with you in 20s, 100s, and so on, to ensure they have worked out the intervals correctly. In question ❷, children are now given number lines that go up to 1,000. Before they begin the question, ask them to consider what is the same about these number lines and what is different. This may help them work out what the number lines go up in. Question ❸ then relates all the work on number lines to weighing scales. Children need to understand that the circular scale is just like a number line. If it helps them to work out the scale, they can imagine stretching it out to form a number line. The same methods can be applied by dividing the difference between two points and the number of intervals.

STRENGTHEN To support children working out what the intervals on a number line increase by, draw a bar model above the number line so it is clear how many sections or intervals the number line is split up into. Then ask children to work out the difference between the start and end points. Ask them how they can work this out by repeatedly adding the amount the number line increases by. For example, if there are five intervals of 20, they should add 20 five times and then be able to tell you that the end point is 100. It is important to get children to check if they are correct at the end. Count aloud with children from the start point to the end point and check the count is correct.

DEEPEN Explore different real-life examples of scales and ask children to work out what the scales measure and what they go up in.

ASSESSMENT CHECKPOINT Use questions ❶ and ❷ to check that children can find what a number line increases by each time.

ANSWERS

Question ❶: The number line goes up in 20s.

Question ❷ a): The number line goes up in 100s.

Question ❷ b): The number line goes up in 200s.

Question ❷ c): The number line goes up in 250s.

Question ❸ a): The scale goes up in 100 g intervals.

Question ❸ b): The scale goes up in 50 g intervals.

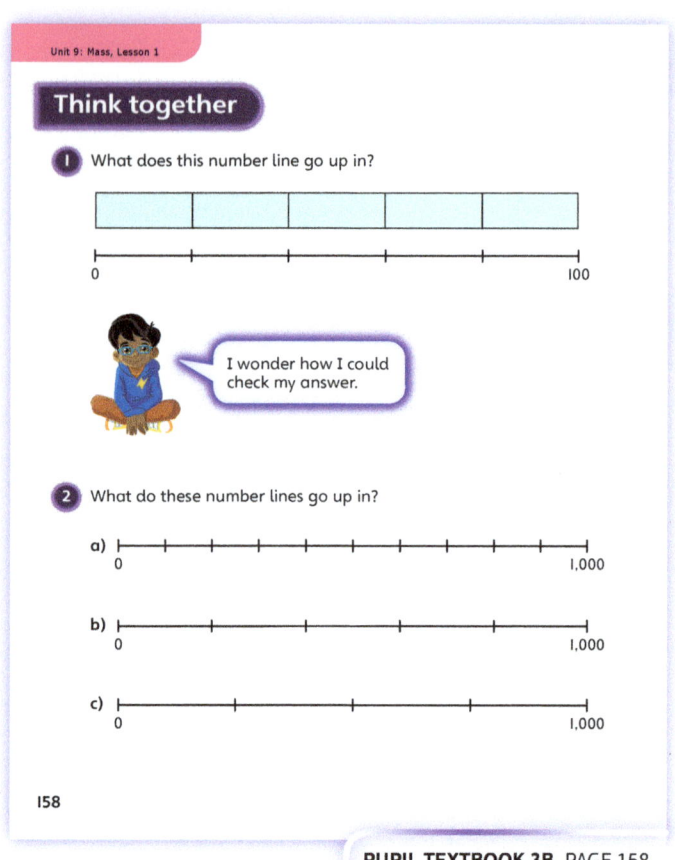

PUPIL TEXTBOOK 3B PAGE 158

PUPIL TEXTBOOK 3B PAGE 159

Unit 9: Mass, Lesson 1

Practice

WAYS OF WORKING Independent thinking

IN FOCUS In question ❶, children use the methods from the main part of the lesson to work out and then check what a number line increases by each time. In question ❷, children count up to check whether the number line goes up in 25s. This helps children explicitly practise the method of checking, which is something that has been encouraged throughout. In question ❸, children are given number lines from 0 to 1,000 and they find and check what the number lines increase by. In question ❹, weighing scales are displayed, and children have to apply what they have learnt using number lines to a circular scale. If they find this challenging, explain that they can try to visualise the scale as a straight number line.

STRENGTHEN To support children as they work out what the intervals on a number line increase by, draw a bar model above the number line so it is clear how many sections or intervals the number line is split into. Then ask children to work out the difference between the start and end points. Ask them how they can use division to work out what the number line increases by each time. It is important that children then check to make sure they are correct. Count aloud with children from the start point to the end point to check the count is correct.

For circular scales, like in question ❹, encourage children to redraw the scale as a straight number line, with the same number of markings.

DEEPEN Explore different real-life examples of scales, similar to those in question ❹, and ask children to work out what the scales measure and what they go up in.

ASSESSMENT CHECKPOINT Use questions ❶, ❷ and ❸ to check that children can find the correct intervals and have an accurate method for checking. Use question ❹ to see if children can apply their knowledge to a set of weighing scales with a circular scale.

ANSWERS Answers for the **Practice** part of the lesson can be found in the *Power Maths* online subscription.

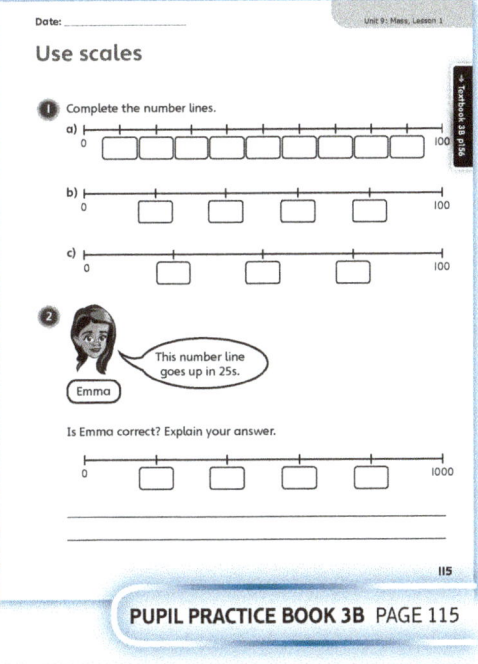

PUPIL PRACTICE BOOK 3B PAGE 115

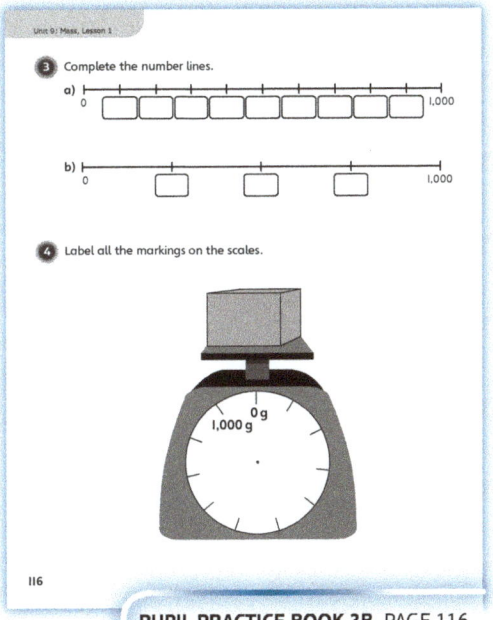

PUPIL PRACTICE BOOK 3B PAGE 116

Reflect

WAYS OF WORKING Independent thinking, pair work

IN FOCUS Children are asked to write their method for working out what a scale goes up in. They should discuss their ideas with a partner. Children may have a variety of different methods and you may want to share them as a class.

ASSESSMENT CHECKPOINT Check that children have a correct method to find what a number line increases by and how they can check their answer.

ANSWERS Answers for the **Reflect** part of the lesson can be found in the *Power Maths* online subscription.

After the lesson

- Can children work out what a number line goes up in?
- Can children check if a number line goes up by a certain amount each time by counting aloud?

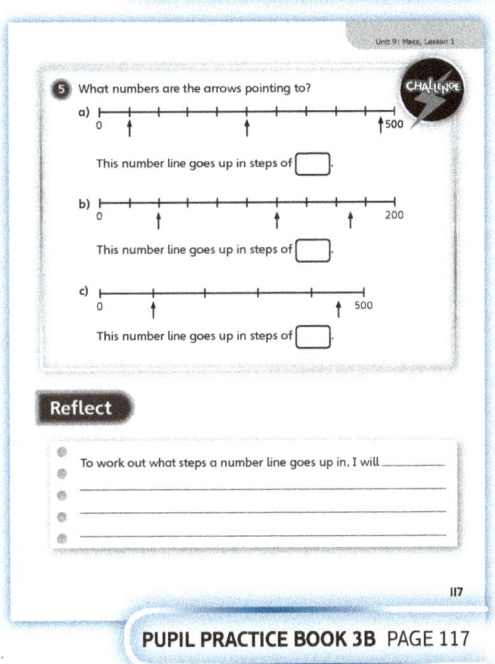

PUPIL PRACTICE BOOK 3B PAGE 117

Unit 9: Mass, Lesson 2

Measure mass

Learning focus

In this lesson, children will learn how to read a range of scales relating to mass, including those with missing intervals.

Before you teach
- What did children learn about mass in Year 2?
- Are all children able to find differences?
- Do children have division strategies they can use?

NATIONAL CURRICULUM LINKS

Year 3 Measurement

Measure, compare, add and subtract: lengths (m/cm/mm); mass (kg/g); volume/capacity (l/ml).

ASSESSING MASTERY

Children can read scales quickly, using the correct method to find a missing interval. Furthermore, they will show a strong understanding of kilograms and grams.

COMMON MISCONCEPTIONS

Children may make errors when working out a missing interval. Ask:
- *What is the method you use to work out a missing interval?*

Children may not understand the magnitudes of kilograms and grams. Ask:
- *How many grams are in a kilogram? Can you think of something that has a mass of 1 gram and something that has a mass of 1 kilogram?*

STRENGTHENING UNDERSTANDING

Children may need support when finding missing intervals. Ask them to write missing interval questions for a partner and to explain the method they followed to find each answer. Summarise the correct methods at the end as a group.

Children may need support when finding the difference between intervals or when dividing. Use number lines to support learning.

GOING DEEPER

Go deeper by doing some practical activities, such as weighing balls of modelling clay on scales. Experiment with balls and parachutes: find out what happens when different masses are attached to the parachutes.

Give children a range of misread intervals. Ask them to reason why the mistakes were made.

KEY LANGUAGE

In lesson: mass, measure, interval, grams (g), kilograms (kg), scale

Other language to be used by the teacher: difference, divide, dial

STRUCTURES AND REPRESENTATIONS

Number line

RESOURCES

Mandatory: weighing scales, modelling clay

Optional: number lines, parachute

 In the eTextbook of this lesson, you will find interactive links to a selection of teaching tools.

Quick recap

Ask children to draw a number line and label the end points 0 and 100. Ask them to mark the points 50, 25 and 90. Then ask them to draw their own arrow and challenge a partner to estimate what number the arrow is pointing to.

Unit 9: Mass, Lesson 2

Discover

WAYS OF WORKING Pair work

ASK

• Question 1 a): *What method did you use to work out the value of each interval?*
• Question 1 b): *What method did you use?*
• Question 1 b): *How can you check your answer is correct?*

IN FOCUS Question 1 a) really gets children thinking about the missing intervals on scales. Encourage them to create some step-by-step instructions to explain how to work missing intervals out.

PRACTICAL TIPS Give children dial scales and modelling clay, so they can measure mass in a practical context.

ANSWERS

Question 1 a): On scale A each interval is 5 g.
On scale B each interval is 10 g.
On scale C each interval is 50 g.

Question 1 b): A: 20 g   B: 130 g   C: 600 g

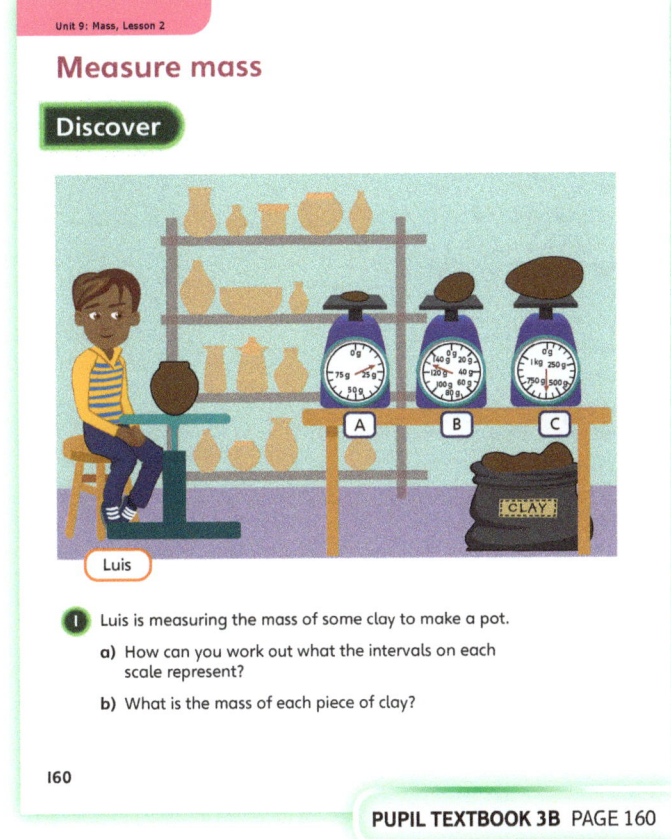

PUPIL TEXTBOOK 3B PAGE 160

Share

WAYS OF WORKING Whole class teacher led

ASK

• Question 1 a): *What does interval mean?*
• Question 1 a): *Let's say the three steps aloud as a class.*
Step 1: Find the difference between the two amounts.
Step 2: Count the number of intervals.
Step 3: Divide the difference by the number of intervals.

IN FOCUS Children will be reminded that 1,000 g equals 1 kg. This is a good opportunity to reinforce this knowledge. Ask: *What is 2 kg in grams? How about 5 kg? 8 kg?*

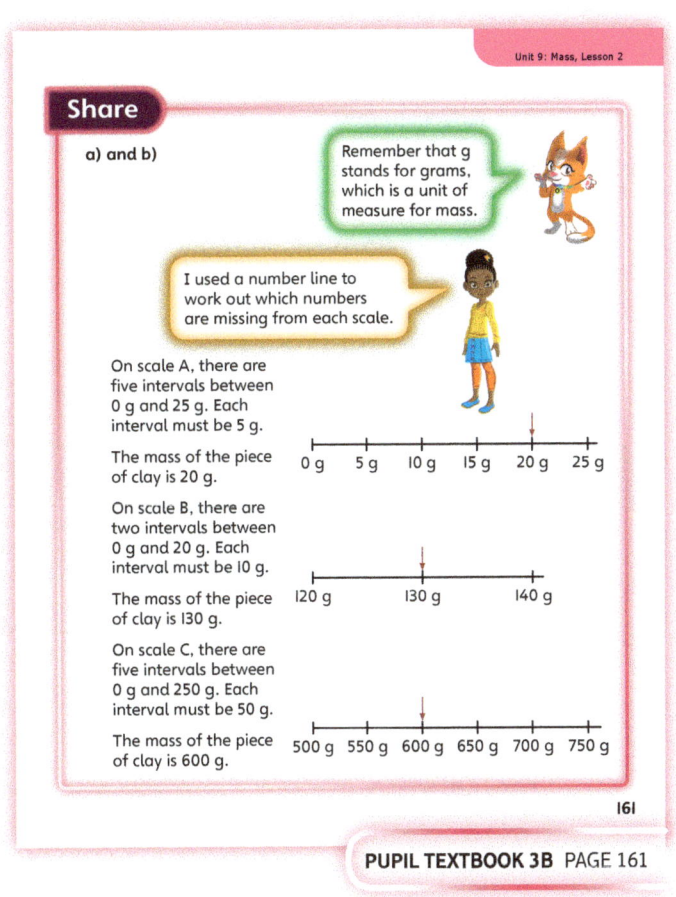

PUPIL TEXTBOOK 3B PAGE 161

Think together

WAYS OF WORKING Whole class teacher led (I do, We do, You do)

ASK
- Question ❷: *What is the value of each interval marked on the number line?*
- Question ❷: *Can you count up in steps of [this interval, for example 250]?*
- *Is there a representation that would help you count?*

IN FOCUS Question ❷ a) shows intervals of 250 g. This may prove a challenge for some children. Ask them to think of an effective strategy to work out what the interval is. Some may see that there is a half-way mark, which has to be 500 g. Children can then halve again to find 250 g. Once children have found the value of each interval, encourage them to count up in steps of this interval until they find the one that matches the given mass.

In question ❸, it might help to show children one of each of the objects listed. This will help them to visually appreciate which items are clearly heavy enough to warrant measuring in kilograms in question ❸ a). It will also help them with the matching activity in question ❸ b).

STRENGTHEN Children will benefit from counting practice. Count in 5s, 10s, 20s, 25s, 50s and 100s.

ASSESSMENT CHECKPOINT Question ❶ will allow you to assess whether children can interpret and read the scales. Look for children who can confidently work out intervals on scales and explain their methodology.

ANSWERS

Question ❶ a): 225 g   b) 375 g   c) 450 g

Question ❷ a):

```
        ↓
|---|---|---|---|
0 kg            1 kg
```

Question ❷ b):

```
              ↓
|-|-|-|-|-|-|-|-|-|-|
0 kg                10 kg
```

Questions ❸ a):
Grams	**Kilograms**
pen	table
t-shirt	bicycle
ring	suitcase
spoon	
mobile phone	

Questions ❸ b):
Pen: 6 g   Table: 50 kg
T-shirt: 140 g   Bicycle: 10 kg
Ring: 6 g   Suitcase: 3 kg
Spoon: 25 g
Mobile phone: 120 g

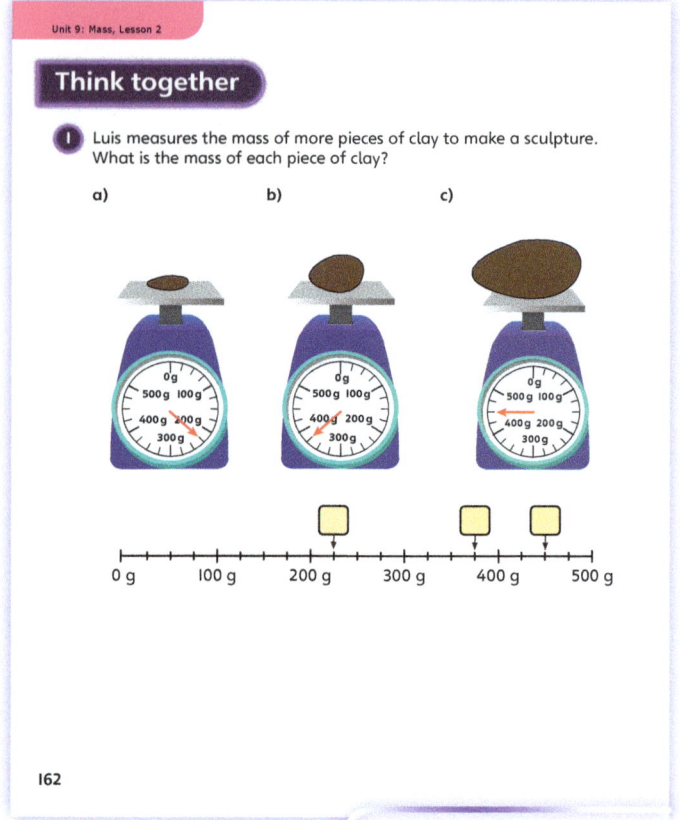

PUPIL TEXTBOOK 3B PAGE 162

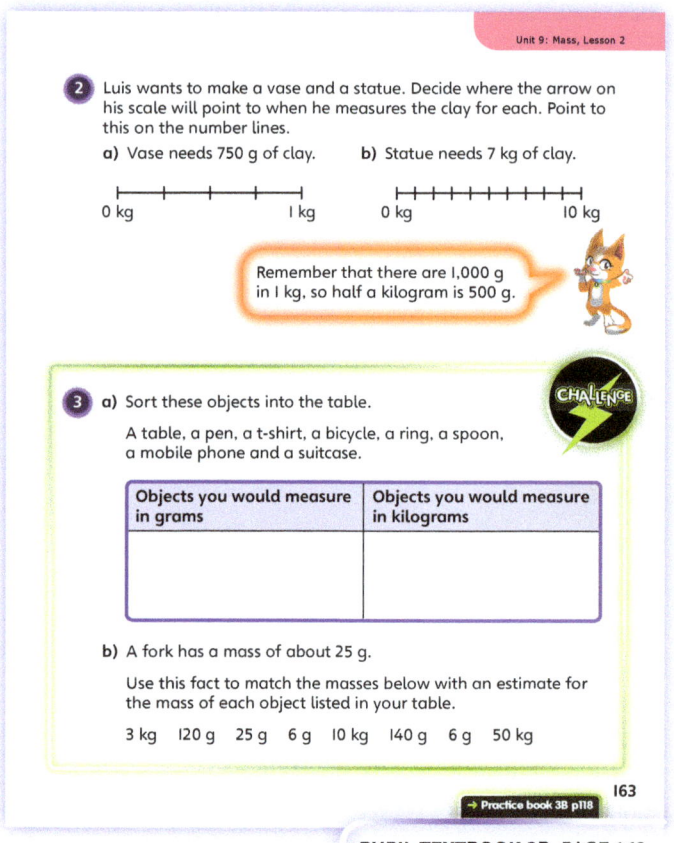

PUPIL TEXTBOOK 3B PAGE 163

Unit 9: Mass, Lesson 2

Practice

WAYS OF WORKING Independent thinking

IN FOCUS Question ❶ requires children to mark the correct mass with an arrow on circular scales. They will need to first work out what the scales increase by each time. In question ❷, children read the scales to determine the correct mass of each object. Question ❹ asks children to consider the correct unit of measurement for different objects. It is important that children know the most likely unit of measurement.

STRENGTHEN To support understanding in this section, give children similar problems to those in question ❶. Use number lines to help children break down the circular scales and intervals.

DEEPEN Ask children to find objects around the classroom and calculate their mass using scales. Ideally, have different scales with different intervals, so children can see the difference when the same object is weighed on different scales.

THINK DIFFERENTLY Question ❸ addresses a common misconception when reading values on scales. In this case, children may think it is 250 g as it lies in the middle of two known points. However, children should realise it is in the middle of 200 g and 400 g, so the missing value is 300 g. If they incorrectly think that the arrow points to 250 g, ask them to check this by counting up by 25 g for each interval. They should notice that they do not reach 400 g at the 400 g marker.

ASSESSMENT CHECKPOINT Use questions ❶ and ❷ to determine whether children can correctly read masses from a variety of different scales.

ANSWERS Answers for the **Practice** part of the lesson can be found in the *Power Maths* online subscription.

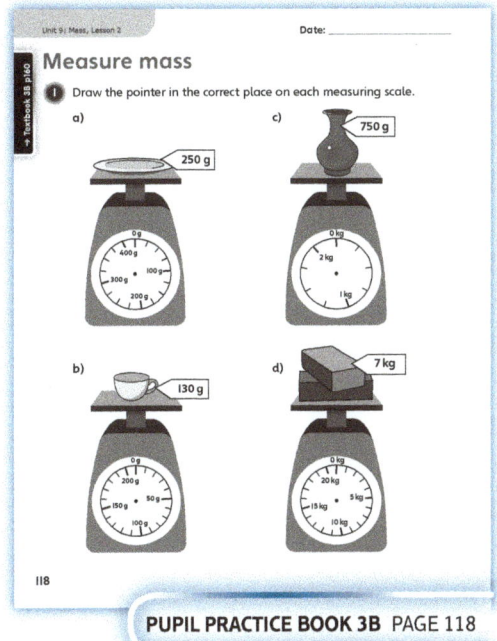

PUPIL PRACTICE BOOK 3B PAGE 118

PUPIL PRACTICE BOOK 3B PAGE 119

Reflect

WAYS OF WORKING Group work

IN FOCUS Children should be encouraged to measure a variety of objects in the classroom using weighing scales. They should read the scales correctly. You may want to ask them to estimate in advance what they think the mass might be, but note that this is a difficult skill. They could compare the mass of one object to another to help them make a prediction.

ASSESSMENT CHECKPOINT Check that children can correctly read off a scale.

ANSWERS Answers for the **Reflect** part of the lesson can be found in the *Power Maths* online subscription.

After the lesson

- Do any children need more practice with reading scales?
- Can children count in 5s, 10s, 20s, 25s, 50s and 100s?

PUPIL PRACTICE BOOK 3B PAGE 120

Unit 9: Mass, Lesson 3

Measure mass in kilograms and grams

Learning focus
In this lesson, children will learn how to read a range of scales in which kilograms and grams are mixed. They will also find midpoints between intervals.

Before you teach
- Are children confident in measuring kilograms and grams separately?
- How can any common misconceptions from Lesson 2 be addressed in this lesson?

NATIONAL CURRICULUM LINKS

Year 3 Measurement

Measure, compare, add and subtract: lengths (m/cm/mm); mass (kg/g); volume/capacity (l/ml).

ASSESSING MASTERY

Children can understand the values of kilograms and grams. Children can represent values using place value counters and work out missing intervals using number lines.

COMMON MISCONCEPTIONS

Children may think that you cannot find anything between values such as 3 kg and 4 kg, because there is no whole number of kilograms between them. Ask:
- *Is there a smaller unit you could use?*

Children may mix up kilograms and grams. Ask:
- *Which is larger? Which should you write first in your answer – the grams or kilograms?*

STRENGTHENING UNDERSTANDING

Children may need support when finding missing intervals between values such as 3 kg and 4 kg. Show them on a number line and practise counting in 100s, 200s and 500s.

You may want to represent kilograms with weights (place value counters would also work), asking children to find different ways to make them; for example, 10 × 100 g = 1 kg.

GOING DEEPER

Go deeper by asking children to find midpoints between intervals. This will call on children's knowledge of finding the values of the intervals and then finding half-way points between two of them.

You may want to continue doing some practical activities such as weighing balls of modelling clay on scales.

KEY LANGUAGE

In lesson: mass, measure, scale, interval, grams (g), kilograms (kg),

Other language to be used by teacher: midpoint, difference, divide, dial

STRUCTURES AND REPRESENTATIONS

Number line

RESOURCES

Mandatory: weighing scales, modelling clay

Optional: number lines, place value counters

 In the eTextbook of this lesson, you will find interactive links to a selection of teaching tools.

Quick recap
On the board, draw a number line that goes from 0 to 1,000 g and ask a child to mark on the midpoint. They should write 500 g. Then draw a number line that goes from 1 kg to 2 kg and ask another child to mark on the midpoint. Discuss why the answer should be 1 kg 500 g.

Discover

WAYS OF WORKING Pair work

ASK

- Questions 1 a) and b): *Could you use a smaller unit to help you?*
- Question 1 a): *Do you need to focus on the kilograms or grams to work out the answer?*

IN FOCUS Question 1 b) will require some deeper thinking from children. Expect to hear some children saying the answer is $6\frac{1}{2}$ kg or $6\frac{3}{4}$ kg (children are likely to call on their knowledge of basic fractions). Focus learning by asking children if there is a smaller unit they could use.

PRACTICAL TIPS Explain the method for working out missing intervals and display the three steps on your maths learning wall:
Step 1: Find the difference between the two amounts.
Step 2: Count the number of intervals.
Step 3: Divide the difference by the number of intervals.

Use number lines to help children count the value of grams between two 1 kg points. Use place value counters to help children visualise 100 and 1,000 values together.

ANSWERS

Question 1 a): The bag of carrots has a mass of 1 kg 300 g.

Question 1 b): The pumpkin has a mass of 6 kg 800 g.

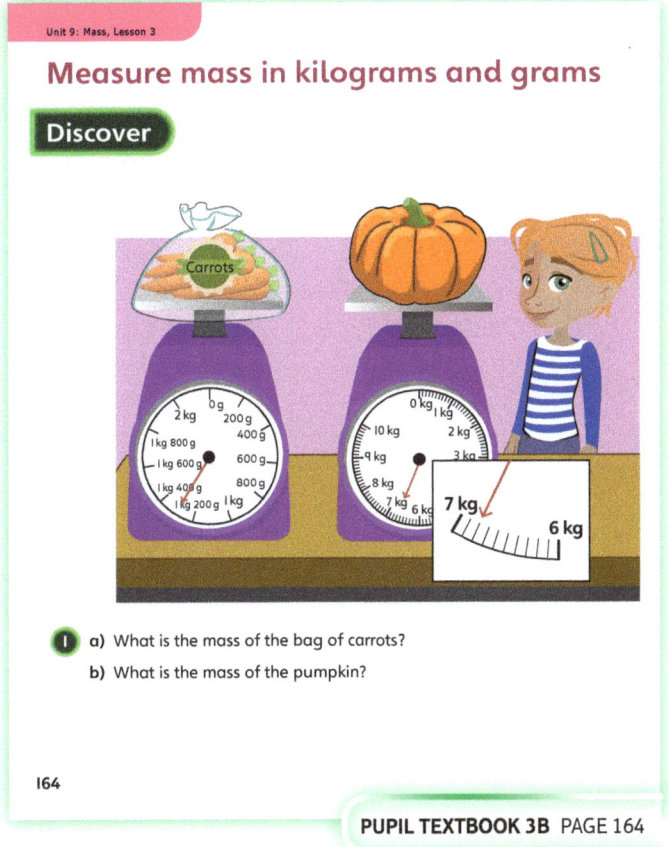

PUPIL TEXTBOOK 3B PAGE 164

Share

WAYS OF WORKING Whole class teacher led

ASK

- Question 1 a): *Can you count aloud in steps of 100 g? What happens when you get to 900 g?*
- Question 1 b): *What does interval mean?*
- Question 1 b): *Let's say the three steps aloud as a class.*
  Step 1: Find the difference between the two amounts.
  Step 2: Count the number of intervals.
  Step 3: Divide the difference by the number of intervals.
- *How can you use a number line to solve question 1 b)?*

IN FOCUS Question 1 introduces children to working with mixed kilograms and grams. This helps them to recognise and record mass within the context of weighing different objects. Help children by modelling how different amounts can be represented: use weights or place value counters on a part-whole model to help with this.

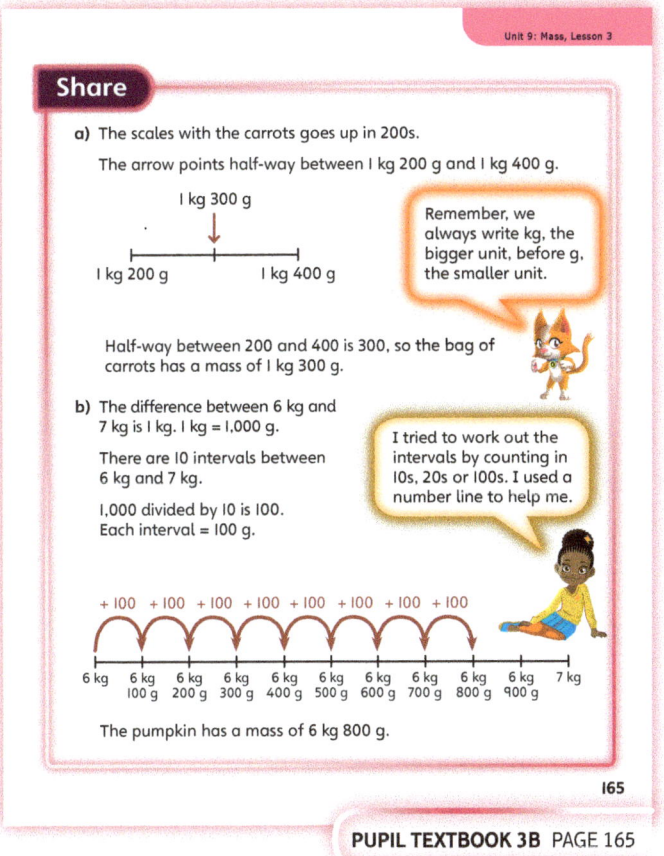

PUPIL TEXTBOOK 3B PAGE 165

Think together

WAYS OF WORKING Whole class teacher led (I do, We do, You do)

ASK
- Question 2: *Can you count the intervals first?*
- Question 3: *How can you work out the midpoint? What calculation do you need to do?*

IN FOCUS Question 2 shows the same mass represented on two different scales. Make learning real here by showing children that this happens in real life: not every scale is the same. You could do some practical measuring using a range of scales.

STRENGTHEN Ask children to count in 5s, 10s, 20s, 25s, 50s, 100s, 200s and 250s up to and past 1 kg. First count in grams (for example, 800 g, 900 g, 1,000 g, 1,100 g, 1,200 g); then count in mixed kilograms and grams (for example, 800 g, 900 g, 1 kg, 1 kg 100 g, 1 kg 200 g).

For question 3, support children by asking them to find midpoints. This will require them to:
- find the interval value
- find half of the interval value
- add it on to the previous interval value.

DEEPEN For question 3, children will have to find midpoints between marked intervals. This will deepen learning in this lesson. Ask children to explain their methods clearly. Show them that they must find the difference between the intervals, halve the amount and then add this onto the previous interval.

ASSESSMENT CHECKPOINT Use question 1 to assess whether children notice that the number of intervals are different on each scale. Listen carefully to their explanations and how they work each mass out.

ANSWERS

Question 1 a): The onions have a mass of 3 kg 400 g.

Question 1 b): The peas have a mass of 400 g.

Question 2 a): Both bags of potatoes weigh 1 kg 500 g.

Question 2 b): Children should notice that the scale intervals are different but the measure is the same on both.

Question 3 a): 4 kg 700 g

Question 3 b): 4 kg 500 g

Question 3 c): 1 kg 950 g

Question 3 d): 10 kg 800 g

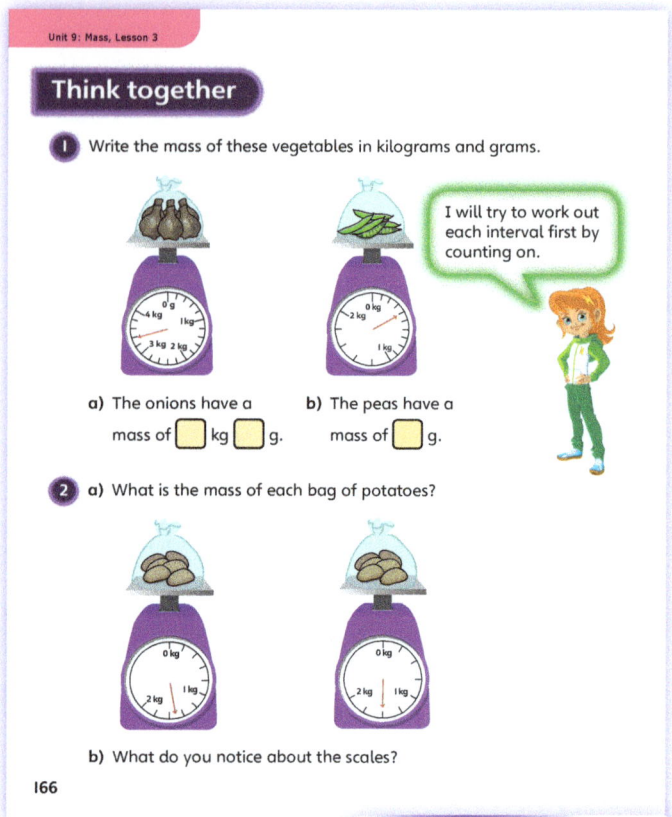

PUPIL TEXTBOOK 3B PAGE 166

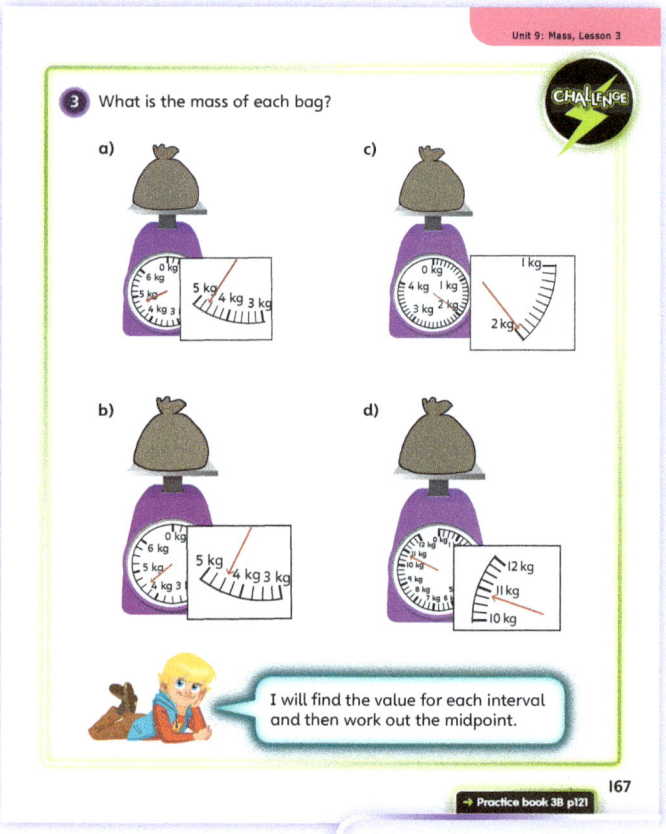

PUPIL TEXTBOOK 3B PAGE 167

Unit 9: Mass, Lesson 3

Practice

WAYS OF WORKING Independent thinking

IN FOCUS Questions 1, 2 and 3 allow children to practise reading different scales. Children should give their answers in kilograms and grams. They should not give their answer in just grams as they have not met 4-digit numbers yet. Use the sentence stems of ☐ kg and ☐ g to support them. Question 4 provides a matching activity, where children match masses to the correct scales. Question 5 allows children to focus their learning by estimating masses on scales with limited interval lines. If children need support with question 5, suggest that they work out what the scale goes up in and what each interval represents, and then mark on all the values.

STRENGTHEN Ask children to work out their answers and then check them by counting around the scale in the interval values they have worked out. For example, in question 1, children should find that the first scale has intervals of 500 g. Ask children to point to each interval and count around in 500s until they get to the answer. Drawing the scale as a number line may also support children.

DEEPEN Question 5 is a great activity for deepening learning; in this question, children are required to estimate the mass of objects where they have limited information. They will need to give their answers in kilograms and grams. Discuss with children why their answers are just estimates.

ASSESSMENT CHECKPOINT Questions 1 and 2 will allow you to see which children can read scales confidently, working out midpoints. For a more thorough assessment, ask children to explain why they matched each scale to the mass it shows in question 4.

ANSWERS Answers for the **Practice** part of the lesson can be found in the *Power Maths* online subscription.

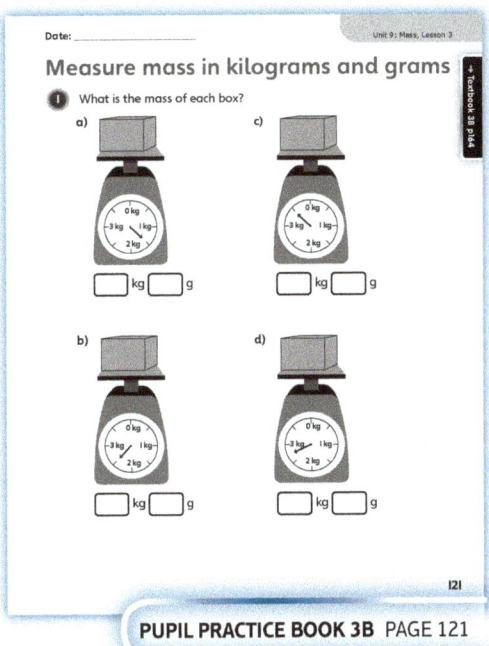

PUPIL PRACTICE BOOK 3B PAGE 121

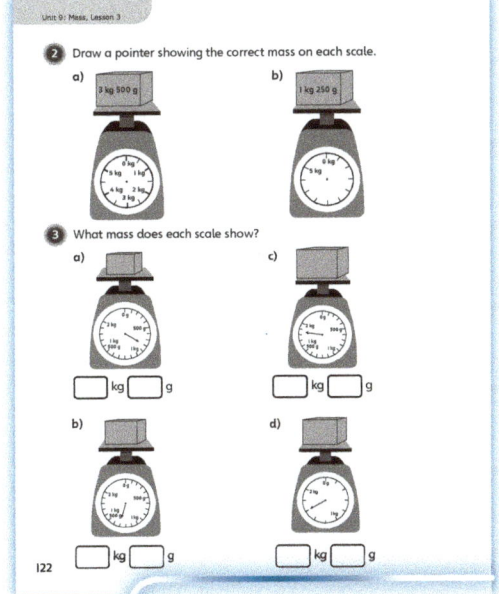

PUPIL PRACTICE BOOK 3B PAGE 122

Reflect

WAYS OF WORKING Pair work

IN FOCUS In this question, children need to start by thinking individually about which question they found most difficult and why. Was it the scale? Was it the numbers? Which questions were easier than others? Ask them to share their answers with a partner.

ASSESSMENT CHECKPOINT Reflect on which questions children found most difficult and potentially ask further similar questions throughout the rest of the unit.

ANSWERS Answers for the **Reflect** part of the lesson can be found in the *Power Maths* online subscription.

After the lesson

- Are children able to measure the mass of an object and give the answer in kilograms and grams?
- Can children mark the mass of an object on a set of scales that is in kilograms and grams?

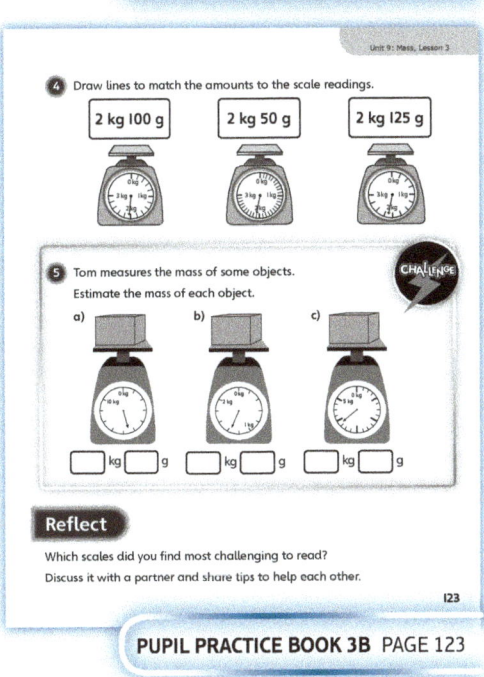

PUPIL PRACTICE BOOK 3B PAGE 123

Unit 9: Mass, Lesson 4

Equivalent masses

Learning focus
In this lesson, children will learn how to convert amounts in grams to values in both kilograms and grams.

Before you teach
- Do all children know that 1 kg = 1,000 g?
- Could conversion examples be made into a display?
- Have you got some weights ready for practical work?

NATIONAL CURRICULUM LINKS

Year 3 Measurement

Measure, compare, add and subtract: lengths (m/cm/mm); mass (kg/g); volume/capacity (l/ml).

ASSESSING MASTERY

Children can quickly convert and represent the amounts using representations such as the part-whole model. Children have a solid understanding of place value and confidently convert amounts such as 1,009 g.

COMMON MISCONCEPTIONS

Children may have some place value misconceptions; for example, they may think 2,011 g = 2 kg 110 g. Ask:
- *Can you represent the amount using weights (1 kg, 100 g, 10 g, 1 g) or place value counters?*

Some children may not be confident with using 0 as a place holder. Ask:
- *Why do you need to include a 0 in 1,022 g?*

STRENGTHENING UNDERSTANDING

Children may need reminding that 1 kg = 1,000 g (it would be useful to display this fact on a learning wall).

Provide children with a range of weights, so they can represent amounts with them and support their conversion mastery.

GOING DEEPER

Go deeper by asking children to represent amounts in different ways. For example: 1,340 g could be: 1 kg + 300 g + 40 g; or 1 kg + 100 g + 100 g + 100 g + 10 g + 10 g + 10 g + 10 g.

KEY LANGUAGE

In lesson: mass, measure, scale, grams (g), kilograms (kg)

Other language used by the teacher: interval, difference, divide, dial, balance

STRUCTURES AND REPRESENTATIONS

Number line, part-whole model

RESOURCES

Mandatory: weighing scales, weights

Optional: modelling clay, number lines, place value counters

 In the eTextbook of this lesson, you will find interactive links to a selection of teaching tools.

Quick recap

Children play a game in pairs. They start with 0–9 digit cards face down on their desk. Children take it in turns to select 3 cards each and then make a 3-digit number with their chosen cards. The child that makes the larger 3-digit number wins.

Play further rounds where the smaller 3-digit number wins. Bonus points can be awarded to the class member who makes the greatest number.

Unit 9: Mass, Lesson 4

Discover

WAYS OF WORKING Pair work

ASK
- Question 1 a): *How can you write your answer?*
- Question 1 b): *How can you work out what each interval on the number line is worth?*

IN FOCUS Question 1 a) introduces children to converting between kilograms and grams. Encourage them to recall the fact that 1 kg = 1,000 g. See if they can identify which digit represents each 1,000 g in 3,400 g. Look at the balance scale together and encourage children to add the kilograms and grams separately. They should work out that 3 kg 400 g is equal to 3,400 g.

PRACTICAL TIPS For this activity, display some conversions on your learning wall, so children have a reference. This will help to scaffold learning. Part-whole models work well when separating larger amounts into kilograms and grams.

ANSWERS

Question 1 a): The bike has a mass of 3 kg 400 g.

Question 1 b):

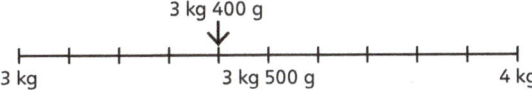

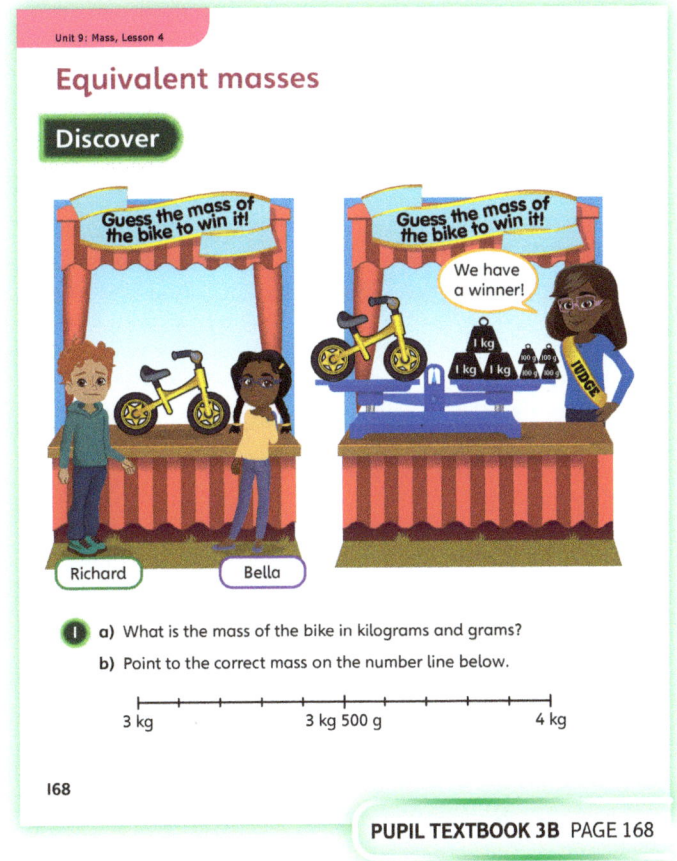

PUPIL TEXTBOOK 3B PAGE 168

Share

WAYS OF WORKING Whole class teacher led

ASK
- Question 1 a): *How does the part-whole model help to show the number of kilograms and grams?*
- Question 1 b): *What intervals does the number line go up in?*

IN FOCUS Question 1 b) allows children to think about the relationship between kilograms and grams on a scale, giving them a context to work out the value of each interval on the number line and also how many grams are in 1 kg.

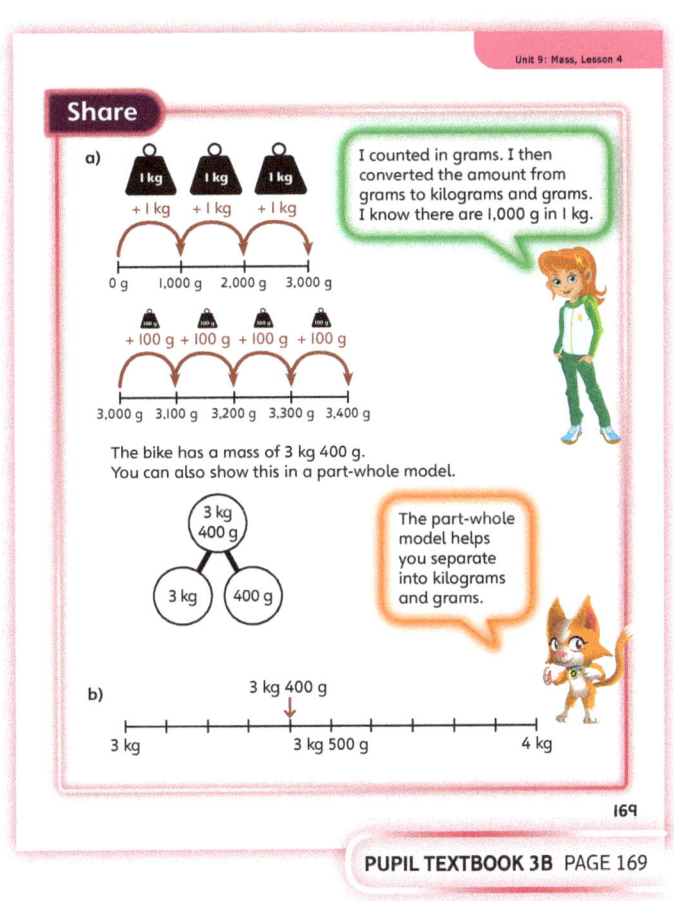

PUPIL TEXTBOOK 3B PAGE 169

207

Think together

WAYS OF WORKING Whole class teacher led (I do, We do, You do)

ASK

- Question ❶: *What do you notice about the masses in the part-whole models? What are the missing parts and wholes?*
- Question ❷: *How many grams are in 1 kg? How do you remember that 1,000 g = 1 kg? How can the bar models help you work out the fractions of a 1 kg?*
- Question ❸: *What different masses are shown?*
- Question ❸ a): *Can you find two masses that add to make 1 kg? What about three or four masses that add to make 1 kg?*

IN FOCUS Note that this lesson is not about converting masses. In question ❶, children use part-whole models to show how a mass can be made up of kilograms and grams. In question ❷, children use bar models to help them work out different fractions of 1 kg. They use their knowledge that 1 kg is made up of 1,000 g to convert amounts such as $\frac{1}{2}$ kg and $\frac{1}{4}$ kg to grams. In question ❸, children further use the fact that 1 kg = 1,000 g to work out which masses can balance the scales. Children should first be encouraged to think about which of the different masses can first be combined to make 1,000 g. This is a good starting point and they should use mental methods to help them do this. They can then progress to thinking about how many additional grams are in 1 kg 700 g.

STRENGTHEN Use bar models to help children understand simple fractions of 1 kg. Use part-whole models to show how a mass can be made up of kilogram and gram parts.

DEEPEN Ask children how many different masses they can make with three of the masses from question ❸. What happens if they use four masses; how many different combinations can they find?

ASSESSMENT CHECKPOINT Use question ❶ to check that children can separate a mass into its kg and g parts. Use question ❷ to check children can find simple fractions of 1 kg in grams.

ANSWERS

Question ❶ a): Parts: 2 kg and 600 g

Question ❶ b): Whole: 3 kg 750 g

Question ❷ a): $\frac{1}{2}$ kg = 500 g

Question ❷ b): $\frac{1}{4}$ kg = 250 g

Question ❸ a): For example: 1 kg or 2 × 500 g or 500 g + 200 g + 200 g + 100 g

Question ❸ b): For example: 1 kg + 500 g + 200 g or 1 kg + 500 g + 100 g + 100 g or 500 g + 500 g + 200 g + 200 g + 200 g + 100 g

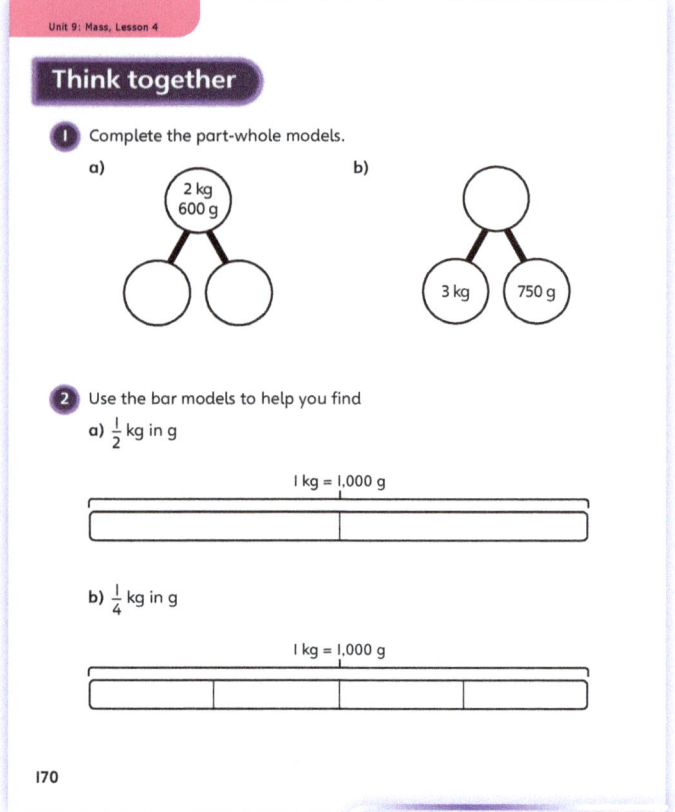

PUPIL TEXTBOOK 3B PAGE 170

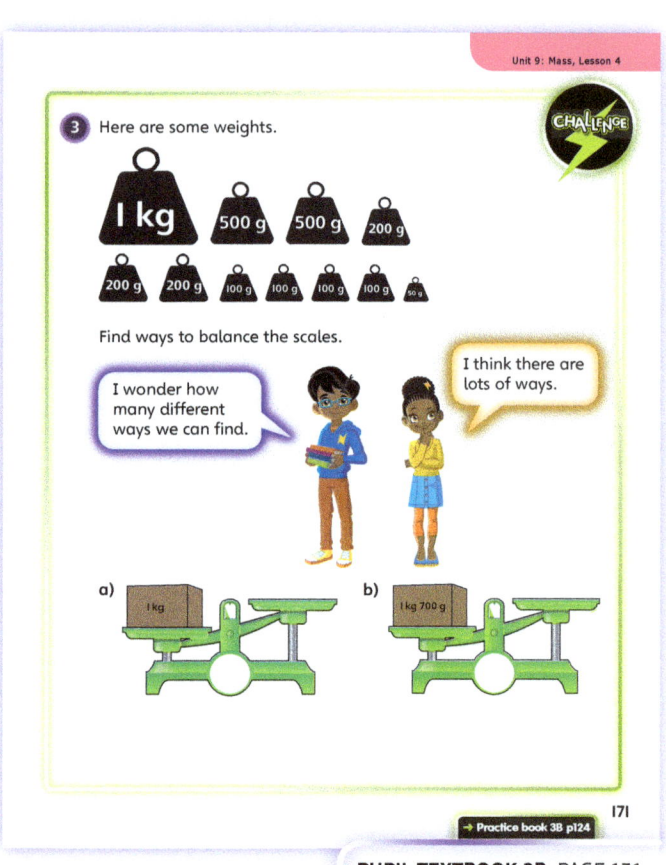

PUPIL TEXTBOOK 3B PAGE 171

Practice

WAYS OF WORKING Independent thinking

IN FOCUS Question ❶ requires children to complete part-whole models. This is an effective way to represent converting between kilograms and grams.

STRENGTHEN Children should practise converting amounts such as 1 kg 200 g, 1 kg 20 g and 1 kg 2 g. Use part-whole models to enable children to gain confidence with this way of modelling mass. Ask them to tell you what the difference is between these amounts. This activity will greatly improve children's understanding of place value.

DEEPEN Question ❻ will deepen learning by making children think about converting kilograms into grams and vice versa. Children will show mastery of the lesson if they can represent amounts in different ways. Repeat the activity for other amounts such as 5 kg 640 g.

THINK DIFFERENTLY Question ❺ is all about place value. Children will have to use their knowledge from the previous two lessons to work out the missing intervals and then reason why Lee is incorrect.

ASSESSMENT CHECKPOINT Question ❹ can be used to assess children's understanding. Children will have to draw on their knowledge of reading scales and apply it to the learning in this lesson.

ANSWERS Answers for the **Practice** part of the lesson can be found in the *Power Maths* online subscription.

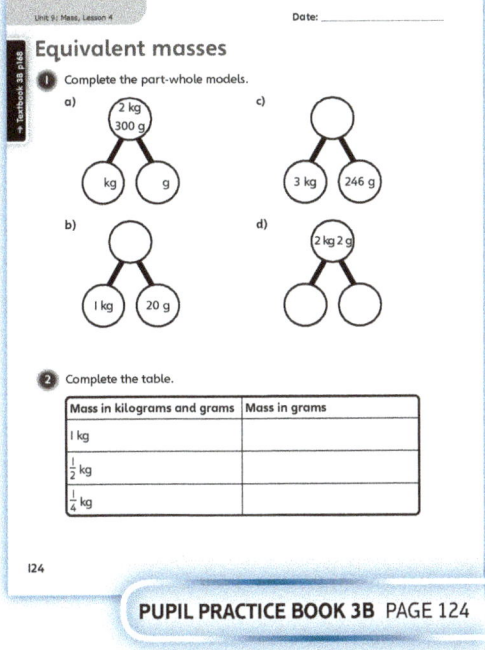

PUPIL PRACTICE BOOK 3B PAGE 124

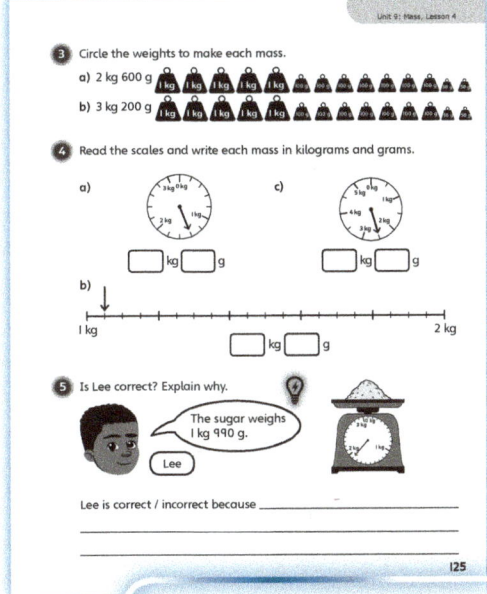

PUPIL PRACTICE BOOK 3B PAGE 125

Reflect

WAYS OF WORKING Independent thinking

IN FOCUS This activity will get children thinking about maths in real-life situations. Prompt children to look at the mass values of products the next time they go to the supermarket (this is a good home-learning activity).

ASSESSMENT CHECKPOINT Reflecting on real-life contexts will allow you to assess whether children can think about the importance of mass value in everyday situations. Children may think that only kilograms are used when we are estimating, or that they do not need to be overly accurate.

ANSWERS Answers for the **Reflect** part of the lesson can be found in the *Power Maths* online subscription.

After the lesson

- Are all children secure at converting kilograms into grams and vice versa?
- Can children represent amounts in different ways?
- How will you link this lesson to the following one (comparing masses) as well as across the curriculum?

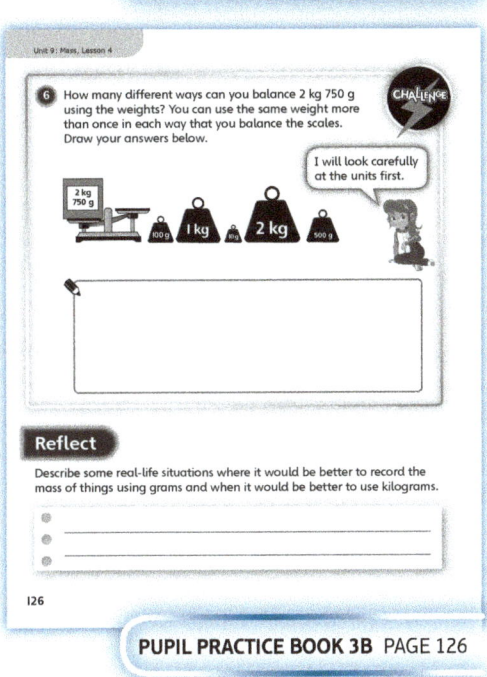

PUPIL PRACTICE BOOK 3B PAGE 126

Unit 9: Mass, Lesson 5

Compare mass

Learning focus
In this lesson, children will compare masses by ordering them on a number line and by using the <, > and = signs.

Before you teach
- Do all children have a sound knowledge of converting between kilograms and grams?
- Is the key vocabulary on display in your classroom?
- Do children know what the <, > and = signs mean?

NATIONAL CURRICULUM LINKS

Year 3 Measurement

Measure, compare, add and subtract: lengths (m/cm/mm); mass (kg/g); volume/capacity (l/ml).

ASSESSING MASTERY

Children can quickly order amounts and use the correct language, such as: greater, more than, less than. Children can solve some problems involving ordering amounts.

COMMON MISCONCEPTIONS

Children may assume an amount in grams is larger than kilograms; for example, 1,245 g > 2 kg 100 g. Ask:
- *Can you convert the grams to kilograms to help you decide?*

Children might confuse the < and > signs. Ask:
- *What is wrong with this sentence: 1 kg < 100 g?*

STRENGTHENING UNDERSTANDING

Recap on the content and skills of converting amounts practised in the previous lesson.

GOING DEEPER

Deepen learning by asking children to order the mass of real-life objects. It is important that children can feel heavier and lighter objects. Ask them to hold different objects and arrange themselves in a line of ascending order of mass. This will deepen learning by allowing children to discuss their object's value with others and will encourage peer correction if they fail to order the objects correctly.

KEY LANGUAGE

In lesson: mass, scale, grams (g), kilograms (kg), weigh

Other language to be used by the teacher: measure, interval, difference, greater than (>), less than (<), equal to (=), compare, divide, dial, more than, order

STRUCTURES AND REPRESENTATIONS

Number line, table

RESOURCES

Mandatory: weighing scales, weights, dice

Optional: modelling clay, number lines, a pineapple, a pumpkin, base 10 equipment, place value counters (different counters to represent kg and g)

 In the eTextbook of this lesson, you will find interactive links to a selection of teaching tools.

Quick recap
Roll a dice three times and ask children to use the numbers to make all the 3-digit numbers they can. Can they make all six possible numbers? Now ask them to put the numbers in order, from smallest to greatest.

Discover

WAYS OF WORKING Pair work

ASK

- Question 1 a): *Would it help to convert grams to kilograms and grams for the pineapple?*
- Question 1 a): *Is the item with the largest mass also the biggest item?*
- Question 1 b): *Did you give your answer in both kilograms and grams? Why?*

IN FOCUS Question 1 a) shows mass in grams and also kilograms and grams, which children will have seen before. In this question, children need to compare these masses in both grams and in kilograms and grams, in a real-life context.

PRACTICAL TIPS Bring a pineapple and a pumpkin into school so children can physically compare them. Discuss why we need scales to measure two objects which are close in mass.

ANSWERS

Question 1 a): The pineapple has the greater mass because 1,243 g is more than 1,230 g.

Question 1 b): The melon has a mass between 1 kg 231 g and 1 kg 242 g.

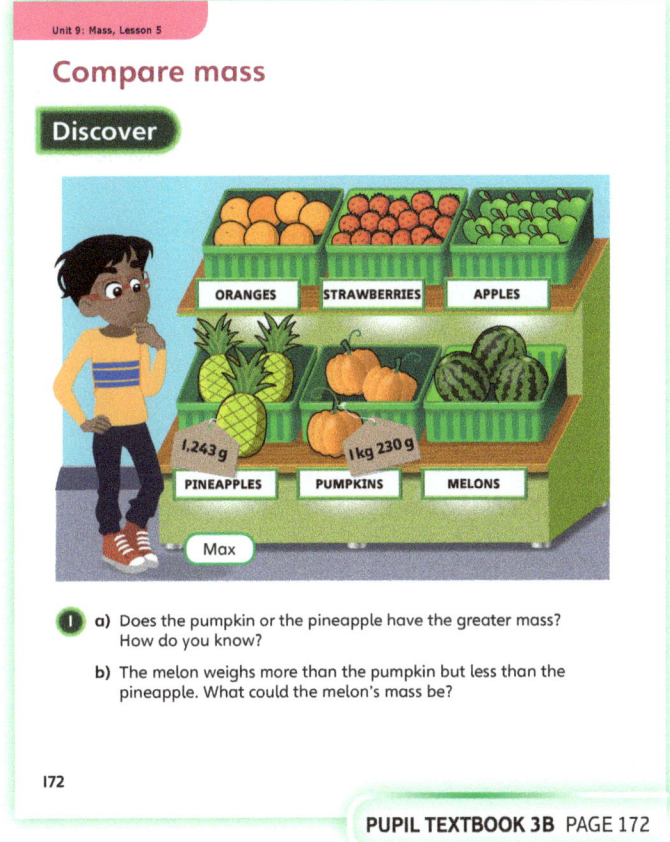

PUPIL TEXTBOOK 3B PAGE 172

Share

WAYS OF WORKING Whole class teacher led

ASK

- Question 1 a): *When comparing the amounts, did you look at the kilograms or grams first?*
- Question 1 b): *Why do you think there is more than one answer?*

IN FOCUS In question 1 a), children may try to convert the mass of the pumpkin into grams. Explain that this is correct but is a less efficient method of working.

In question 1 b), children are presented with a number line. Explain that number lines can be useful when asked to find the possible numbers between two numbers.

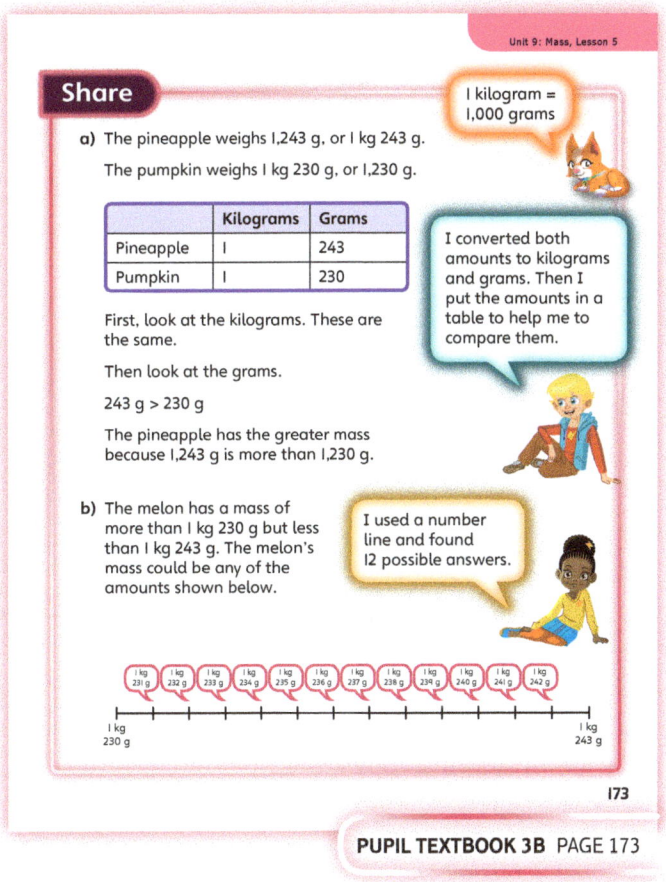

PUPIL TEXTBOOK 3B PAGE 173

Think together

WAYS OF WORKING Whole class teacher led (I do, We do, You do)

ASK
- Question ①: *What do the signs < and > mean?*
- Question ②: *Why are the balance scales incorrect?*

IN FOCUS Question ② allows children to convert the amounts and then work out which balance scales are incorrect. There are some amounts that will require a good knowledge of place value: for example, in question ② D 1 kg 3 g is very similar to 1,001 g.

STRENGTHEN Children may find it helpful to use a table to convert, compare and order masses, as shown below:

Kilograms	Grams

DEEPEN In question ②, ask children to explain why the balance scales are incorrect. Can they suggest an amount which would make each scale correct? Some children may point out that although ② D is correct, in reality the two dishes would be almost balanced because the difference between each side is only 2 g; this shows a good understanding of place value in the context of mass.

ASSESSMENT CHECKPOINT Question ③ can be used to assess whether children can compare more than two amounts. Assess children's ability to find ways to compare the amounts and to explain their answers clearly and concisely.

ANSWERS

Question ①: 1 kg 456 g < 1 kg 500 g
1 kg 211 g < 1 kg 215 g
1 kg 90 g > 1 kg 9 g
2 kg 211 g > 2 kg 210 g

Question ②: B and C are not working correctly. On scale B, 499 g should be lighter than (not heavier than) 1 kg 402 g. On scale C, 3 kg 200 g should be heavier than (not equal to) 3 kg 20 g.

Question ③: Lightest to heaviest: 754 g, 1 kg 9 g, 1,090 g, 1,098 g, 1,432 g, 1 kg 900 g.

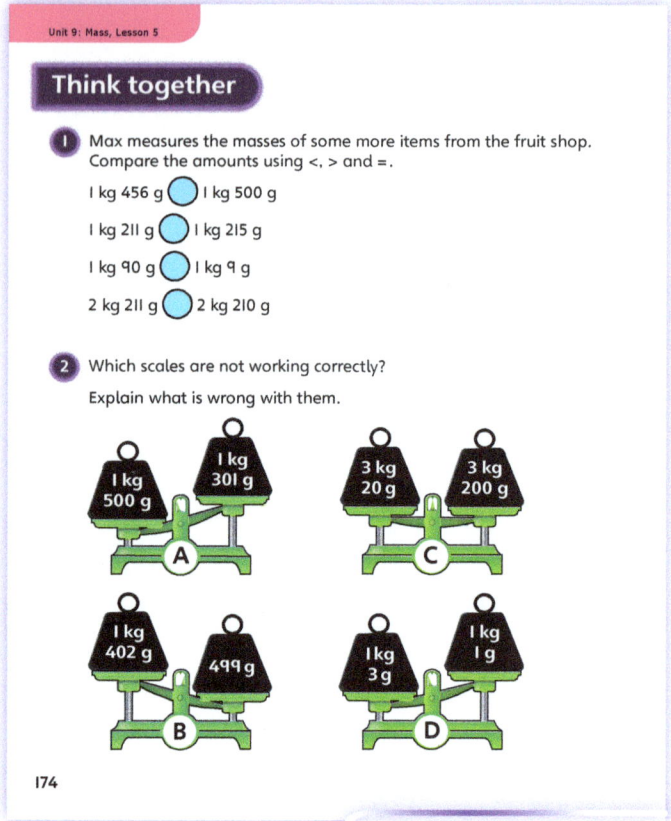

PUPIL TEXTBOOK 3B PAGE 174

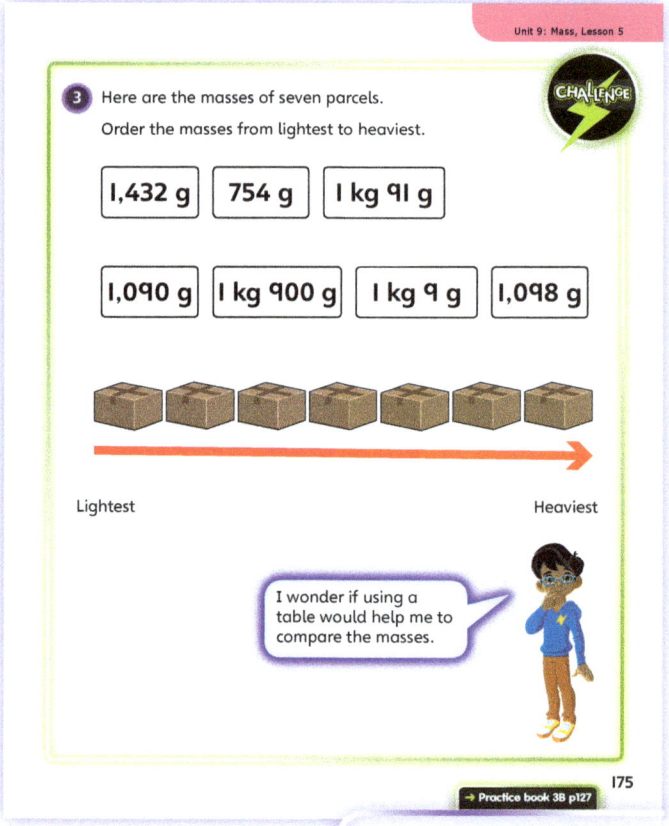

PUPIL TEXTBOOK 3B PAGE 175

Unit 9: Mass, Lesson 5

Practice

WAYS OF WORKING Independent thinking

IN FOCUS In questions ➊ and ➋, children are asked to compare masses and then tick either the heavier or lighter mass. Examples include where the masses are all in grams and all in kilograms, but later examples include where the masses are in kilograms and grams. Children start to see that they can compare the kilograms first and, if they are the same, they then need to compare the grams. Question ➌ provides further practice, with children using the inequality signs to compare two masses. In question ➎, children extend their knowledge of comparing numbers to ordering a set of masses.

STRENGTHEN Use a number line to support children when comparing and ordering masses. For those that struggle with positioning masses on a number line to compare them, you may want to make the masses using base 10 equipment. Children will be able to see the smallest and largest numbers, helping them work out the lighter and heavier masses.

DEEPEN Question ➏ requires children to work out the possible masses of different boxes by looking at the given information. As a further challenge, ask children to work out the possible masses between the boxes, once ordered.

ASSESSMENT CHECKPOINT Questions ➊, ➋ and ➌ will help you assess if children can compare two masses. Question ➎ will show whether children can order a set of masses.

ANSWERS Answers for the **Practice** part of the lesson can be found in the *Power Maths* online subscription.

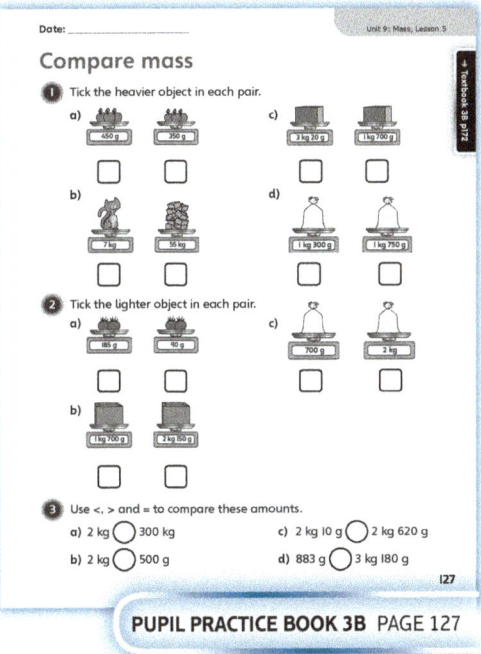

PUPIL PRACTICE BOOK 3B PAGE 127

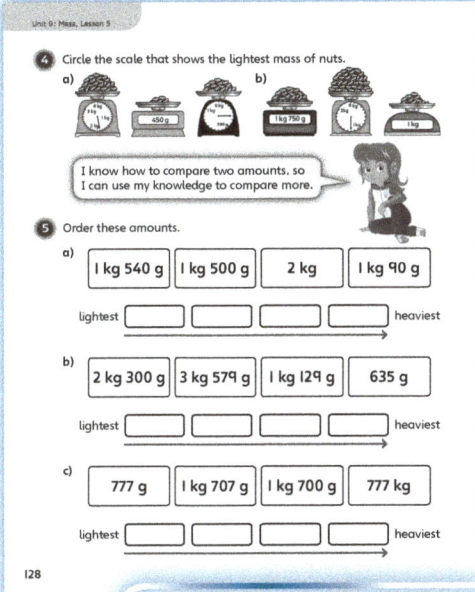

PUPIL PRACTICE BOOK 3B PAGE 128

Reflect

WAYS OF WORKING Independent thinking

IN FOCUS This activity will get children thinking about efficient methods of ordering. Children should realise that they always need to look at the kilograms first, before the grams.

ASSESSMENT CHECKPOINT Listen carefully to children's reasoning: they should explain that, once Max has checked the kilograms and realised that they are not equal, he does not need to look at the number of grams.

ANSWERS Answers for the **Reflect** part of the lesson can be found in the *Power Maths* online subscription.

After the lesson

- Are all children secure in their knowledge of comparing and converting mass values?
- How can you provide further opportunities across the curriculum for comparing a range of objects?

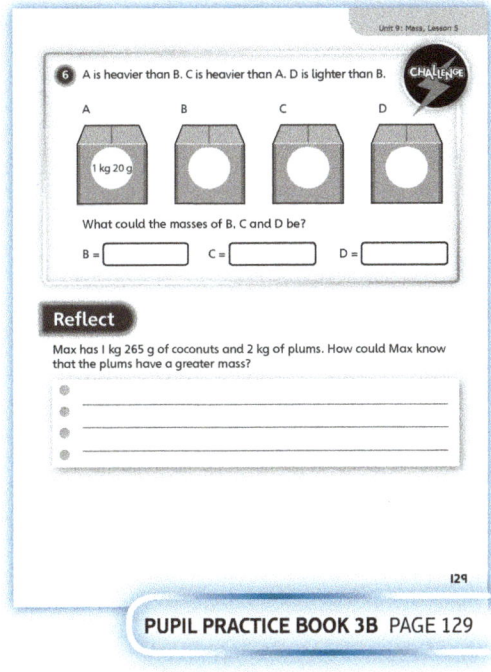

PUPIL PRACTICE BOOK 3B PAGE 129

Unit 9: Mass, Lesson 6

Add and subtract mass

Learning focus
In this lesson, children will add and subtract masses, which include mixed units, using a range of strategies. They will continue to convert between kilograms and grams.

Before you teach
- Are all children secure in addition and subtraction operations?
- How will you support children who find counting on a number line difficult?

NATIONAL CURRICULUM LINKS

Year 3 Measurement

Measure, compare, add and subtract: lengths (m/cm/mm); mass (kg/g); volume/capacity (l/ml).

ASSESSING MASTERY

Children can use efficient strategies to add and subtract mixed-value masses. They should be able to explain their methods and give reasons for why they have used a particular method.

COMMON MISCONCEPTIONS

Children may not add the kilograms and grams separately. Ask:
- *Can you add kilograms if they cross the 1,000 g barrier?*

Children may not convert the grams to kilograms. For example, they may write 1,200 g. Ask:
- *Do you need to convert the grams? Why not?*

STRENGTHENING UNDERSTANDING

To highlight the importance of conversion before carrying out an operation, ask children if they can add 1 kg 100 g to 1,500 g without converting. Then remind children how to convert grams to kilograms if needed.

Use representations such as the number line to support understanding of addition and subtraction.

GOING DEEPER

Deepen learning by challenging children to solve missing number problems. Ask:
- *On a balance scale, if a vase weighs 1 kg 500 g on the heavier side and there are three toy cars weighing 375 g on the lighter side, how many more cars do you need to make the scales balance?*

KEY LANGUAGE

In lesson: mass, weigh, scale, grams (g), kilograms (kg)

Other language to be used by the teacher: measure, interval, difference, divide, dial, add, subtract, more than, difference, take away, plus, minus, sum, total, inverse

STRUCTURES AND REPRESENTATIONS

Number line, bar model, column addition

RESOURCES

Mandatory: weighing scales

Optional: modelling clay, number lines, base 10 equipment, counters

 In the eTextbook of this lesson, you will find interactive links to a selection of teaching tools.

Quick recap
Write the digits 5, 4, 3, 2, 1 and 0 on the board. Ask children to make two 3-digit numbers. Then ask them to add the numbers together and then find the difference between the numbers. A partner should check their answers.

Unit 9: Mass, Lesson 6

Discover

WAYS OF WORKING Pair work

ASK
- Question 1 a): *What are the key words that help you identify which operation to use?*
- Question 1 b): *What strategies could you use to work this out?*

IN FOCUS Question 1 a) requires children to focus their attention on the language of the question and correctly identify the words that indicate which operation they should use to work out the answer.

In question 1 b), some children may think 'more … than' means they should add. Explain this question in the context of 'more than' as finding a difference between two values.

PRACTICAL TIPS Addition and subtraction in a real-life context is important for learning. Do some practical measuring of mass in the classroom and then ask children to add more mass, or take some away. This could be done in a cooking and nutrition lesson.

ANSWERS

Question 1 a): Zac and Alex buy 3 kg 750 g of flour altogether.

Question 1 b): Alex buys 750 g more flour than Zac.

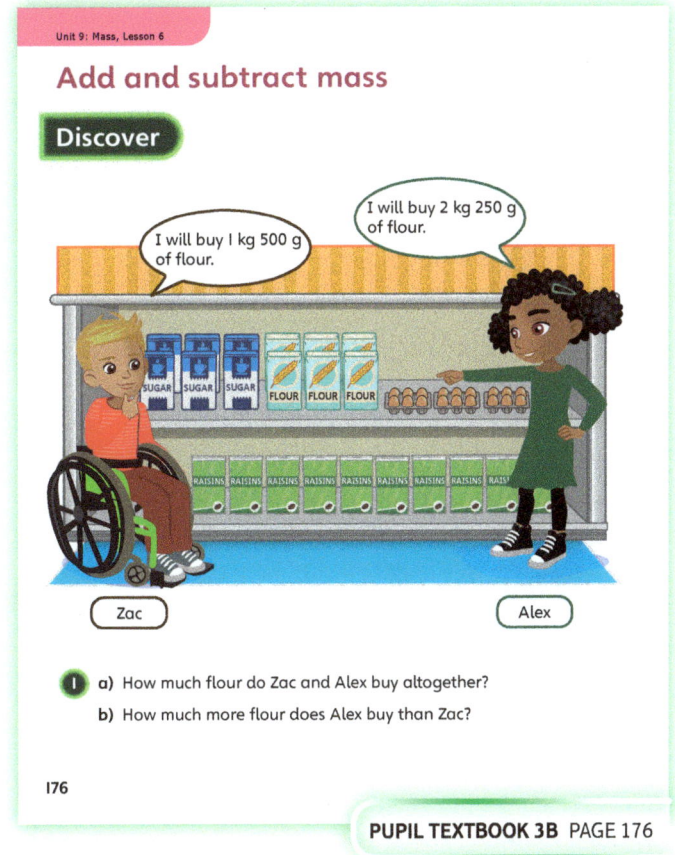

PUPIL TEXTBOOK 3B PAGE 176

Share

WAYS OF WORKING Whole class teacher led

ASK
- Question 1 a): *Did you remember to add the kilograms and grams separately?*
- Question 1 b): *Which strategy did you find the most efficient: using a number line or using the column method?*

IN FOCUS Question 1 encourages children to use different strategies to add and subtract, in particular the column method and the number line method. Discuss how Flo has used the number line to find the difference between both numbers and the nearest kilogram. Ask: *Is this an efficient method to find the total difference?* Ask children which strategy they find more efficient (this is a great whole class talking point).

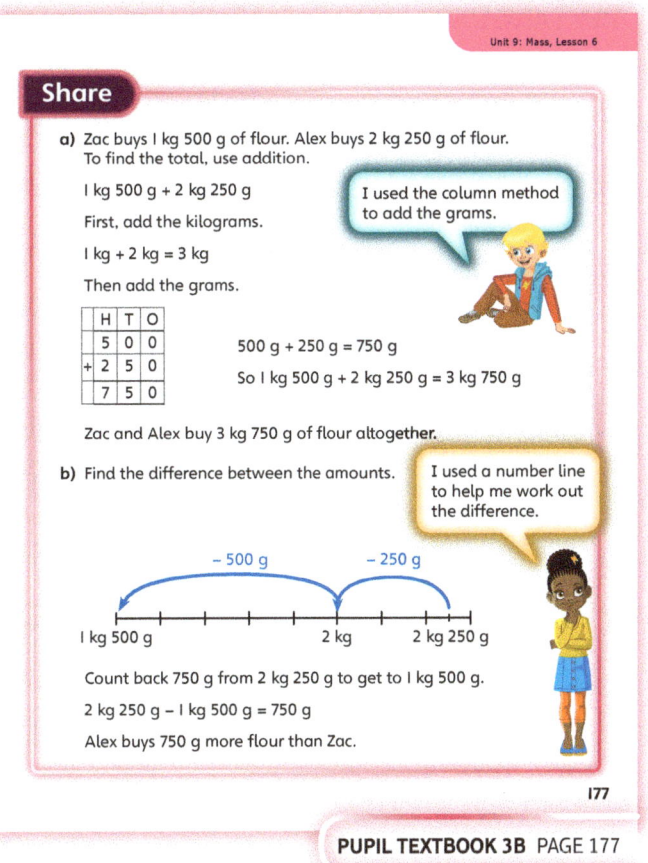

PUPIL TEXTBOOK 3B PAGE 177

215

Think together

WAYS OF WORKING Whole class teacher led (I do, We do, You do)

ASK

- Question ①: *Can you explain your chosen method?*
- Question ③: *How might the column method be useful for solving these problems?*
- Question ③: *How can you use the inverse operation to check your answers?*

IN FOCUS In question ①, point out that the answers could be written in grams or kilograms (depending on which way children convert). Encourage children to convert to kilograms: it makes the amounts more manageable (also children are not expected to add 4-digit numbers in Year 3).

STRENGTHEN Provide children with more activities similar to question ②. Practising with the number line is an important skill for children; at first, they may need to make multiple jumps.

DEEPEN In question ②, ask children to explain what operation is needed for each number line. They will have to think carefully about the missing number problems. Challenge children to draw their own number lines with missing numbers for a partner to solve.

ASSESSMENT CHECKPOINT Question ① can be used to assess whether children can add amounts that require a conversion first.

ANSWERS

Question ① a): 348 g + 295 g = 643 g

Question ① b): 2 kg 423 g + 1 kg 221 g = 3 kg 644 g

Question ① c): 2 kg 800 g + 200 g = 3 kg

Question ① d): 1,950 g + 5 kg 100 g = 7 kg 50 g

Question ② a): 750 g

Question ② b): 1 kg 300 g

Question ③ a): 500 g

Question ③ b): 1 kg 150 g

Question ③ c): 3 kg 800 g

Question ③ d): Whole: 3 kg 500 g
            Parts: 2 kg 600 g and 900 g

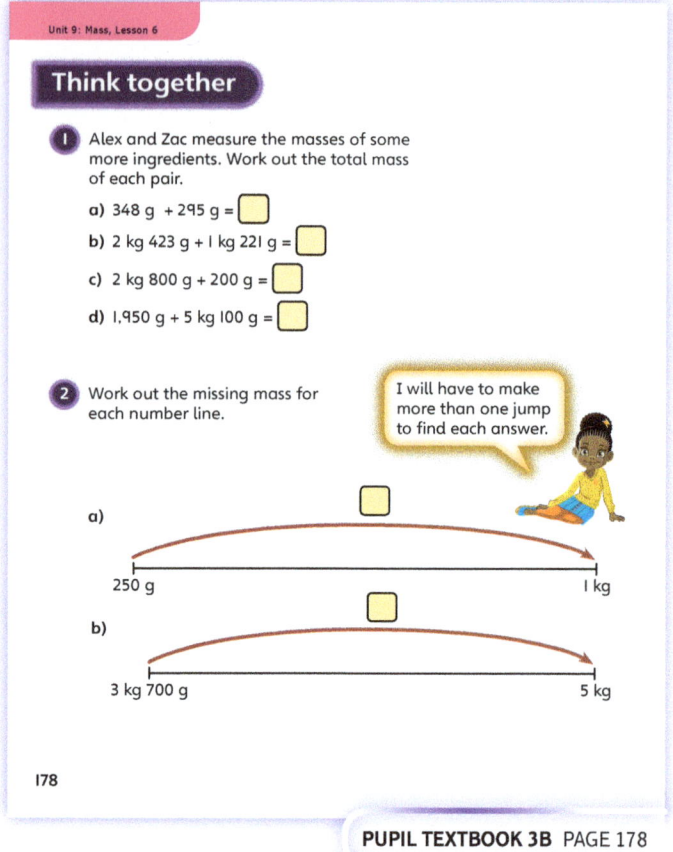

PUPIL TEXTBOOK 3B PAGE 178

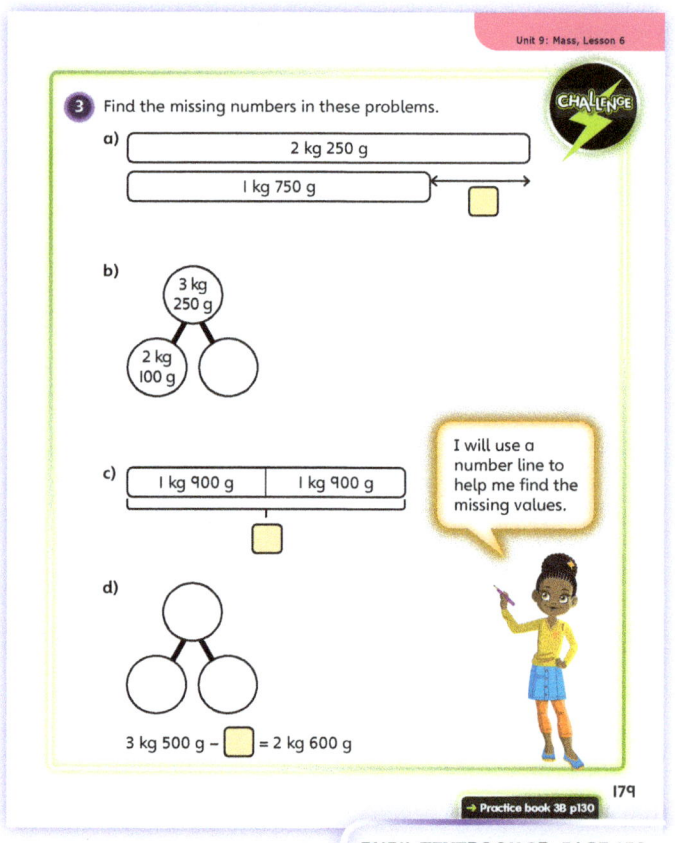

PUPIL TEXTBOOK 3B PAGE 179

Unit 9: Mass, Lesson 6

Practice

WAYS OF WORKING Independent thinking

IN FOCUS In question ①, children are required to add two masses together. In questions ① b), c) and d), they should realise they need to add the kilograms first and then the grams. They should not be converting to grams before adding. In question ②, children need to work out what they need to add to make the next whole kilogram. Questions ③ and ④ provide further examples of adding and subtracting masses. In question ③, the calculations are presented as bar models and then, in question ④, they are given as number sentences. Children are likely to use column addition and subtraction to find the answers, although some children may be able to work them out mentally.

STRENGTHEN If children are struggling to add or subtract the masses, use base 10 equipment or counters and remind them of the concepts. When adding, you could allow children to practise using number lines. Showing the addition or subtraction on a number line may help children as opposed to the column method.

DEEPEN In question ⑤, children are adding and subtracting where they need to cross into the next kilogram. They need to work out how many they need to make the next kilogram and then subtract this amount to work out how many above that kilogram they need to go. To support children's understanding of this harder concept, you might want to provide more questions like question ③, where children first have to work out how many grams make the next kilogram.

ASSESSMENT CHECKPOINT Questions ① to ④ will help you assess whether children can confidently add and subtract masses. Question ⑤ will help you assess if children can add and subtract masses where they need to cross a kilogram.

ANSWERS Answers for the **Practice** part of the lesson can be found in the *Power Maths* online subscription.

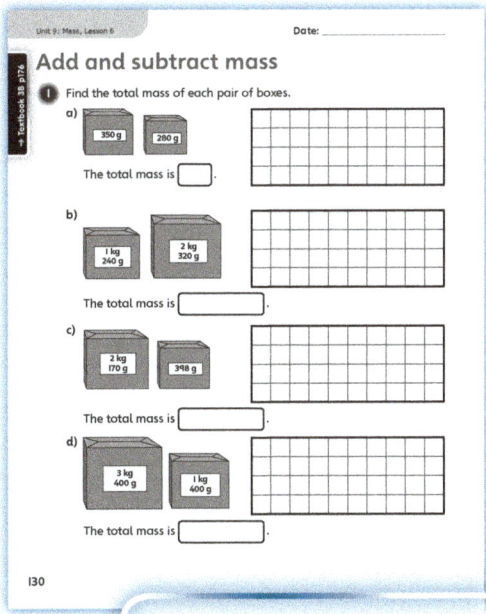

PUPIL PRACTICE BOOK 3B PAGE 130

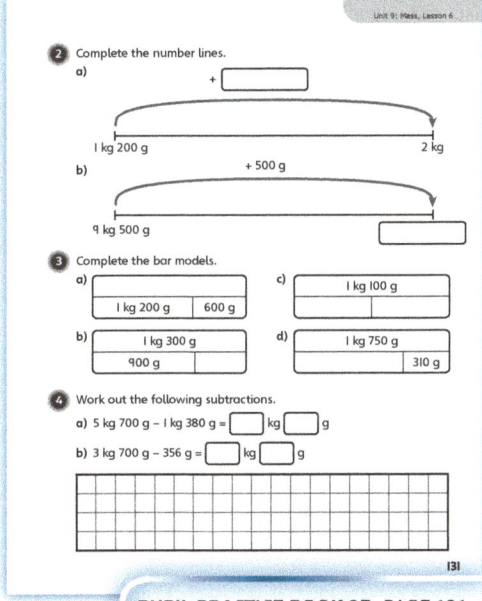

PUPIL PRACTICE BOOK 3B PAGE 131

Reflect

WAYS OF WORKING Pair work

IN FOCUS This activity allows the sharing of ideas and methods. Children will have to explain their method: a great learning point in itself.

ASSESSMENT CHECKPOINT Look carefully at children's explanations of methods. Do they understand why some methods are more efficient than others?

ANSWERS Answers for the **Reflect** part of the lesson can be found in the *Power Maths* online subscription.

After the lesson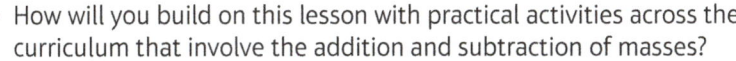

- How will you build on this lesson with practical activities across the curriculum that involve the addition and subtraction of masses?
- Are all children secure in solving missing number problems? Do some need further support?
- Do children know how to add and find the difference between two masses given as a combination of grams, and kilograms and grams?

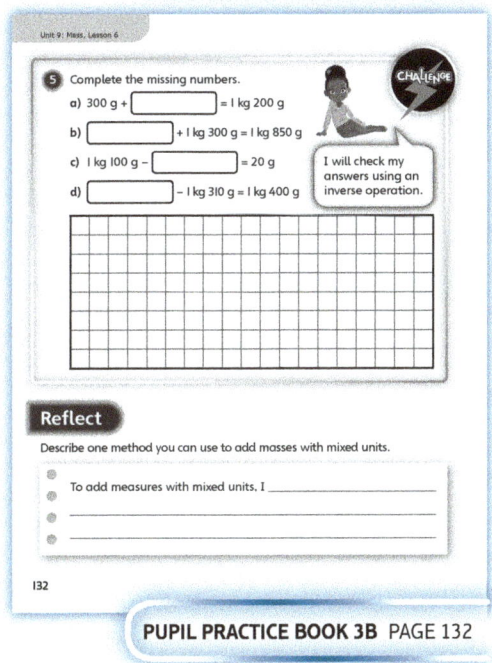

PUPIL PRACTICE BOOK 3B PAGE 132

217

Unit 9: Mass, Lesson 7

Problem solving – mass

Learning focus
In this lesson, children will use all of the knowledge and strategies they have learnt in this unit to solve problems involving mass.

Before you teach
- Can a classroom display support problem solving with mass?
- Are all children secure in operations such as multiplication and division?

NATIONAL CURRICULUM LINKS

Year 3 Measurement

Measure, compare, add and subtract: lengths (m/cm/mm); mass (kg/g); volume/capacity (l/ml).

ASSESSING MASTERY

Children can solve problems effectively, using suitable strategies such as choosing the correct method of addition or subtraction. Children can explain how they solved the problem clearly and using the correct vocabulary.

COMMON MISCONCEPTIONS

Children may misinterpret the question and carry out an incorrect calculation. Ask:
- *How can you check your answer?*

Children may not be able to show the question using a visual representation. Ask:
- *Can you draw a bar model to help you solve the problem?*

STRENGTHENING UNDERSTANDING

Ask children to construct bar models to represent the problems. This should help them understand what is required to find a solution.

GOING DEEPER

To deepen learning, give children some answers, for example, 4 kg 20 g, and ask them to create multi-step problems to match them. Asking children to explain their method for constructing and solving their problems will help to consolidate their understanding of the different ways to solve problems.

KEY LANGUAGE

In lesson: mass, scale, interval, grams (g), kilograms (kg)

Other language to be used by the teacher: measure, difference, divide, dial, problem

STRUCTURES AND REPRESENTATIONS

Number line, bar model, part-whole model, column addition

RESOURCES

Optional: weighing scales, modelling clay, number lines

 In the eTextbook of this lesson, you will find interactive links to a selection of teaching tools.

Quick recap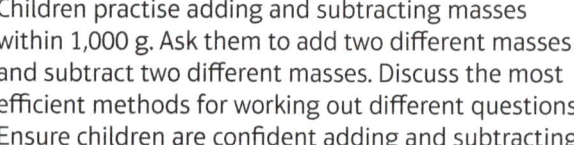

Children practise adding and subtracting masses within 1,000 g. Ask them to add two different masses and subtract two different masses. Discuss the most efficient methods for working out different questions. Ensure children are confident adding and subtracting 3-digit numbers.

Unit 9: Mass, Lesson 7

Discover

WAYS OF WORKING Pair work

ASK
- Question 1 a): *Which method could you use?*
- Questions 1 a) and b): *How many steps are in this problem?*

IN FOCUS Question 1 b) focuses on subtraction. Children may have used the column method to find the solution to question 1 a). You may want to model finding the difference on a number line as a more efficient mental strategy.

PRACTICAL TIPS For this activity, encourage children to use a range of strategies to work out the answers, and use bar models to highlight how you can find the difference between two values. Afterwards, share the different methods.

ANSWERS

Question 1 a): Zac has 200 g of flour left.

Question 1 b): There is 75 g of flour left.

PUPIL TEXTBOOK 3B PAGE 180

Share

WAYS OF WORKING Whole class teacher led

ASK
- Question 1 a): *How does the bar model help?*
- Question 1 b): *How does the number line help?*
- Questions 1 a) and b): *Are there any other methods you could use?*

IN FOCUS For both parts of the question, encourage children to work methodically and in clear steps. Children should record their workings in a series of number sentences. Some children may need the bar model and number line to be explained to them, to help them understand the problems.

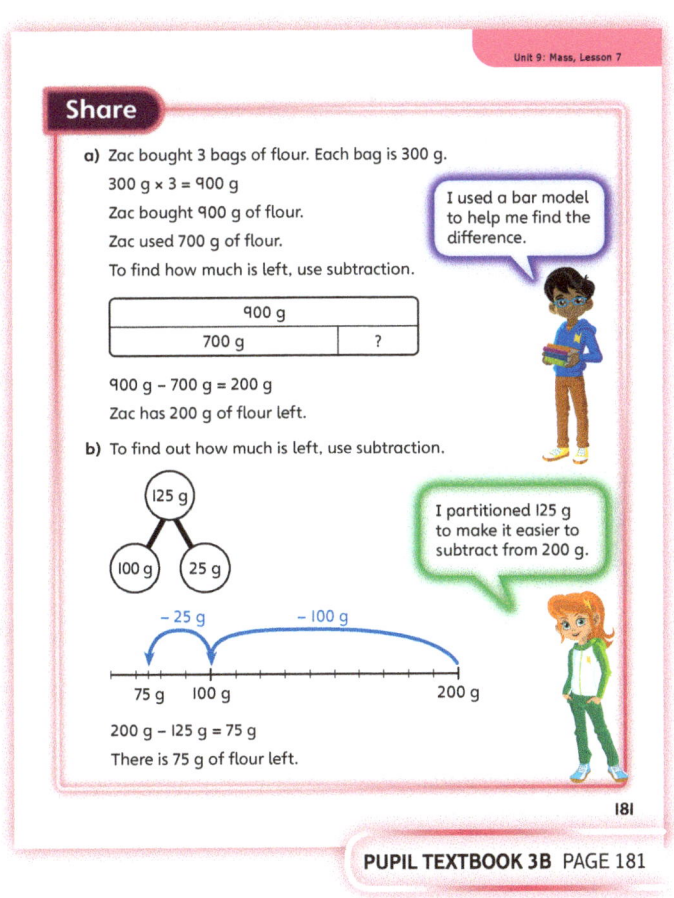

PUPIL TEXTBOOK 3B PAGE 181

Unit 9: Mass, Lesson 7

Think together

WAYS OF WORKING Whole class teacher led (I do, We do, You do)

ASK

- Question ❶: *How can the scale help you work out the problem?*
- Question ❷: *Which object do you think you should work out first? Why?*
- Question ❸: *How could you find out the mass of one robot?*

IN FOCUS In questions ❶ and ❷, children may work out the intervals on the scale and then use it as a number line to help them work out the answer. This is a good strategy and should be encouraged.

STRENGTHEN Use question ❸ to explore the various calculation steps with learners who need more support and scaffolding. Ask children what they should find out first. If they are unsure, you may want to point at the two robots on the second scale and ask how they could work out the mass of one robot.

In question ❷, remind children that they can work out 40 × 3 by using place value knowledge (some may start with a written method).

DEEPEN Question ❸ can be used to deepen learning, because it allows children to break down a problem clearly. Ask children to create their own similar problem.

ASSESSMENT CHECKPOINT Question ❸ is a good indication of which children have mastered the lesson. Those who have will demonstrate a clear understanding of the question and will be able to use appropriate strategies to work out the answers.

ANSWERS

Question ❶: The mass of the flour is 475 g.

Question ❷: The arrow will be pointing to 150 g (the third interval after 0 on the scales).

Question ❸: The mass of one empty wagon is 500 g.

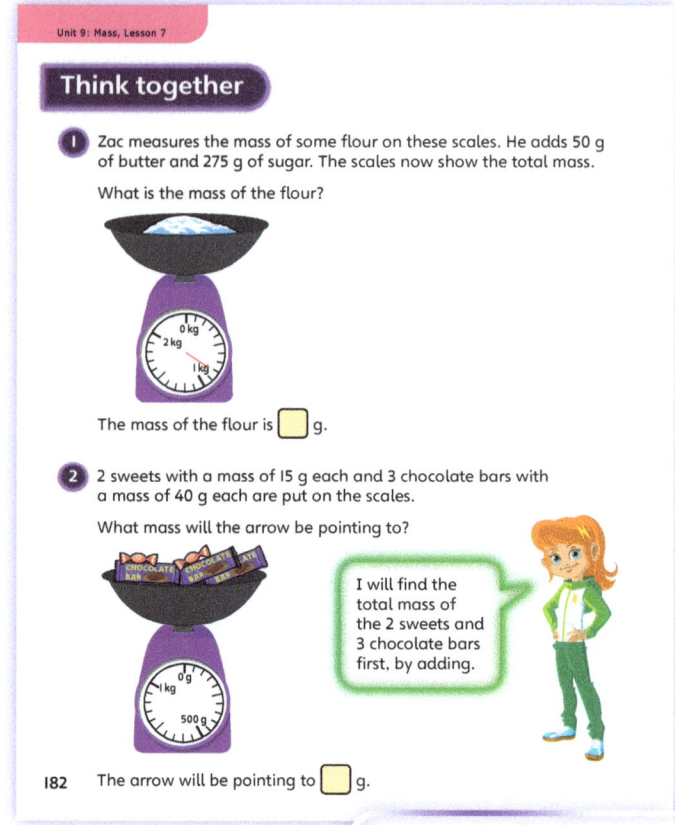

PUPIL TEXTBOOK 3B PAGE 182

PUPIL TEXTBOOK 3B PAGE 183

Unit 9: Mass, Lesson 7

Practice

WAYS OF WORKING Independent thinking

IN FOCUS Question ❶ allows children to practise adding masses. For question ❷, encourage children to explore the problem using a representation such as a bar model. This is important as it gives children a visual understanding of the question.

STRENGTHEN Run intervention sessions in which children practise calculation strategies. Apply these strategies to word problems and use a bar model to represent them.

DEEPEN After completing question ❹, challenge children to create similar problems for a partner to solve. This will deepen learning as children will need to think carefully about their own problems – this will involve both creativity and logic.

ASSESSMENT CHECKPOINT Question ❸ can be used to see if children can extract the right information from the word problem in order to find the answer.

ANSWERS Answers for the **Practice** part of the lesson can be found in the *Power Maths* online subscription.

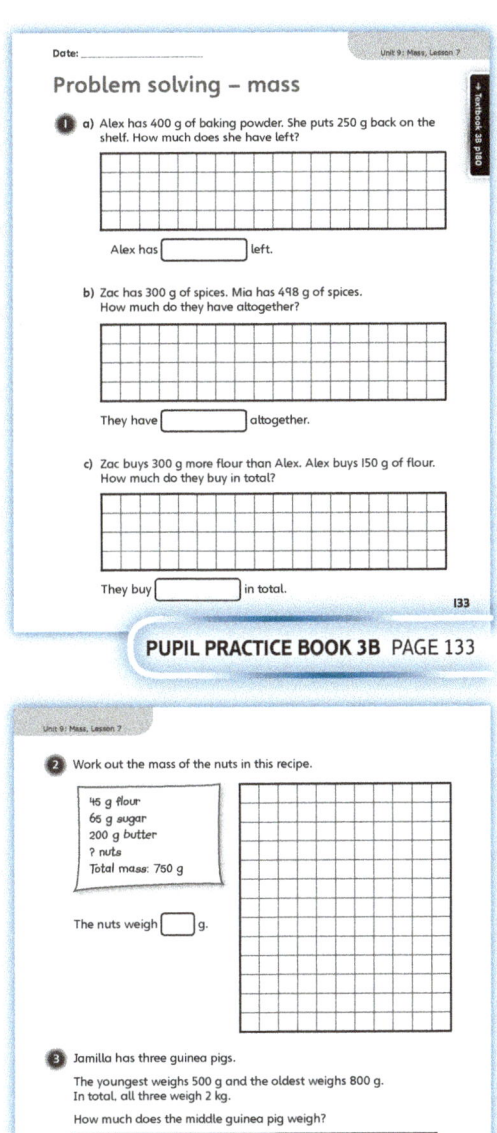

Reflect

WAYS OF WORKING Pair work

IN FOCUS This activity consolidates learning in the lesson. Children will have to use their knowledge of problem solving in order to create a question to match a given answer. Encourage children to think of a multi-step question.

ASSESSMENT CHECKPOINT Ask children to swap their books with a partner and solve each other's questions. Look for multi-step questions which are correct.

ANSWERS Answers for the **Reflect** part of the lesson can be found in the *Power Maths* online subscription.

After the lesson ⏸

- Are all children secure in the aims for each lesson and ready for the **End of unit check**?
- Could children recognise and explain what approach they took to solve each problem?
- What were the main misconceptions children had during this lesson? How did you resolve them?

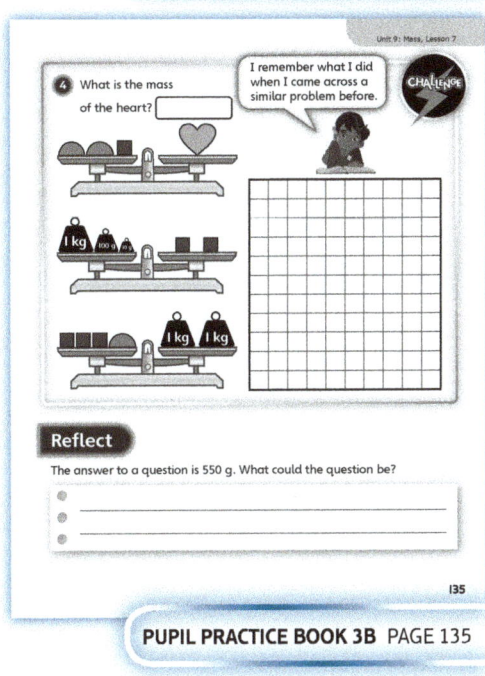

End of unit check

Don't forget the unit assessment grid in your *Power Maths* online subscription.

WAYS OF WORKING Group work adult led

IN FOCUS

This **End of unit check** will allow you to focus on children's understanding of measuring mass and whether they can apply their knowledge in order to solve problems.

Look carefully at the answer that is given for question ❺. It will tell you if children can read two scales and then find the sum of the amounts.

Question ❻ is a SATs-style question. Talk through the question and, after children have answered it, explore answers.

ANSWERS AND COMMENTARY

Children should be secure with reading a range of scales and finding missing intervals. They should also be able to convert between grams and kilograms. Children should be confident when calculating answers to addition and subtraction questions involving mass, and should be able to check their strategies. Finally, they should apply their learning when solving problems.

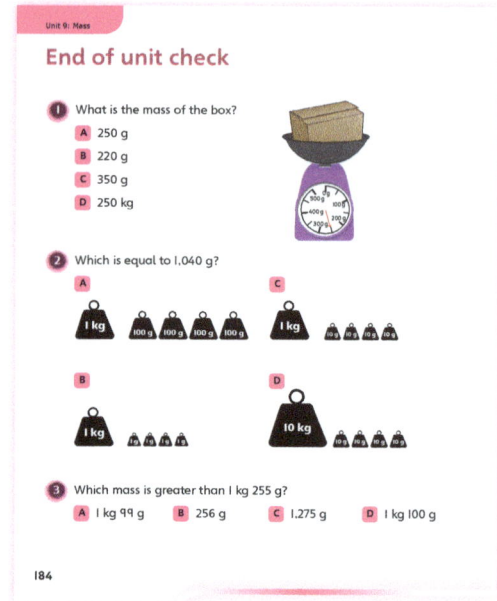

PUPIL TEXTBOOK 3B PAGE 184

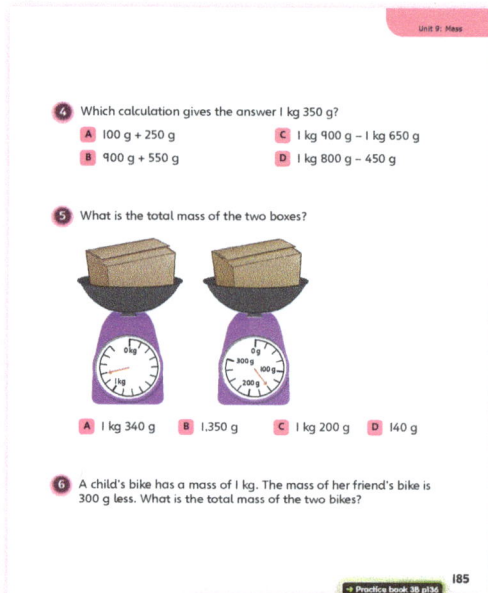

PUPIL TEXTBOOK 3B PAGE 185

Q	A	WRONG ANSWERS AND MISCONCEPTIONS	STRENGTHENING UNDERSTANDING
1	A	D suggests that the child has not read the units at the top of the scale.	Give children support with real-life contexts for measuring mass by: • asking children to convert measurements from grams to kilograms and vice versa • displaying the key vocabulary of the unit in your classroom • asking children to match word problems with representations such as bar models and number lines.
2	C	Any other answer suggests that children are not secure with the fact that 1 kg = 1,000 g or that they do not fully understand place value.	
3	C	B may suggest that children are struggling to convert and then compare amounts.	
4	D	B suggests that children have some difficulty crossing the kilogram barrier.	
5	B	Any other answer suggests that children are not secure in reading two different scales and adding the amounts together.	
6	1 kg 700 g	Have children followed both steps of the problem?	

Unit 9: Mass

My journal

WAYS OF WORKING Independent thinking

ANSWERS AND COMMENTARY

Children should populate the table with accurate objects that they would measure the mass of in kilograms or grams. They should explain their reasoning.

For the second part, they may choose to use balance scales or weighing scales to check if their estimates are correct.

Power check

WAYS OF WORKING Independent thinking

ASK
- *What visual representation helped you in this unit?*
- *What do you know now that you did not at the start of the unit?*
- *Can you write down the new words you have learnt and what they mean?*

PUPIL PRACTICE BOOK 3B PAGE 136

PUPIL PRACTICE BOOK 3B PAGE 137

Power puzzle

WAYS OF WORKING Pair work

IN FOCUS Children should realise that they have to look at the balance scales first to work out the mass of the pineapple. They can then take this away from the value on the first scale to find the mass of the melon.

ANSWERS AND COMMENTARY The calculations are as follows:

500 g + 200 g + 50 g + 5 g = 755 g

1 kg 300 g − 755 g = 545 g

Children are most likely to use the column method for the addition and a number line for the subtraction. If children are finding it difficult to write an explanation, ask: *Should you find the total amounts of each side of the balance scales first?*

After the unit

- Are all children secure in converting mass units, finding an interval value and applying the correct method when working out multi-step problems?
- How will you link this unit to other areas of the curriculum, for example weighing materials in art and design?

Strengthen and **Deepen** activities for this unit can be found in the *Power Maths* online subscription.

Unit 10
Capacity

Mastery Expert tip! 'I found it valuable to make the links between this unit and the work children have done in number and calculating. Thinking of the scale as a number line helped the children to apply their understanding of number to the context of capacity.'

Don't forget to watch the Unit 10 video!

WHY THIS UNIT IS IMPORTANT

This unit explores capacity and comes after other units about measure. It asks children to interpret a range of scales and apply their knowledge of place value and the number system. Children will learn to compare and order measurements, and convert between millilitres and mixed units of litres and millilitres. They will then use knowledge of all four operations to solve problems involving capacity.

WHERE THIS UNIT FITS

→ Unit 9: Mass
→ **Unit 10: Capacity**
→ Unit 11: Fractions (2)

This unit builds on from children's previous work in measures involving length and mass. Children should already have experience in reading and interpreting a range of scales and converting between units of measure, which will help them in this unit. Children will learn to compare, calculate and solve problems in the context of capacity. Children will need to apply their knowledge of the number system and calculating, in order to solve capacity word problems.

Before they start this unit, it is expected that children:
- understand place value in 3-digit numbers
- know how to add and subtract 3-digit numbers
- know multiplication facts for the 2, 5 and 10 times-tables.

ASSESSING MASTERY

Children will be able to use and interpret a variety of scales in order to measure amounts. They will be able to work with mixed units and do simple conversions between litres and millilitres. Children will be able to add and subtract capacities that cross the litre boundary and they will be able to solve problems using addition, subtraction, multiplication and division. Children will understand that the capacity of a container is how much it holds when full.

COMMON MISCONCEPTIONS	STRENGTHENING UNDERSTANDING	GOING DEEPER
Children may struggle to interpret scales where not all divisions are labelled. They may make errors in place value, particularly when measuring and calculating in mixed units and converting between litres and millilitres.	Make the links to the work on calculations and the number system clear to children. Encourage children to think of the scales as number lines, and to represent them horizontally as well as vertically. Allow children to explore with capacity equipment by measuring, comparing and calculating capacities practically.	Encourage children to write their own capacity word problems that they can try on a partner. You could challenge them by specifying what operations their problem should include. Ask: *Can you write a problem with three or four steps?*

Unit 10: Capacity

UNIT STARTER PAGES

Use these pages to introduce what children will be covering in this unit. Ask children if they have done anything similar in other units. Children may be able to make links between this unit and their work on length and mass.

Talk through the bar model to ensure children understand what it represents. You could start with easier examples, using number bonds to 100. Ask: *How does 35 + ☐ = 100 relate to 350 + ☐ = 1,000?* Discuss how a subtraction such as 100 − ☐ = 45 would be represented and how that relates to 1,000 − ☐ = 450. Practise counting up and back in steps of 100 and 1,000.

Talk through the part-whole model, using it to practise number bonds to 100, such as: 35 + ☐ = 100; 100 − ☐ = 45. Ask: *How many multiples of 5 pairs can you write in 3 minutes? Can you then relate this to multiples of 10 bonds to 1,000?*

Finally, go through the key vocabulary with children. Ask: *Do you recognise any words? Can you explain what they mean? Which words are new?*

STRUCTURES AND REPRESENTATIONS

Bar model:

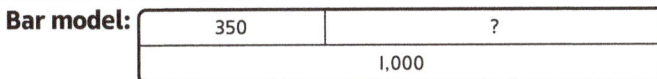

Part-whole model: **Scales:**

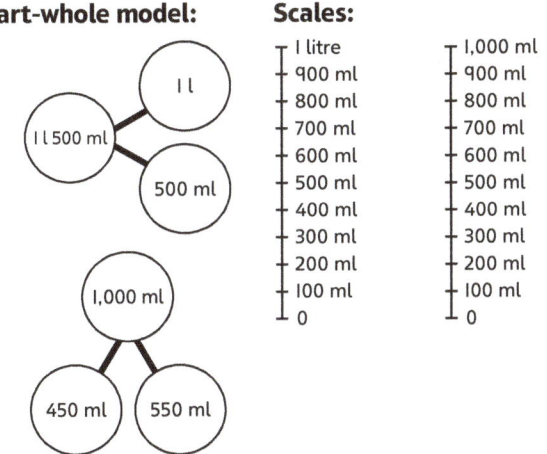

Number line:

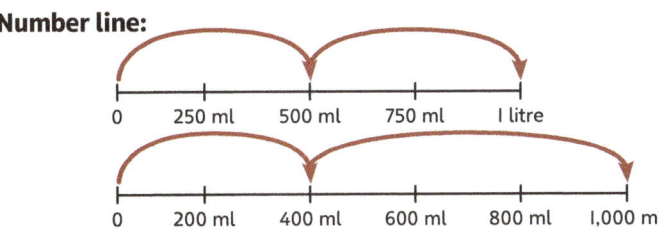

KEY LANGUAGE

There is some key language that children will need to know as part of the learning in this unit:

→ capacity, amount, measurement
→ litre (l), millilitre (ml)
→ scale, number line, interval
→ compare, convert, order

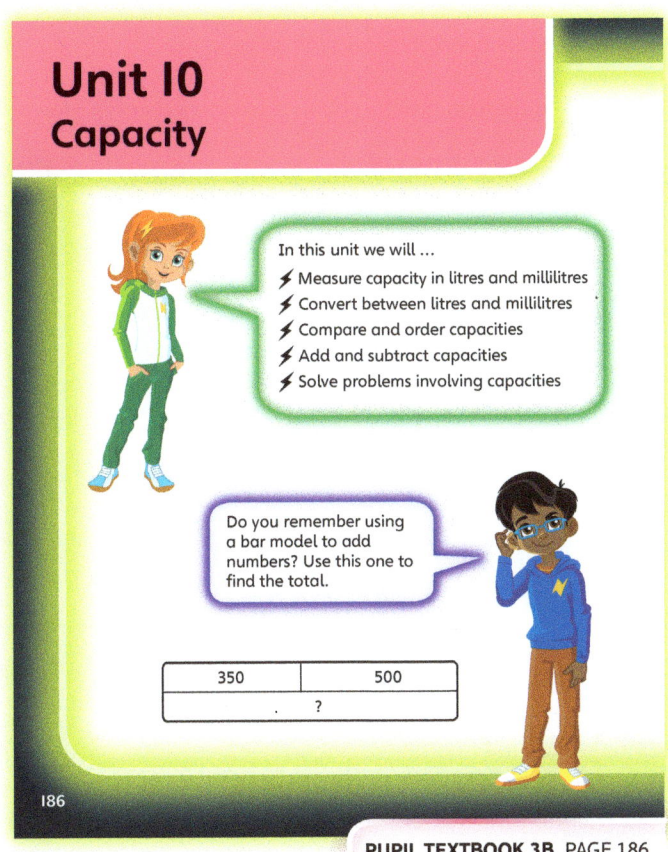

PUPIL TEXTBOOK 3B PAGE 186

PUPIL TEXTBOOK 3B PAGE 187

Unit 10: Capacity, Lesson 1

Measure capacity and volume in litres and millilitres

Learning focus

In this lesson, children will learn to measure volume in litres and in millilitres. They will learn how to read a variety of scales where only some of the divisions are labelled, drawing on their understanding of number, division and multiplication.

Before you teach

- What practical activities could you use to improve learning?
- What other areas of the curriculum could you link this lesson to (for example, science)?
- Can children count in 10s, 2s and 5s?

NATIONAL CURRICULUM LINKS

Year 3 Measurement

Measure, compare, add and subtract: lengths (m/cm/mm); mass (kg/g); volume/capacity (l/ml).

ASSESSING MASTERY

Children can read a scale in either litres or millilitres in order to determine volume. They will be able to work out the value of intervals not labelled, by applying known multiplication and division facts.

COMMON MISCONCEPTIONS

Children may struggle to determine how the scale is divided up. This could be due to not applying division facts or miscounting the intervals between two labelled measurements. Encourage children to look at the scale as a number line. Draw the scales horizontally, as this may be a more familiar orientation for them. Ask:

- *What is the total amount between these two labelled markers? How many jumps are there between these two labelled markers? How can you work out how much each jump is worth?*

Children may not understand that capacity is the amount a container holds when full, whereas the amount actually in the container is often not its full capacity. Ask:

- *What is the capacity of this jug when it is full? What is the amount of liquid in the jug? How much more will fill it to its capacity?*

STRENGTHENING UNDERSTANDING

Encourage children to use a range of different containers and a variety of scales. Ask children to fill the containers to a given level, discussing how they would find this on the scale and encouraging them to identify the container's capacity.

GOING DEEPER

Children could create their own capacity problems by coming up with their own scales. They could think of different ways to divide multiples of 100 ml or 1,000 ml on their scales.

KEY LANGUAGE

In lesson: capacity, litres (l), millilitres (ml), scale, interval, amount, number line, divided by, half-way between, least, most, approximate, measure, gauge

Other language to be used by the teacher: volume, add, subtract, divide, multiply, unit of measure

STRUCTURES AND REPRESENTATIONS

Number line, bar model

RESOURCES

Mandatory: rulers

Optional: blank number lines, capacity measuring equipment (selection of 100 ml, 500 ml and litre containers), water, poster paint, empty milk and other drinks bottles

 In the eTextbook of this lesson, you will find interactive links to a selection of teaching tools.

Quick recap

Ask children to draw a number line from 0 to 100. Divide the line into different amounts. For example, ask children to put three marks equally spaced along the number line. Ask them what the number line is going up in. Repeat for lines going up in 10s and 20s.

Discover

WAYS OF WORKING Pair work

ASK
- Question 1 a): *How many intervals are there between 0 litres and 100 litres?*
- Question 1 a): *Is the interval the number of markers or the number of jumps between the markers?*
- Question 1 a): *How could you work out how much each interval is worth?*
- Question 1 a): *If the tank was full at the start, how many litres has the elephant drunk?*

IN FOCUS The focus of questions 1 a) and b) is on identifying the intervals between unlabelled markers on a scale. Modelling the scale as a number line (with arrowed jumps between markers) will help children to see how to work out what each interval is worth. Remind children that, when counting on, you do not count the start number, which means that the number of intervals is the same as the number of jumps between marked points. Some children will count just the unlabelled markers and others may include the ones given as well.

PRACTICAL TIPS Give children number lines modelled on the scale in the scenario to practise counting the number of intervals between marked points. Practise counting in 25s to 300 and in 20s to 300, then count in 10s, 20s, 25s and 50s to see which works. Discuss how you know you have the right interval.

ANSWERS

Question 1 a): There are 150 litres of water left in the tank.

Question 1 b):

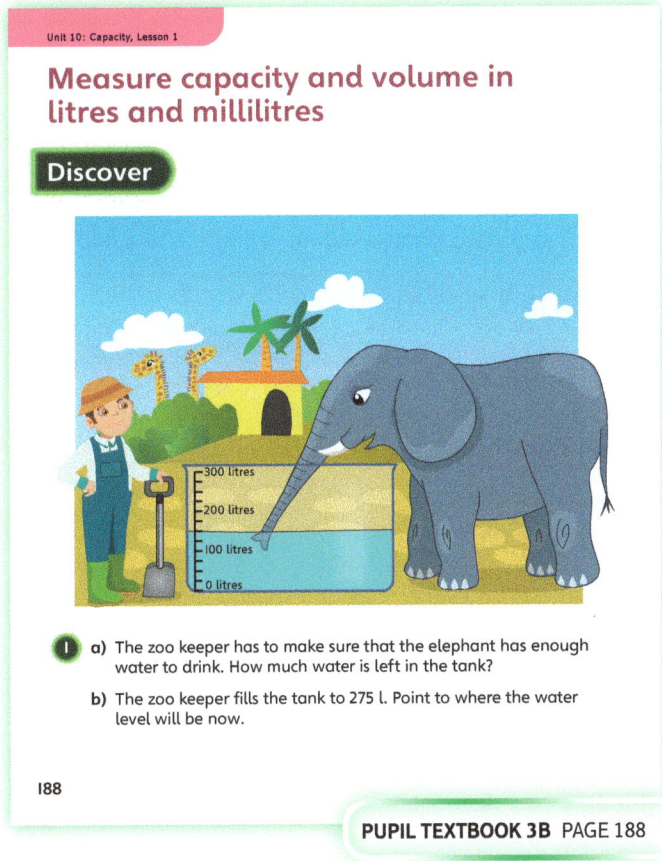

Share

WAYS OF WORKING Whole class teacher led

ASK
- Question 1 a): *How many jumps are there from 0 to 100 litres?*
- Question 1 a): *How does Astrid's comment help you to work out how much is left in the tank?*
- Question 1 b): *What does the scale go up in? How does this help you know where 275 ml would be on the scale?*

IN FOCUS In question 1 a), the scale is divided into four equal sections, each worth 25 litres. It is important for children to see that they can use their knowledge of counting and calculations within this context.

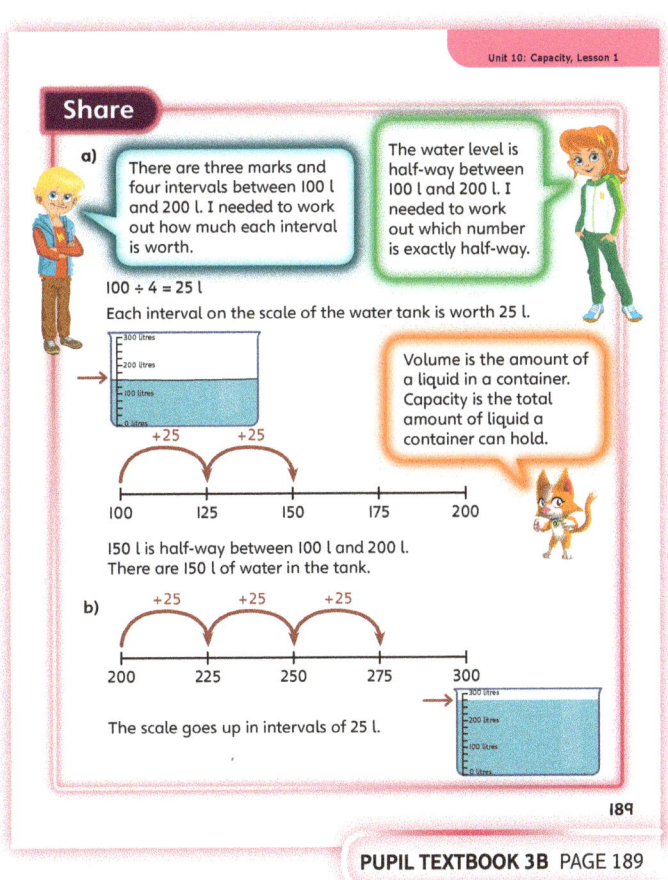

PUPIL TEXTBOOK 3B PAGE 188

PUPIL TEXTBOOK 3B PAGE 189

Think together

WAYS OF WORKING Whole class teacher led (I do, We do, You do)

ASK
- Questions ① and ③: *How can you calculate what each interval is worth?*
- Questions ① and ③: *What number facts can you draw on to help you work out what each interval is worth?*
- Question ②: *What does capacity mean? What knowledge can you use to match the amounts?*
- Question ③: *What number is half-way between 150 and 200?*

IN FOCUS In question ②, children will not know the actual capacities of these items, but encourage logical reasoning to work out the matches, even though the items are not shown to scale.

Question ③ involves three different scales and an amount that is half-way between two intervals. Once children have worked out what each interval represents, encourage discussion about what the values half-way between markers would be and how to work this out.

STRENGTHEN Allow children to experiment with 500 ml and 1 l containers with scales going up in 100 ml, 50 ml and 25 ml. They can first work out the in-between values to identify the intervals, then partially fill the containers to read the amounts. Help children to read the quantities by letting them use coloured water (add a little poster paint to the water). Let children look at a variety of milk and other drink bottles to see what each capacity is.

DEEPEN Ask children to create different scales for 1 litre and to identify one(s) that make it easy to measure 250 ml and 200 ml.

ASSESSMENT CHECKPOINT Can children use division or counting strategies to work out the intervals on a millilitre or litre scale?

Children should know that amounts of liquids are measured in millilitres or litres. They should be able to give examples of capacities measured in litres and capacities measured in millilitres. Children are beginning to understand that capacity is the amount a container can hold when filled up.

ANSWERS

Question ① a): There are 35 litres of fuel left in the tank.

Question ① b): There are 140 ml of water in the bottle

Question ②:

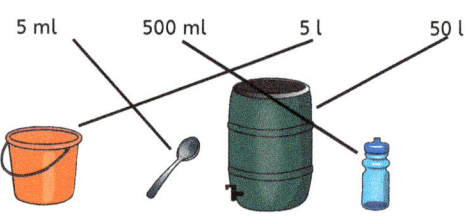

Question ③ a): A = 80 ml   B = 175 ml   C = 125 ml

Question ③ b): Smallest to greatest: A, C, B.

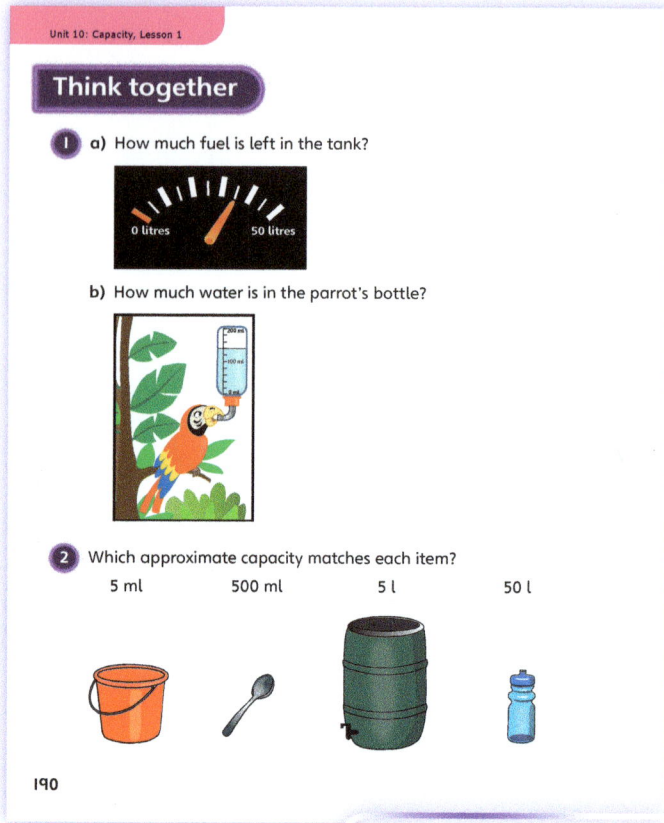

PUPIL TEXTBOOK 3B PAGE 190

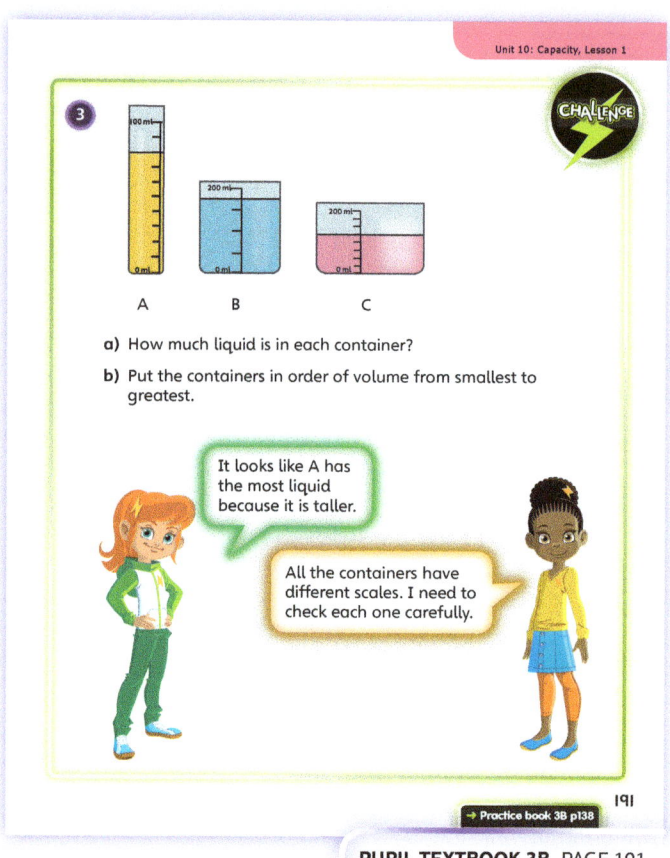

PUPIL TEXTBOOK 3B PAGE 191

Unit 10: Capacity, Lesson 1

Practice

WAYS OF WORKING Independent thinking

IN FOCUS The focus in questions 1, 3 and 4 is on reading scales and working out what each interval is worth. A variety of different types of measuring scales are used, including gauges. There is plenty of opportunity to work out the value of each interval, which is a key skill in this unit.

STRENGTHEN Look at scales that children may be more familiar with, such as scales on a ruler. Talk about how the scales have been divided up. Children could use a ruler to draw a line of 10 cm to create a scale that goes up to 100 millilitres in intervals of 10 ml, or a scale that goes up to 1,000 ml in intervals of 100 ml. Ask them to number each marker. Discuss different ways that a 10 cm scale could be divided. Give children the opportunity to explore what 5 ml, 50 ml, 500 ml and 1 litre look like on a 10 cm number line.

DEEPEN Ask children to draw three vertical lines, measuring 20 cm, 50 cm and 80 cm in length. Then ask them to label the top of each line 100 ml. Can children divide these scales so that each has 10 ml divisions? What strategies did they use to work it out? Can children draw an accurate scale to show 0–100 ml with 20 ml or 25 ml intervals?

THINK DIFFERENTLY Question 2 involves children using their general knowledge of the world to decide whether something will be measured in litres or millilitres. An understanding that 1 litre is quite a lot and 1 millilitre is a very small drop will help.

ASSESSMENT CHECKPOINT Questions 1, 3 and 4 will determine whether children are able to read a variety of scales. Question 2 will determine whether children understand the relative sizes of litres and millilitres.

ANSWERS Answers to the **Practice** part of the lesson can be found in the *Power Maths* online subscription.

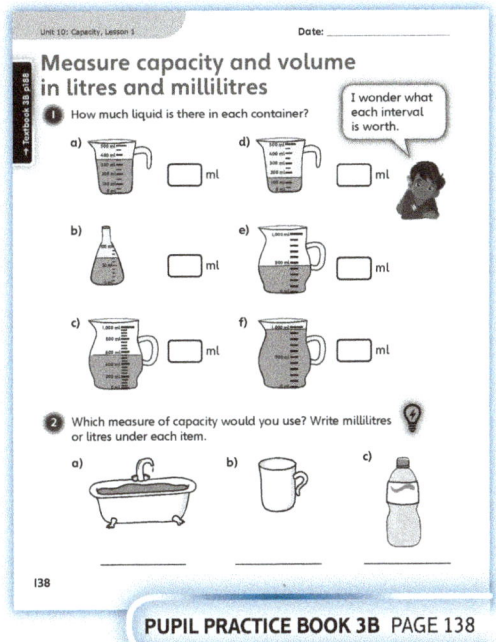

PUPIL PRACTICE BOOK 3B PAGE 138

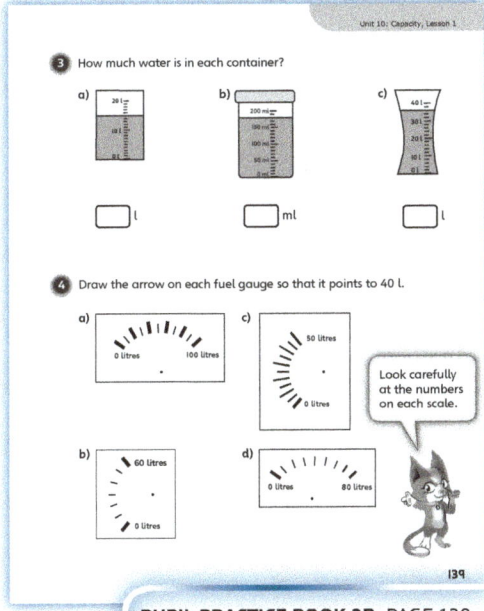

PUPIL PRACTICE BOOK 3B PAGE 139

Reflect

WAYS OF WORKING Independent thinking

IN FOCUS This question encourages children to think about what information is important when reading scales. They have to consider the value of the labelled divisions and the number of divisions between each labelled division. Children should reflect on the types of operations they need to carry out, in order to determine how much each division is worth.

ASSESSMENT CHECKPOINT Do children understand the importance of the scale on measuring equipment? Can children explain how to work out the unlabelled parts of the scale?

ANSWERS Answers to the **Reflect** part of the lesson can be found in the *Power Maths* online subscription.

PUPIL PRACTICE BOOK 3B PAGE 140

After the lesson

- Can you provide opportunities for children to practise these skills in other curriculum areas?
- Are children confident with how to read a scale where not every interval is labelled?
- Did you draw links with other areas of maths, such as number, calculations and other measures?

Unit 10: Capacity, Lesson 2

Measure in litres and millilitres

Learning focus
In this lesson, children will learn to read mixed units of capacity given in litres and millilitres and as $\frac{1}{2}$ litres, and convert them to millilitres. They will also read scales showing amounts over 1 litre.

Before you teach
- Are there any children who may need support reading scales?
- Can you provide practical activities to support children's understanding of measuring capacity?
- How will you make the links between measuring capacity and their understanding of number patterns clear to children?

NATIONAL CURRICULUM LINKS

Year 3 Measurement

Measure, compare, add and subtract: lengths (m/cm/mm); mass (kg/g); volume/capacity (l/ml).

ASSESSING MASTERY

Children can measure amounts greater than 1 litre. They will understand that capacity between whole litres is measured in a mixture of litres and millilitres. Children understand that $\frac{1}{2}$ litre is the same as 500 ml. Children will begin to do simple conversions between mixed units and millilitres.

COMMON MISCONCEPTIONS

Children may read only the millilitres and overlook any litres when measuring with mixed units. Ask:
- *Is this amount more than 1 litre? How many whole litres are there?*

STRENGTHENING UNDERSTANDING

Encourage children to create their own amounts with mixed units. First, measure 1 litre of water and pour it into a larger container then add 500 millilitres of water. Ask: *How much water is in the container altogether?*

Add different amounts to the litre and record as, for example, 1 l 500 ml.

Make links with the previous lesson clear. Encourage children to think of a scale as a number line. Draw the scale horizontally, if that helps children to understand it.

GOING DEEPER

Give children a measurement with mixed units. Ask them to create their own scale to record the measurement.

KEY LANGUAGE

In lesson: capacity, litre (l), millilitre (ml), scale, intervals, half ($\frac{1}{2}$), partition, amount, whole, half-way, number line

Other language to be used by the teacher: mixed unit, greater than (>), less than (<)

STRUCTURES AND REPRESENTATIONS

Number line, bar model, column addition

RESOURCES

Optional: capacity measuring equipment

 In the eTextbook of this lesson, you will find interactive links to a selection of teaching tools.

Quick recap
Ask children to draw a number line from 0 to 1,000, going up in 100s. Then ask them to draw a number line going up in 250s to 1,000 and finally to draw a number line going up in 200s to 1,000.

Discover

WAYS OF WORKING Pair work

ASK

- Question 1 a): *Look at bucket A. What does the scale go up in? Litres or millilitres? How do you know the amount of water is not a whole number of litres? How far between 3 l and 4 l is the water?*
- Question 1 b): *Look at bucket B. Where is 1 litre on the scale? Will 1 l 200 ml be above or below this? How many ml do the markings represent?*

IN FOCUS In questions 1 a) and b), children show they can measure in both litres and millilitres. Children should not go into 1,000s as they have not met these numbers yet. Encourage children to draw a number line representing the buckets. They may need support to understand that there are two parts to the answers, in that they need a whole number of litres first and then millilitres.

PRACTICAL TIPS Use a jug or similar measuring device and water to recreate the scenario in the classroom.

ANSWERS

Question 1 a): There are 3 l and 500 ml of water in bucket A.

Question 1 b):

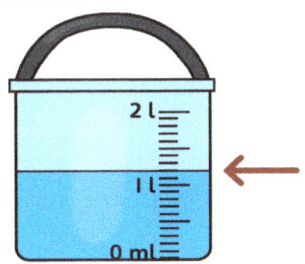

Share

WAYS OF WORKING Whole class teacher led

ASK

- Question 1 a): *How do you know the water is not a whole number of litres? How far is the water between 3 l and 4 l?*
- Question 1 b): *How many ml does each marking represent? How did you work that out?*

IN FOCUS

In question 1 a), discuss with children that the water is not an exact amount of litres. Use what Dexter says to help explain that the water is half-way between 3 l and 4 l. Sparks's comment reminds children that there are 1,000 ml in 1 litre and so the marking must be pointing to 3 litres 500 ml. Some children may want to write down $3\frac{1}{2}$ l and then convert to ml, but encourage them instead to first measure the whole litres and measure whatever is left in ml. In question 1 b), children need to understand that each interval is worth 100 ml. They can work this out by using their knowledge of number lines or by realising that there are 10 intervals and so they need to divide 1,000 by 10.

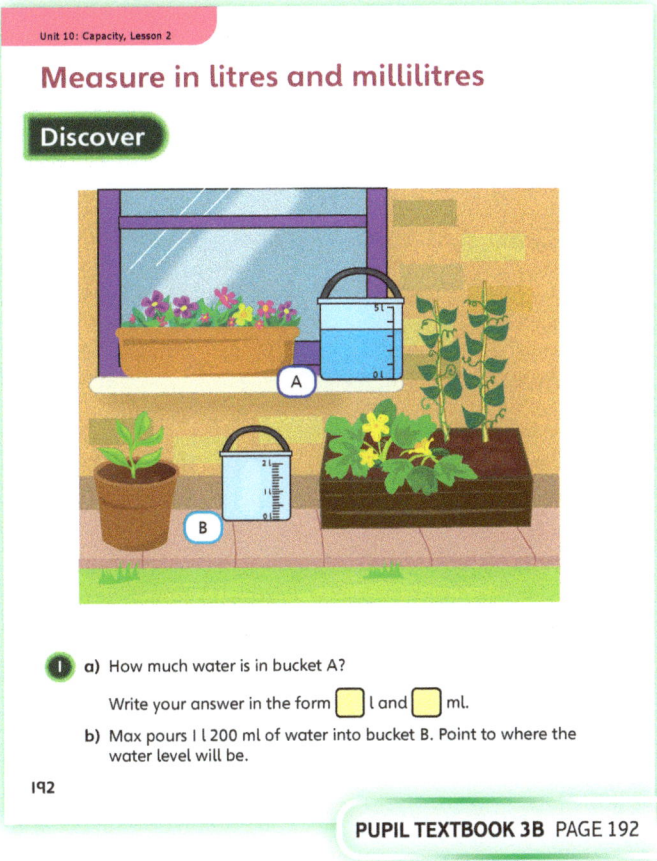

PUPIL TEXTBOOK 3B PAGE 192

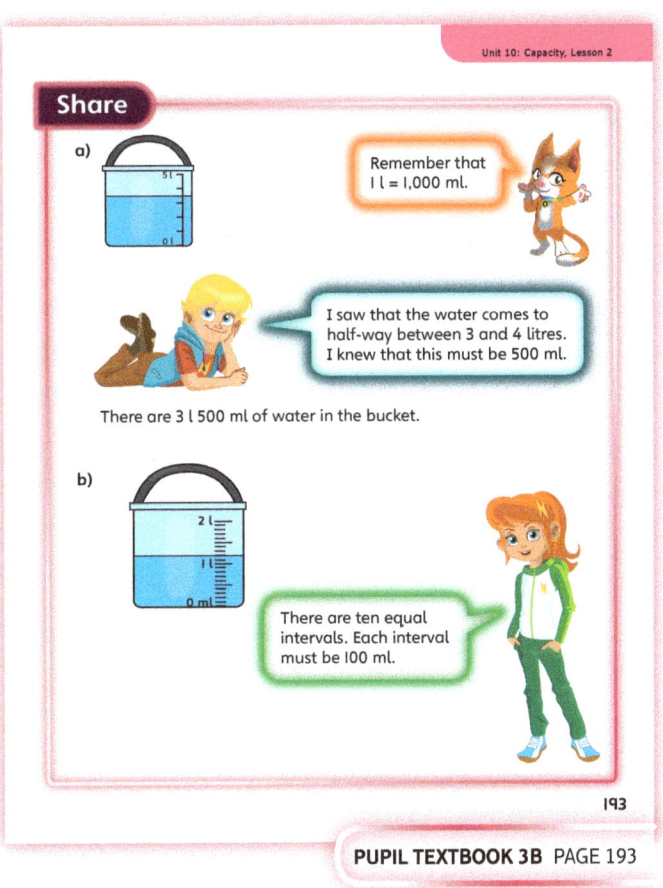

PUPIL TEXTBOOK 3B PAGE 193

Think together

WAYS OF WORKING Whole class teacher led (I do, We do, You do)

ASK

- Question ❶: *What do you need to know in order to work out the value of the intervals?*
- Question ❷: *How do you work out amounts when the liquid is between intervals?*
- Question ❸: *How are the scales the same? How are they different?*

IN FOCUS In question ❸, children are given only 0 ml and various maximum values. Encourage children to start by working out what the intervals represent on each scale. On the first scale, each interval represents 500 ml so there is 1 l of water in the jug. On the second scale, the top mark is 1 l, so it is clear that this jug does not contain 1 l 400 ml. On the third scale, each interval represents 200 ml. This one contains 1 l 400 ml.

STRENGTHEN Children could make the amounts using a range of measuring jugs and cylinders.

DEEPEN Look at question ❸. Can children draw other scales that will show 1 l 400 ml? Encourage them to make each scale increase by a different amount.

ASSESSMENT CHECKPOINT Question ❶ assesses whether children can accurately read a scale using mixed litres and millilitres. Use question ❷ to see if children can read scales where the liquid is between intervals. Question ❸ allows you to assess whether children can interpret a range of scales and reason about them in order to estimate a capacity.

ANSWERS

Question ❶ a): 1 l 500 ml

Question ❶ b): 1 l 800 ml

Question ❷ a): 1 l 200 ml

Question ❷ b): 3 l 400 ml

Question ❷ c): 1 l 50 ml

Question ❸: Children should explain that each interval is 200 ml, so the seventh interval is 1,400 ml which equals 1 litre 400 ml.

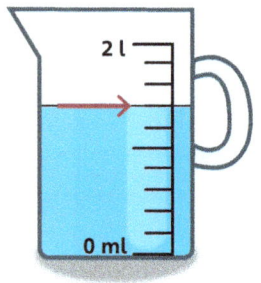

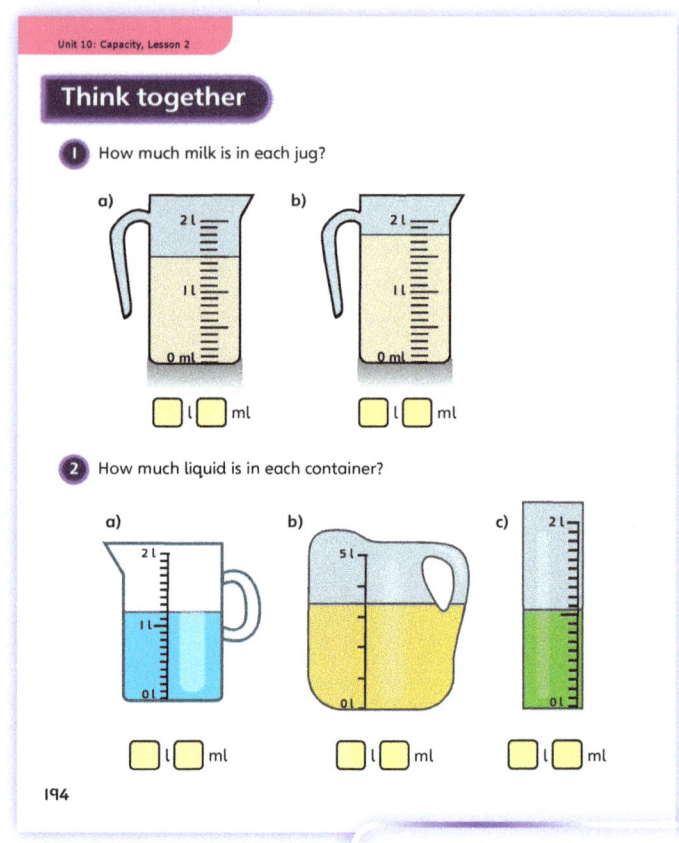

PUPIL TEXTBOOK 3B PAGE 194

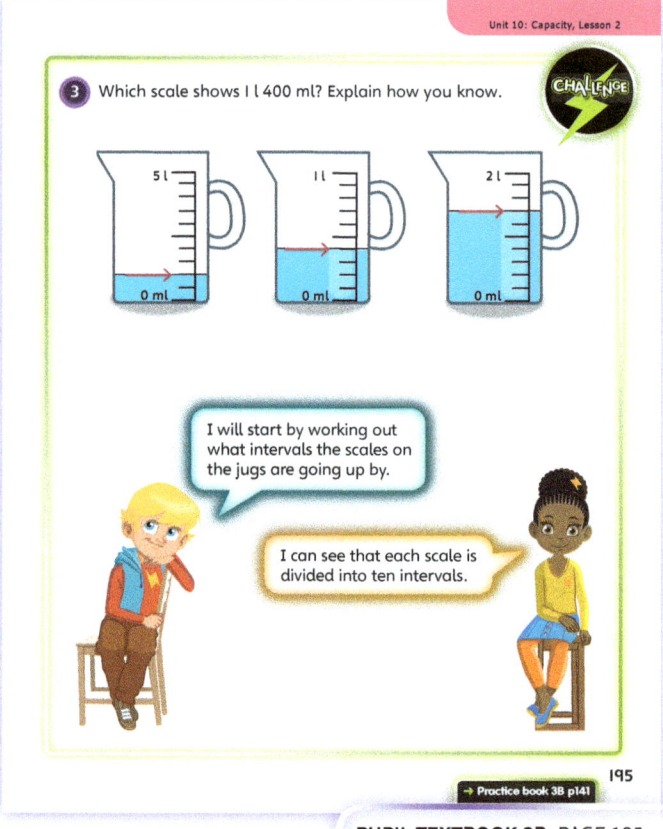

PUPIL TEXTBOOK 3B PAGE 195

Unit 10: Capacity, Lesson 2

Practice

WAYS OF WORKING Independent thinking

IN FOCUS The focus in questions ❶, ❷ and ❸ is on determining the intervals on a variety of scales and writing the amounts shown in mixed units and also in millilitres. Encourage children to realise that the intervals are likely to be 100 ml, 200 ml, 250 ml or 500 ml and that the number of unlabelled intervals (jumps between markers rather than the number of markers) will determine which it is (10 for 100 ml, 5 for 200 ml, 4 for 250 ml and 2 for 500 ml as 10 × 100 ml, 5 × 200 ml, 4 × 250 ml and 2 × 500 ml all equal 1,000 ml = 1 l).

In question ❶ c), the amount is 500 ml. Encourage children to write this as 0 l 500 ml.

STRENGTHEN Ask children to label some of the unlabelled intervals in between the whole litre intervals, to make the scales easier to read. They can practise counting in 100s, 200s, 250s and 500s to see which works for each scale. Ask: *Does the count result in the next recorded value?*

DEEPEN Ask children to create their own versions of question ❺ to try out on a partner, and to discuss how accurate they can be when there is no scale shown.

THINK DIFFERENTLY The focus of question ❹ is on measuring an amount in a slightly different context. Encourage children to think about what they need to do to answer the question (use Flo's comment: measure the amount in the large jug). They can do this by working out where the whole litre intervals are on the large jug, and what each small interval represents. Then, they need to interpret the scale on each of the three smaller jugs and identify which contains the same amount as the big jug.

ASSESSMENT CHECKPOINT Questions ❶, ❷ and ❸ will determine whether children are able to read a variety of scales to measure capacity in mixed units. Question ❹ will determine whether children can solve a word problem involving scales and mixed units. Question ❺ will determine whether children can apply their knowledge of number to look at the two end-points on a scale and make an estimation of a value between them.

ANSWERS Answers to the **Practice** part of the lesson can be found in the *Power Maths* online subscription.

Reflect

WAYS OF WORKING Independent thinking

IN FOCUS Children should be aware that, as the number of intervals changes, the amount that they represent also changes. Responses should also include the unit of measurement.

ASSESSMENT CHECKPOINT Is it clear from children's responses that they understand the use of intervals on a scale? Have they also included 0 l and 1 l at either ends of the scale?

ANSWERS Answers for the **Reflect** part of the lesson can be found in the *Power Maths* online subscription.

After the lesson

- Are children confident in reading a range of scales?
- How could you further support those children who are not secure in measuring amounts using mixed units?
- Can you provide practical opportunities to reinforce children's learning in other curriculum areas?

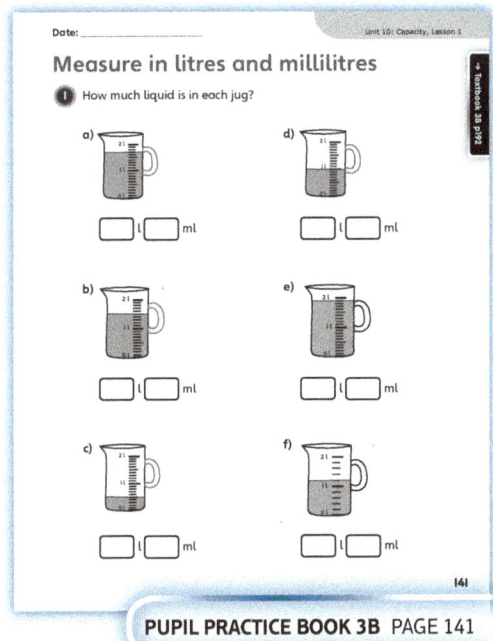

PUPIL PRACTICE BOOK 3B PAGE 141

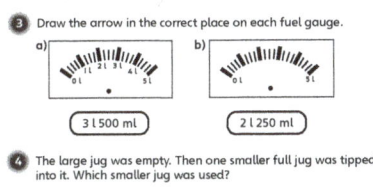

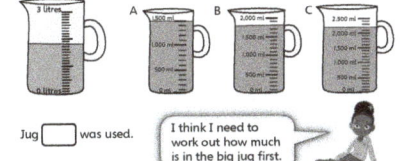

PUPIL PRACTICE BOOK 3B PAGE 142

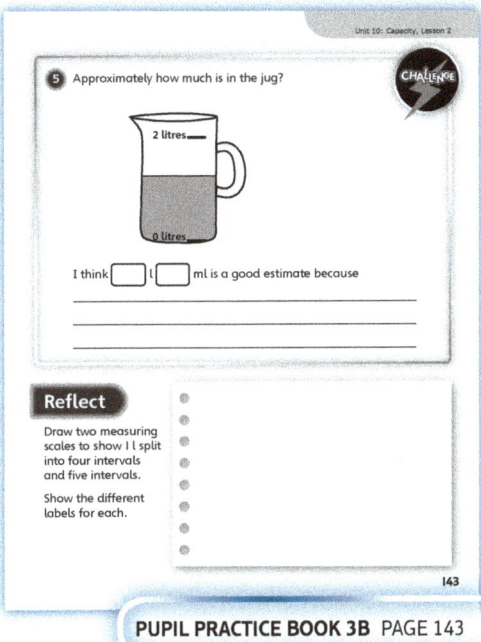

PUPIL PRACTICE BOOK 3B PAGE 143

233

Unit 10: Capacity, Lesson 3

Equivalent capacities and volumes (litres and millilitres)

Learning focus

In this lesson, children will continue to learn how to convert between litres and millilitres, including mixed units, in the context of real-life scenarios.

Before you teach

- Can you make links to previous learning on converting measurements?
- Are there any children who may need support in applying their knowledge of place value?
- Do children know that 1,000 ml = 1 litre?

NATIONAL CURRICULUM LINKS

Year 3 Measurement

Measure, compare, add and subtract: lengths (m/cm/mm); mass (kg/g); volume/capacity (l/ml).

ASSESSING MASTERY

Children will know that 1 litre is the same as 1,000 millilitres and that $\frac{1}{2}$ litre equals 500 millilitres. Children will be able to do simple conversions of measurements between litres and millilitres, and millilitres and litres in the context of real-life word problems.

COMMON MISCONCEPTIONS

If children are not fully secure in their understanding of the place value of 3-digit numbers, they may make errors when converting. Ask:

- *How do you write the number one hundred and twenty? Three hundred and five?*

STRENGTHENING UNDERSTANDING

Support children by using bar models or fraction strips to support fraction conversions of 1 litre.

GOING DEEPER

Provide children with a blank scale, with the bottom labelled 0 l and the top labelled 10 l. Write a variety of mixed measurements greater than 1 l on the board and ask them to plot the measurements on their scale.

KEY LANGUAGE

In lesson: capacity, litres (l), millilitres (ml), scale, bar model, number line, place value, convert, approximately, thousand

Other language to be used by the teacher: mixed unit, measure, measurement, compare, interval, partition

STRUCTURES AND REPRESENTATIONS

Bar model, number line, part-whole model, fraction strips

RESOURCES

Optional: base 10 equipment, capacity measuring equipment, place value columns, number lines, milk containers with a capacity greater than 1 litre

 In the eTextbook of this lesson, you will find interactive links to a selection of teaching tools.

Quick recap

Write several capacities greater than 1 litre on the board, such as: 5 l 350 ml, 2 l 895 ml, 4 l 49 ml. Ask children to draw part-whole models to partition these capacities into litres and millilitres.

Unit 10: Capacity, Lesson 3

Discover

WAYS OF WORKING Pair work

ASK

- Question 1 a): *What do you notice about the scale on the jug? Where would 1,000 ml be on the scale?*
- Question 1 b): *Can you use a bar model to help you work this out?*

IN FOCUS In question 1 a), encourage children to look carefully at the scale on the jug to remind themselves that 1 l is equivalent to 1,000 ml. In question 1 b), encourage children to draw a bar model to help them.

PRACTICAL TIPS Use large measuring containers to partially fill and read the levels. Ask children to give the measurements in mixed units.

ANSWERS

Question 1 a):

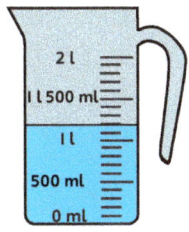

Question 1 b): $\frac{3}{4}$ litre = 750 ml

PUPIL TEXTBOOK 3B PAGE 196

Share

WAYS OF WORKING Whole class teacher led

ASK

- Question 1 a): *What do you think the prefix 'milli-' in millilitres means?*
- Question 1 b): *What does the bar model show? How does this help you work out how many millilitres are in $\frac{3}{4}$ l?*

IN FOCUS The focus of question 1 b) is on partitioning the measurement using a bar model. Children should be able to do the simple conversion of $\frac{3}{4}$ l to millilitres with pictorial support.

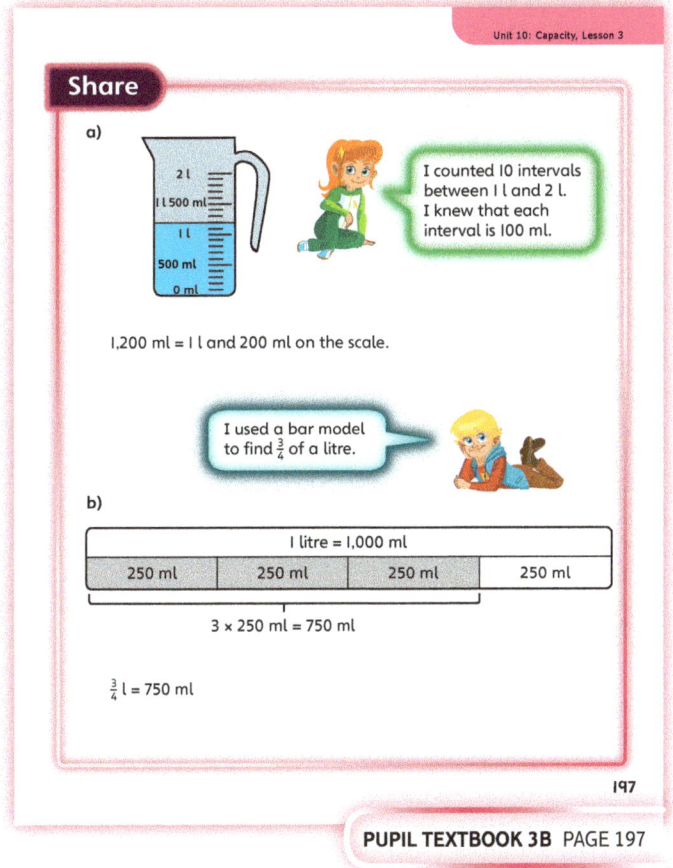

PUPIL TEXTBOOK 3B PAGE 197

235

Think together

WAYS OF WORKING Whole class teacher led (I do, We do, You do)

ASK
- Question ❶ a): *How many millilitres are in 1 litre?*
- Question ❶ b): *The bar model helps you partition 1 litre into five equal parts. How many millilitres are in each part? How many parts do you need?*
- Question ❷: *Would drawing bar models help you work out these conversions?*

IN FOCUS Question ❸ has a measurement of 1 l 50 ml. Some children may be confused about how to represent this in millilitres, as it requires a zero placeholder in the 100 ml column. Children could write the 1,000 ml and the 50 ml as a column addition, making sure to align the digits in the correct columns. This may help them to identify the need for the zero placeholder in the hundreds column.

STRENGTHEN In question ❷, encourage children to draw bar models to help them with their conversions from litres to millilitres. For all questions, some children may find it helpful to draw fraction strips to help with their fraction conversions of 1 litre.

DEEPEN Children roll dice to create a 3-digit number. Explain that this number represents a measurement in millilitres. Ask: *Can you convert it to litres without the support of a bar model? How would you work it out?*

ASSESSMENT CHECKPOINT Question ❶ shows whether children are able to find a fraction of a litre in millilitres using a bar model. Question ❷ shows whether children are able to convert simple fractions of a litre to millilitres without supporting bar models. Question ❸ shows whether children can securely use their knowledge of place value when converting capacity.

ANSWERS

Question ❶ a): 1,000 ml of orange juice are needed.

Question ❶ b): 200 ml of apple juice are needed.

Question ❶ c): 900 ml of lemonade are needed.

Question ❷ a): 250 ml

Question ❷ b): 500 ml

Question ❷ c): 600 ml

Question ❸: The baker will have 1,050 ml of liquid altogether.

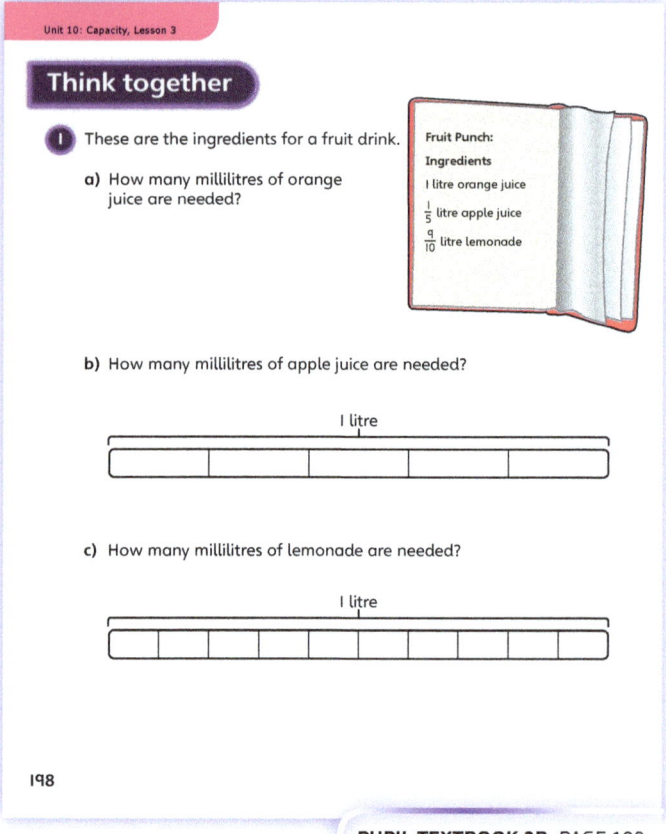

PUPIL TEXTBOOK 3B PAGE 198

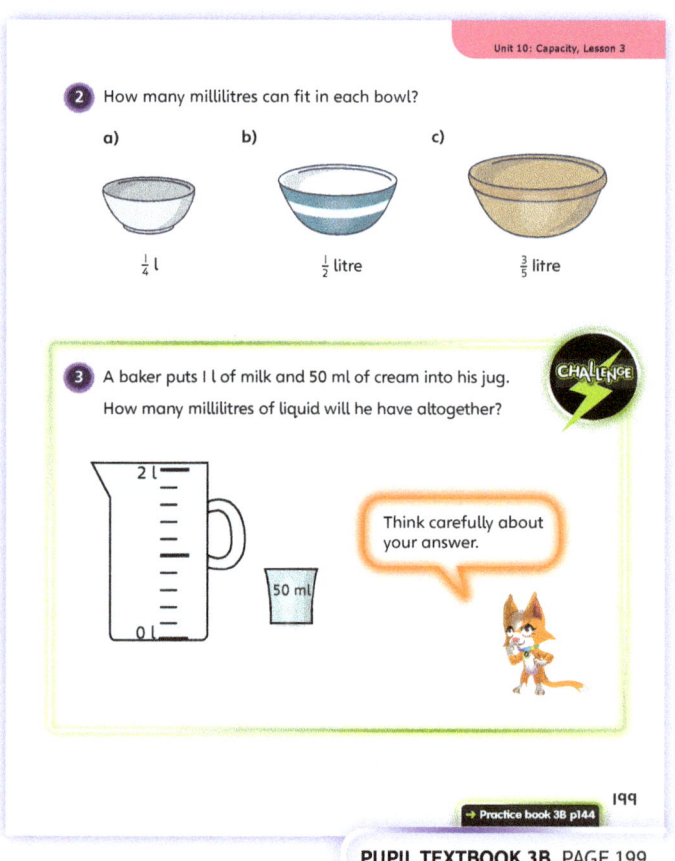

PUPIL TEXTBOOK 3B PAGE 199

Unit 10: Capacity, Lesson 3

Practice

WAYS OF WORKING Independent thinking

IN FOCUS This section consolidates understanding of key facts about millilitres and litres. Remind children that 1,000 ml = 1 l. Question ❶ uses a bar model to help children convert $\frac{1}{2}$ l into millilitres. Question ❷ focuses on children working out $\frac{1}{10}$ l as being equal to 100 ml. They should draw a bar model or fraction strip to help them. Question ❸ builds on their understanding from Question ❷, with children explaining that there are ten lots of 100 ml in 1 l. They should then then use this knowledge to work out how many 50 ml glasses there are. Encourage children to count up in 50s. Question ❹ asks children to write amounts of liquid in litres and millilitres. Question ❻ focuses on children finding a more complicated fraction of an amount.

STRENGTHEN Use bar models and number lines throughout to support children's understanding of common fraction conversions for 1 l. Drawing a number line that goes up in 100s and a bar model underneath will help children see that $\frac{1}{10}$ l is equal to 100 ml.

DEEPEN Can children create their own conversion problems set in a real-life context? Children could also draw and label their own containers, shaded to show a level, and swap with a partner for them to write the amount. Children should be encouraged to use fractions, too.

THINK DIFFERENTLY Question ❺ requires children to use their reasoning when given a measurement that is more than 1 litre as a fraction.

ASSESSMENT CHECKPOINT Questions ❶ and ❷ will help you see if children know common fractions of 1 litre. Other questions build on this understanding.

ANSWERS Answers to the **Practice** part of the lesson can be found in the *Power Maths* online subscription.

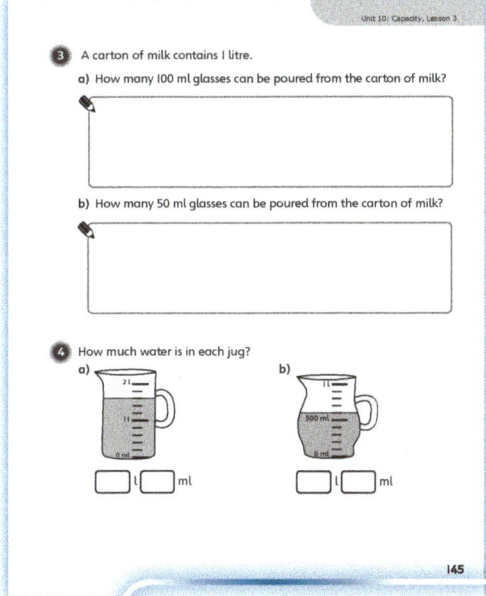

PUPIL PRACTICE BOOK 3B PAGE 144

PUPIL PRACTICE BOOK 3B PAGE 145

Reflect

WAYS OF WORKING Independent thinking

IN FOCUS Children write down the fractions of a litre that they know. For example, they may know that $\frac{1}{2}$ l is equal to 500 ml and $\frac{1}{4}$ l is equal to 250 ml. Encourage them to look back in the **Textbook** and also the **Practice** part of this lesson to remind them of other ones. Discuss with children how they might find more complicated fractions of amounts.

ASSESSMENT CHECKPOINT Check that children know $\frac{1}{2}$ l, $\frac{1}{4}$ l and $\frac{1}{10}$ l conversions to ml.

ANSWERS Answers for the **Reflect** part of the lesson can be found in the *Power Maths* online subscription.

PUPIL PRACTICE BOOK 3B PAGE 146

After the lesson ⏸

- Children should be confident that 1 litre = 1,000 ml.
- Children know simple fractions of a litre conversions (for example, $\frac{1}{2}$ l = 500 ml, $\frac{1}{4}$ l = 250 ml, $\frac{1}{5}$ l = 200 ml and $\frac{1}{10}$ l = 100 ml).

237

Unit 10: Capacity, Lesson 4

Compare capacity and volume

Learning focus

In this lesson, children will learn to compare capacities by first comparing the number of litres, then the number of millilitres. Children will also apply their knowledge of converting when comparing capacities given in different units.

Before you teach

- Are children secure at reading capacities of mixed units on a range of scales?
- What practical opportunities can you incorporate in the lesson?
- Are there any children who may need support in place value in order to make comparisons?

NATIONAL CURRICULUM LINKS

Year 3 Measurement

Measure, compare, add and subtract: lengths (m/cm/mm); mass (kg/g); volume/capacity (l/ml).

ASSESSING MASTERY

Children can compare and order a range of capacities. Children will be able to make simple conversions of measures so that they are all in the same units before making comparisons.

COMMON MISCONCEPTIONS

Children may not consider the number of litres before looking at the number of millilitres. Be sure to emphasise that they must first look at how many litres there are in each measure. Make links to comparing numbers, where you look at the largest place value first. Ask:

- *How many litres is in this measure? And in this one?*

STRENGTHENING UNDERSTANDING

Start by comparing measures that are just in litres, to ensure that children focus on litres first. Then move to comparing two capacities, where one is in whole litres and the other is in mixed litres and millilitres. This will help children see the importance of looking at the litres first.

GOING DEEPER

Children play a game in pairs or small groups. Each child rolls four dice to generate the largest 4-digit number. This gives them a capacity in millilitres. They each convert their capacity to mixed units and order them from largest to smallest. The child who generates the largest capacity in each round wins a point. Children could have other rules: generate a capacity between 2 l and 4 l, or generate capacities as close to 3 l as possible.

KEY LANGUAGE

In lesson: order, compare, litres (l), millilitres (ml), capacity, more, less, most, least, greater, greatest, greater than (>), less than (<), half ($\frac{1}{2}$)

Other language to be used by the teacher: place value

STRUCTURES AND REPRESENTATIONS

Bar model, number line

RESOURCES

Optional: dice, capacity measuring equipment

 In the eTextbook of this lesson, you will find interactive links to a selection of teaching tools.

Quick recap

Play a game where a child writes down a 3-digit number and the class get ten goes to try to work out the number using less than or greater than questions. For example, 'Is the number less than 200?'. After each turn, children can have a guess. Use a number line on the board to record the strategy and guesses.

Unit 10: Capacity, Lesson 4

Discover

WAYS OF WORKING Pair work

ASK

- Question 1 a): *What should you look at first when comparing these measurements?*
- Question 1 b): *What do you need to do before you can compare all these measurements?*
- Question 1 b): *Which holds the most and which holds the least of all four jugs?*

IN FOCUS Questions 1 a) and b) cover the key aspects of comparing two or more capacities. As with place value, the litres have the highest value, so this is the first number to consider. If the litres are the same, then the number of millilitres becomes important. When capacities are in different forms, then one or more of the measures will have to be converted so that they are all the same. Encourage children to verbalise their reasoning when tackling these questions to ensure that they go through the correct process.

PRACTICAL TIPS Practical experience of using pairs of similar sized containers (where one has measurements in mixed measures and the other just millilitres) would be a helpful way for children to see that, for example, $\frac{1}{2}$ l is the same as 500 ml.

ANSWERS

Question 1 a): 950 ml > 450 ml
There is more orange juice.

Question 1 b): Greatest to smallest: water, orange juice, apple juice, lemon squash

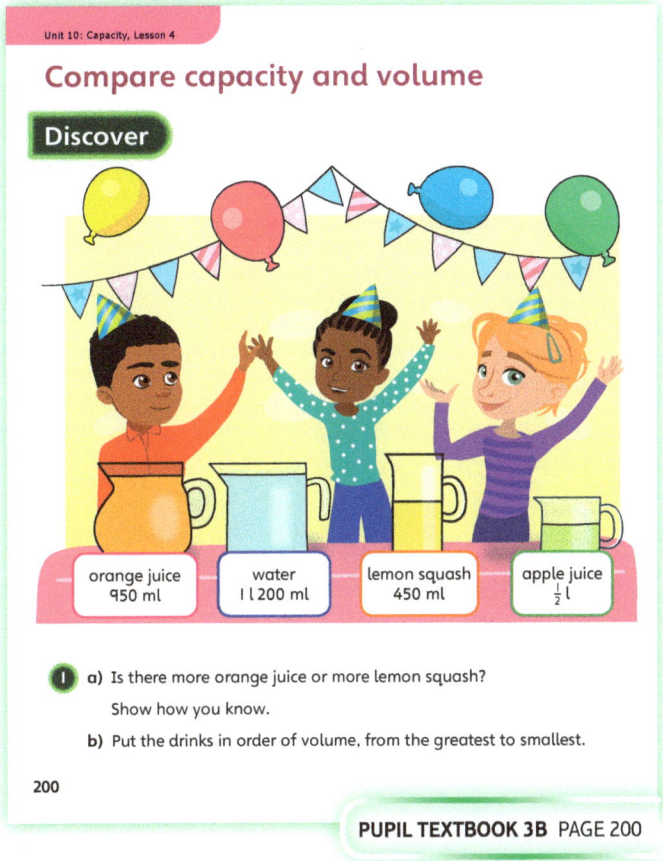

PUPIL TEXTBOOK 3B PAGE 200

Share

WAYS OF WORKING Whole class teacher led

ASK

- Question 1 b): *How can you easily tell which drink has the greatest volume?*
- Question 1 b): *Which is easier, to convert millilitres into mixed litres and millilitres or to convert mixed litres and millilitres into millilitres?*

IN FOCUS These questions highlight the importance of comparing the number of litres first, because they are the larger unit. Question 1 b) encourages children to convert $\frac{1}{2}$ l to 500 ml, making a direct comparison easier.

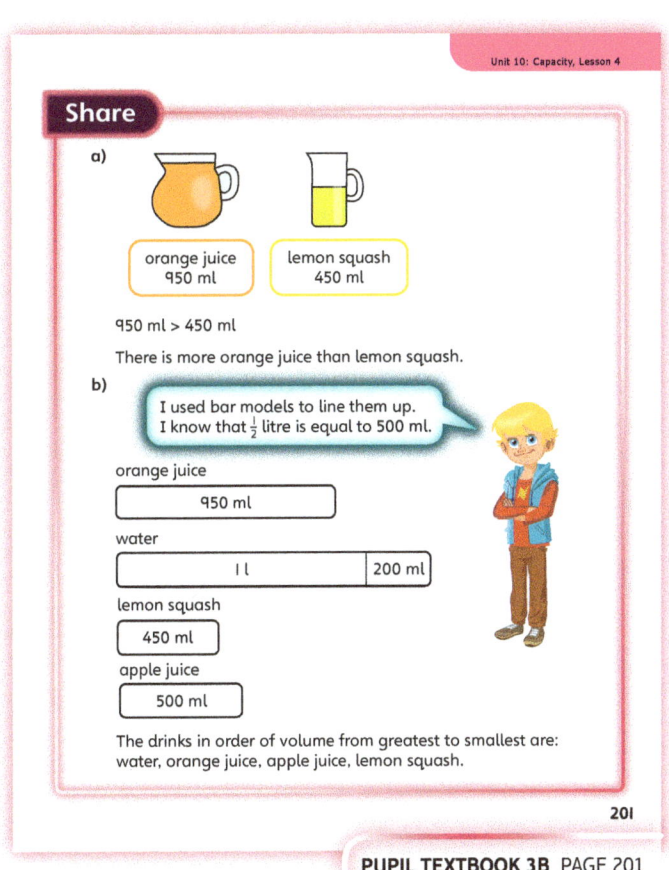

PUPIL TEXTBOOK 3B PAGE 201

Unit 10: Capacity, Lesson 4

Think together

WAYS OF WORKING Whole class teacher led (I do, We do, You do)

ASK
- Question ❶: *What do you have to do before you can compare these capacities?*
- Question ❷: *Is it easy to see which one will hold the least? Why?*
- Question ❸: *Can you draw a picture to help you solve this?*

IN FOCUS Question ❸ requires children to carry out a multiplication to solve the problem. They could use repeated addition in order to do this. They will then need to carry out a conversion to compare the capacities. Encourage children to draw or make notes to keep track of their thinking. The fact that the capacities are the same may make some children question whether they calculated correctly.

STRENGTHEN Use capacity equipment so children can physically measure and compare the capacities in the questions.

DEEPEN For question ❸, challenge children to think of other ways to divide 4 l of liquid into smaller containers. Children might suggest, for example, ten cups of 400 ml each, or five jugs with 800 ml each.

ASSESSMENT CHECKPOINT Question ❶ will determine whether children can make simple conversions from litres to millilitres to compare two capacities. Question ❷ will determine whether children can convert, compare and order capacities that have different numbers of litres. Question ❸ will determine whether children can apply their knowledge of multiplication in order to solve a multi-step problem comparing capacities.

ANSWERS

Question ❶ a): Car A has the greater fuel capacity.

Question ❶ b): 2 l 650 ml is less than 2 l 900 ml because 2,650 < 2,900 or 650 < 900.

Question ❷: Smallest to greatest: C, A, D, B

Question ❸: Reena and Max both buy 4 litres. They have the same amount of fizzy drinks.

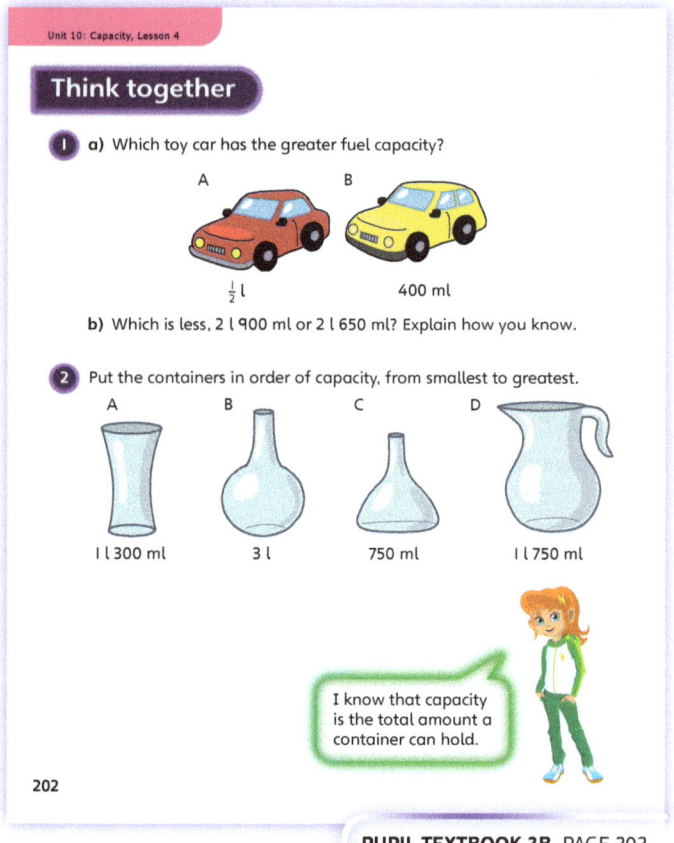

PUPIL TEXTBOOK 3B PAGE 202

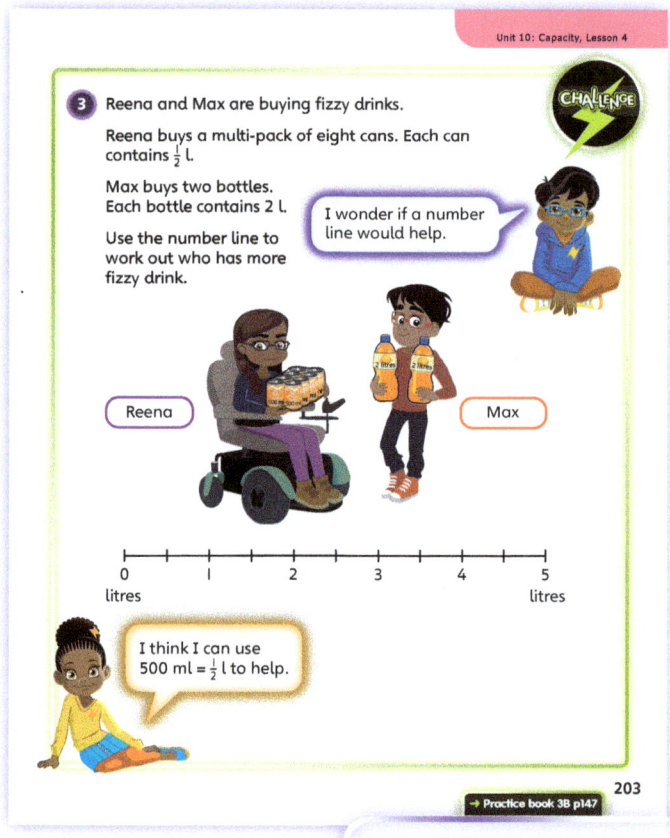

PUPIL TEXTBOOK 3B PAGE 203

Practice

WAYS OF WORKING Independent thinking

IN FOCUS In question ❻, some children may just guess, thinking that the tall, thin container is fuller than the short wide container. But the total capacity of the wide one holds 2 l and it is at least $\frac{3}{4}$ full which is 1 l 500 ml (1,500 ml); whereas the total capacity of the taller container is only 1,500 ml (1 l 500 ml) and it is not full.

STRENGTHEN Children can make comparisons by using capacity equipment. First, ask children to work in pairs and give each pair two different containers. Ask them to pour 1 or 2 litres of water into them. Ask: *Which looks as if it has more liquid?* Provide two 2 litre jugs, one scale should be in in litres and the other in millilitres. Ask children to pour the same amount into each jug and read and record the amounts. Children then should then pour more water into one container and record each amount as a comparison: $a > b$.

DEEPEN Children work in pairs. Give each pair a scale from 0 l to 7 l, marked in 100 millilitre intervals with whole litres labelled. Children then take it in turns to roll two dice to generate two numbers. They choose one for the litres and the other for the hundreds of millilitres (for example, 3 and 4 could give 3 l 400 ml or 4 l 300 ml) and plot their capacity on the scale. The first child to get three numbers in a row wins.

THINK DIFFERENTLY Question ❹ combines learning from this and previous lessons. Children need to read the amounts from a variety of vertical line scales and order the amounts. Some may need support reading the intervals on each scale.

ASSESSMENT CHECKPOINT Children should compare by first looking at the number of litres in mixed units, or by using place value to 4 digits for millilitres. Children should realise that they may need to convert units, so that all the measurements are in the same unit of measure. Children can read off a scale where not all intervals are labelled.

ANSWERS Answers to the **Practice** part of the lesson can be found in the *Power Maths* online subscription.

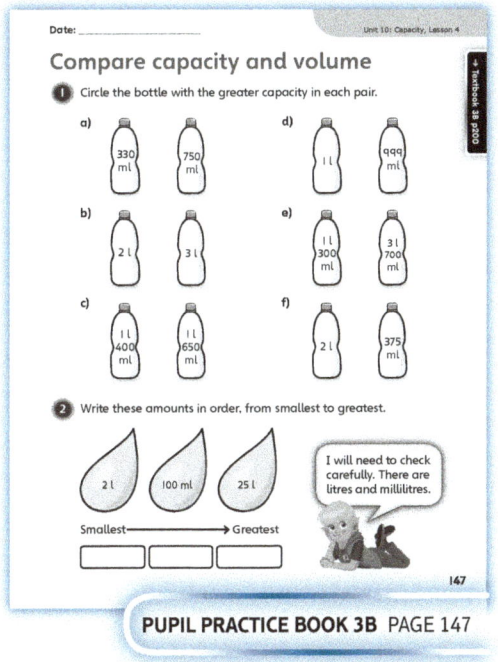

PUPIL PRACTICE BOOK 3B PAGE 147

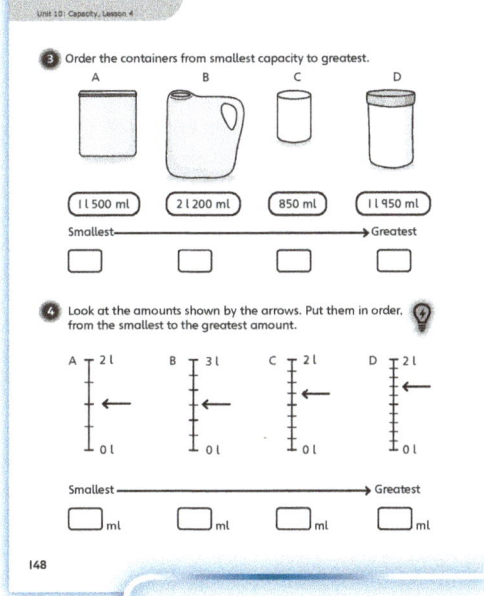

PUPIL PRACTICE BOOK 3B PAGE 148

Reflect

WAYS OF WORKING Independent thinking

IN FOCUS This question focuses on comparing an amount in litres and an amount in millilitres. It addresses the misconception where children just look at the numbers but not the units of measure. Discuss answers from children to make sure everyone understands the reason that 1 l is greater than 750 ml. You might want to convert 1 l to 1,000 ml to further explain this fact, or look at measuring jugs showing 1 l to be higher than 750 ml.

ASSESSMENT CHECKPOINT Check that children are not comparing amounts just using the numbers, but instead paying attention to the units of measurement.

ANSWERS Answers for the **Reflect** part of the lesson can be found in the *Power Maths* online subscription.

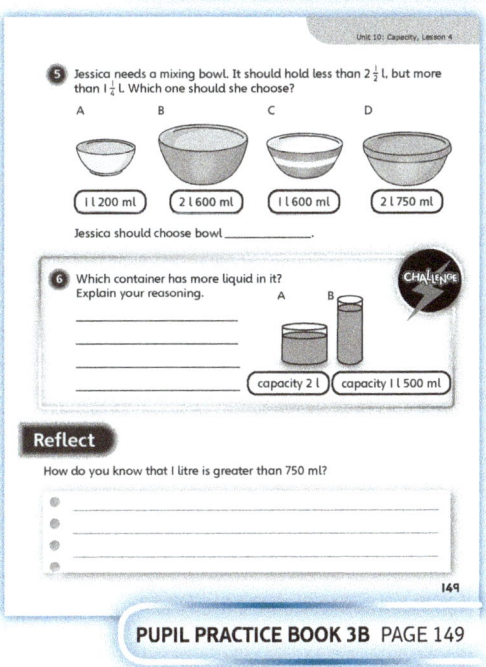

PUPIL PRACTICE BOOK 3B PAGE 149

After the lesson

- Were there any gaps in understanding that hindered children's ability to compare capacities?
- Were children able to convert to the same units in order to compare capacities?
- Were children able to draw on their knowledge from other units on measure to help them in this lesson?

Unit 10: Capacity, Lesson 5

Add and subtract capacity and volume

Learning focus

In this lesson, children will know that 1 litre is the same as 1,000 millilitres and that $\frac{1}{2}$ litre equals 500 millilitres. Children will be able to convert measurements between litres and millilitres within a litre, but focusing on fraction of litre conversions.

Before you teach

- Do children know that 1,000 ml = 1 litre?
- Can children find a simple fraction of an amount?
- Can children draw simple bar models?

NATIONAL CURRICULUM LINKS

Year 3 Measurement

Measure, compare, add and subtract: lengths (m/cm/mm); mass (kg/g); volume/capacity (l/ml).

ASSESSING MASTERY

Children can convert simple fractions of a litre into equivalent amounts in millilitres. As a minimum, children should be able to convert $\frac{1}{2}$, $\frac{1}{4}$ and $\frac{1}{10}$ of a litre confidently.

COMMON MISCONCEPTIONS

Children may think that l litre = 100 ml. Ask:
- *What do you think the prefix 'milli-' means in 'millilitre'? How many millillitres are in 1 litre?*

Children may think that $\frac{1}{2}$ of a litre = 200 ml as children focus on the numbers as opposed to the actual fraction of an amount. Ask:
- *Can you draw a bar model or a number line to show the fraction in millilitres?*

STRENGTHENING UNDERSTANDING

Children may struggle to find fractions of 1,000 ml. Use a bar model and a number line to support children with their understanding. For example, draw a number line with two intervals between 0 and 1,000 and ask them what is in the middle. They often find this easier than dividing 1,000 by 2. Explain how the number line helps them find $\frac{1}{2}$ of 1,000. Repeat for fractions such as $\frac{1}{4}$ and $\frac{1}{10}$ by dividing a number line into 4 and 10 intervals.

GOING DEEPER

Depending on children's confidence, you may want to find different fractions of an amount or convert measures such as 1 litre 200 ml into 1,200 ml. Note that children have not explored 4-digit numbers in depth and so, if doing this, keep to simple amounts.

KEY LANGUAGE

In lesson: add, subtract, total, difference, capacity, millilitres (ml), litres (l), scale, column addition, counting on, subtraction, greater than (>), less than (<), total, number bonds to 1,000

Other language to be used by the teacher: number bonds to 10 or 100, mixed measure, pattern

STRUCTURES AND REPRESENTATIONS

Bar model, column addition, column subtraction, number line

RESOURCES

Optional: capacity measuring equipment, base 10 equipment

 In the eTextbook of this lesson, you will find interactive links to a selection of teaching tools.

Quick recap

Write two additions (175 + 362 and 240 + 300) and two subtractions (700 − 175 and 500 − 200) on the board. Ask children to work out the answers. Discuss and compare the methods that children used (for example, column addition, number line, mental methods).

Discover

WAYS OF WORKING Pair work

ASK
- Question 1 a): *Do you need to do an addition or subtraction? How do you know? How much liquid is in each of the beakers?*
- Question 1 b): *Do you need to do an addition or subtraction? What methods do you know?*

IN FOCUS In question 1 a), bar models should help children to answer the question, rather than adding the amounts. As both of Professor Smith's amounts are less than 500 ml, their total must be less than 1,000 ml. Children may use column methods or mental methods to find the answer. Question 1 b) requires children to realise they are finding the difference and so have to do a subtraction. As both numbers are in the 100s, encourage children to find the difference using a mental method (for example, by subtracting 100s or counting on) rather than using a more inefficient column method for this question.

PRACTICAL TIPS Allow children to act out the scenario. They can physically measure out the amounts and pour them into a clearly labelled container to find the totals. Children could also use the scale as a counting-on number line for the subtraction.

ANSWERS

Question 1 a): Professor Smith has 600 ml in total.

Question 1 b): There is 200 ml more liquid in A than B.

PUPIL TEXTBOOK 3B PAGE 204

Share

WAYS OF WORKING Whole class teacher led

ASK
- Question 1 a): *How much liquid is in each measuring beaker? Can you work this out without doing a column addition? Could you use a number line or do this in your head?*
- Question 1 b): *How much liquid is in each measuring beaker? What methods can you use to subtract?*

IN FOCUS The focus of questions 1 a) and b) is on discussing the language (for example, total and difference) and how these words help children work out whether they need to add or subtract. Once decided, the next focus is on looking at different methods that children could use to find the answers. In question 1 a), a column addition has been used, but ask children if any of them could do it in their head without the need to set it out. In question 1 b), children discuss Flo's explanation and focus on non-column methods to find the answer. Share with the class different methods that can be used to get the answer, asking which is the quickest and perhaps most efficient.

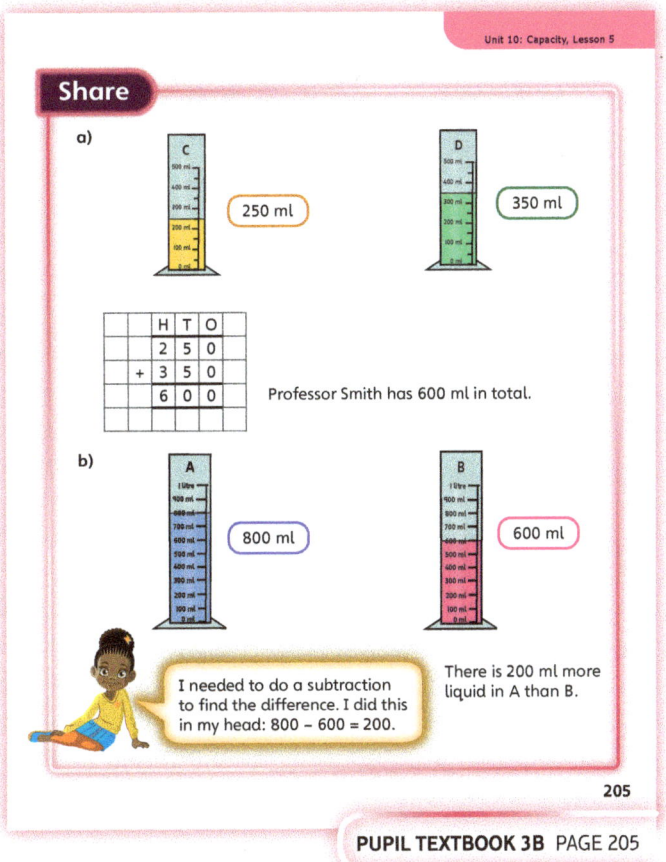

PUPIL TEXTBOOK 3B PAGE 205

Unit 10: Capacity, Lesson 5

Think together

WAYS OF WORKING Whole class teacher led (I do, We do, You do)

ASK
- Question ❶: *Which addition facts could help you?*
- Question ❷: *Do you need to convert between units? Why not?*

IN FOCUS Question ❸ involves crossing the litre boundary when adding and also when subtracting. Discuss some strategies for adding the amounts, including using number bonds to 1,000 then adding on, or adding the litres and millilitres separately and converting the units. Ask: *Which do you find easier?* Steps for this question are modelled using bar models, but number lines marked as scales could also be used.

STRENGTHEN Model the additions and subtractions using a part-whole model and a bar model. It may support some children's understanding of adding and subtracting capacities to explore the questions practically.

DEEPEN Set children word problems to deepen learning. Ask: *What three quantities could you add together to make a total of 2 l 200 ml? Each quantity must be a multiple of 100 ml. How many different combinations can you find?*

ASSESSMENT CHECKPOINT Question ❶ shows whether children are able to use number bonds to 1,000 in the context of capacity. Question ❷ shows whether children can add mixed units by partitioning the litres and millilitres. Question ❸ shows whether children can add and subtract capacities that involve crossing litre boundaries.

ANSWERS

Question ❶ a): 400 ml

Question ❶ b): 800 ml

Question ❶ c): 450 ml

Question ❶ d): 650 ml

Question ❷ a): 1 l + 2 l = 3 l
400 ml + 250 ml = 650 ml
Total capacity = 3 l 650 ml

Question ❷ b): 750 ml is left in the jug.

Question ❸: 400 ml + 800 ml = 1 l 200 ml
1 l + 1 l 200 ml = 2 l 200 ml
3 l = 2 l 200 ml + 800 ml
800 ml is needed to fill the jug.

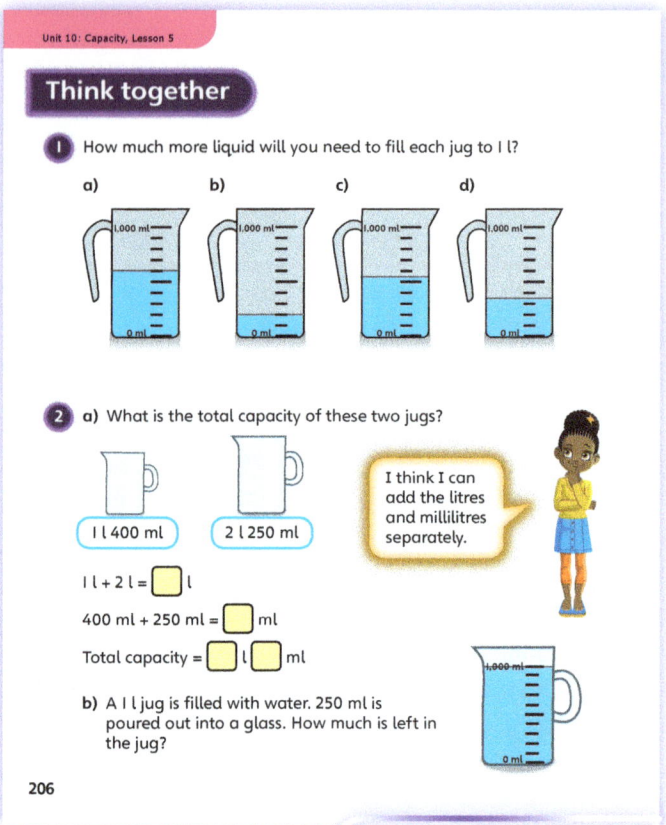

PUPIL TEXTBOOK 3B PAGE 206

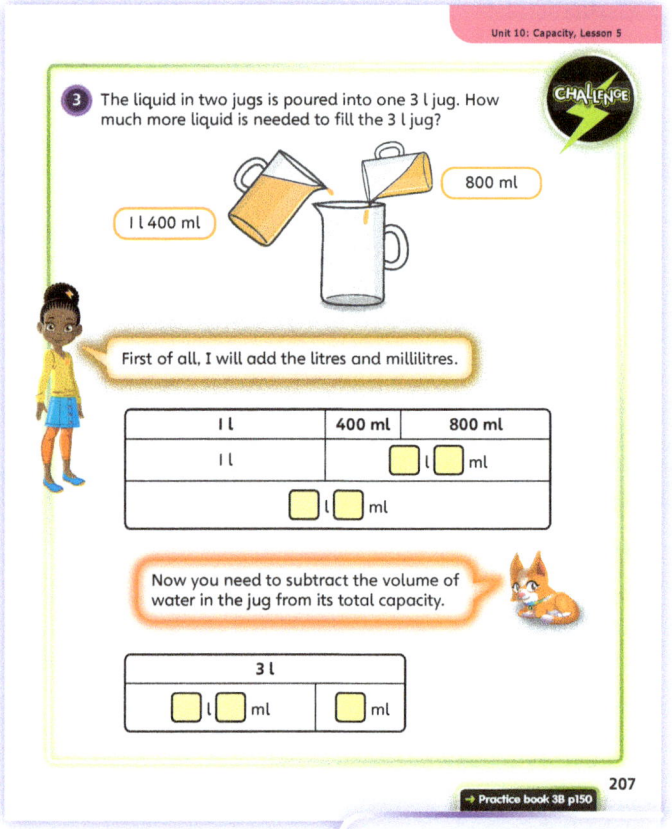

PUPIL TEXTBOOK 3B PAGE 207

Unit 10: Capacity, Lesson 5

Practice

WAYS OF WORKING Independent thinking

IN FOCUS Question ❺, the challenge question, is a multi-step problem. Some children may need to physically act it out, then draw bar models or number lines to realise they need to add the first two amounts and subtract that from 2 l (2,000 ml) to then work out what must be in cylinder C.

STRENGTHEN For support, children should use bar models or number lines to represent each problem. They can use base 10 equipment to help calculate with the 1,000 cubes representing litres. Practise multiple of 10 and multiple of 100 number bonds to 1,000.

DEEPEN Ask children to develop their own addition and subtraction two-step problems. Their problems need to require the crossing of a litre boundary.

THINK DIFFERENTLY Question ❹ is an abstract word problem, in which children need to extract the key information that will tell them whether they need to add or subtract. The addition that they need to do crosses the litre boundary.

ASSESSMENT CHECKPOINT Questions ❶, ❷ and ❸ assess whether children can add or subtract amounts without crossing a litre boundary. Questions ❹ and ❺ assess whether children can solve problems that involve crossing litre boundaries.

ANSWERS Answers to the **Practice** part of the lesson can be found in the *Power Maths* online subscription.

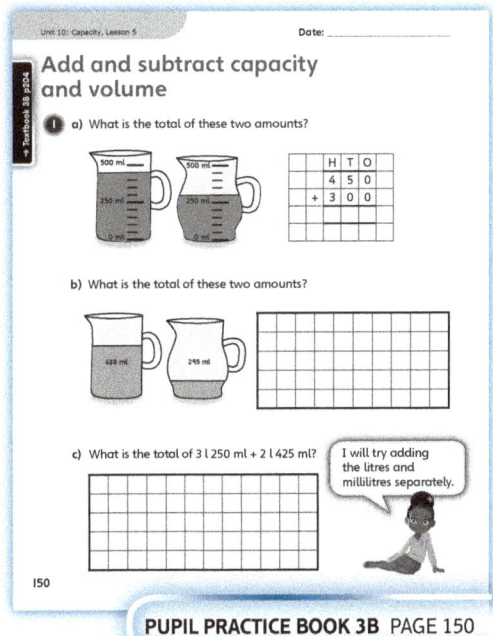

PUPIL PRACTICE BOOK 3B PAGE 150

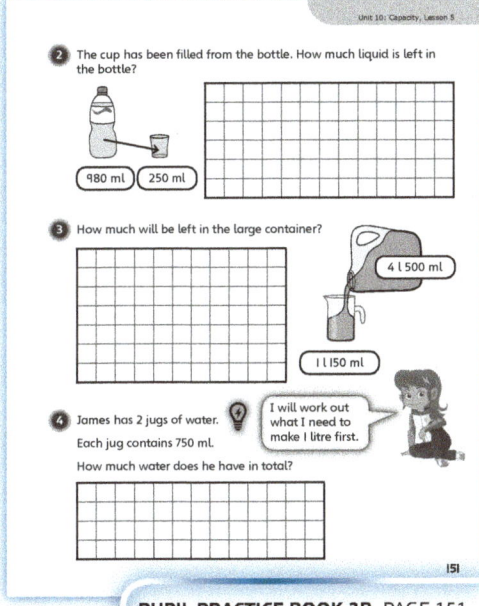

PUPIL PRACTICE BOOK 3B PAGE 151

Reflect

WAYS OF WORKING Independent thinking

IN FOCUS Children are asked to explain how they would add together two amounts, given in different units of measure.

ASSESSMENT CHECKPOINT Can children identify strategies such as converting between units, adding the litres and millilitres separately, or using knowledge of number bonds to 1,000?

ANSWERS Answers to the **Reflect** part of the lesson can be found in the *Power Maths* online subscription.

After the lesson

- Were there any gaps in children's knowledge and application of place value, number bonds or partitioning that acted as barriers to their learning?
- Were you able to make the links between this lesson and previous lessons in this unit?

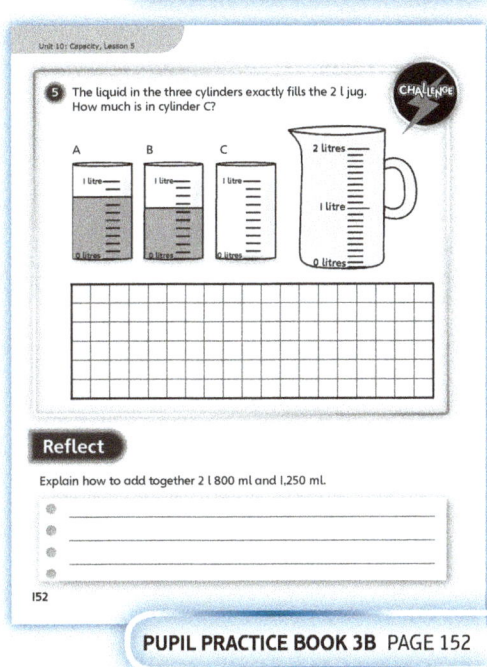

PUPIL PRACTICE BOOK 3B PAGE 152

Unit 10: Capacity, Lesson 6

Problem solving – capacity

Learning focus
In this lesson, children will apply their learning from this unit to solve problems involving all four operations.

Before you teach
- Are there any children who may need support with calculating?
- How can you model the problems for children to help them identify the calculations required?

NATIONAL CURRICULUM LINKS

Year 3 Measurement

Measure, compare, add and subtract: lengths (m/cm/mm); mass (kg/g); volume/capacity (l/ml).

ASSESSING MASTERY

Children can carry out multi-step capacity problems involving all four operations. Children will be able to apply their knowledge of multiplication, division, subtraction and addition, as well as converting between litres and millilitres.

COMMON MISCONCEPTIONS

Children may need support to identify what calculations they need to do, particularly if it is a multi-step problem. Encourage children to draw a picture of the problem and use bar models in order to identify what operations are required. Ask:
- *What is the question asking you to do? What do you need to do first? What could you draw to show what the problem is asking?*

STRENGTHENING UNDERSTANDING

Where possible, represent the problem using concrete apparatus. Encourage children to talk the problem through with a partner or an adult and to say the problem in their own words. It can help to model children's thinking process by reasoning aloud whilst solving the problem with them.

GOING DEEPER

Ask children to adapt some of the problems in order to change the type of operations required. For example, they could adapt a problem that requires addition, so that it requires subtraction instead.

KEY LANGUAGE

In lesson: capacity, millilitres (ml), litres (l), add (+), subtract (–), multiply (×), divide (÷), difference, total, convert, equivalent

Other language to be used by the teacher: calculation, scale, division

STRUCTURES AND REPRESENTATIONS

Bar model, number line, part-whole model

RESOURCES

Optional: capacity measuring equipment, base 10 equipment

 In the eTextbook of this lesson, you will find interactive links to a selection of teaching tools.

Quick recap

Check children's understanding of fractions of a litre, using a number line or scale to support them. Ensure children know the facts $\frac{1}{2}$ l = 500 ml, $\frac{1}{4}$ l = 250 ml, $\frac{1}{5}$ l = 200 ml and $\frac{1}{10}$ l = 100 ml.

Discover

WAYS OF WORKING Pair work

ASK
- Question 1 a): *What do you need to do to solve this?*
- Question 1 b): *What could you draw to show a picture of the problem?*
- Question 1 b): *What calculation do you need to do to solve the problem?*

IN FOCUS Both questions involve division, but question 1 a) is best modelled as a repeated addition, whereas question 1 b) is better suited to being modelled as a division. It is useful to draw bar models for both questions. Children need to draw on their knowledge of converting 1 litre to millilitres.

PRACTICAL TIPS You can support children by drawing a scale for 1 litre with five divisions. This will help children apply their learning from reading scales to solve the problem. Children could physically enact the questions by using a 200 ml measure and a litre jug.

ANSWERS

Question 1 a): Holly uses three 200 ml cartons of milk for the pancake recipe.

Question 1 b): There will be 200 ml of juice in each cup.

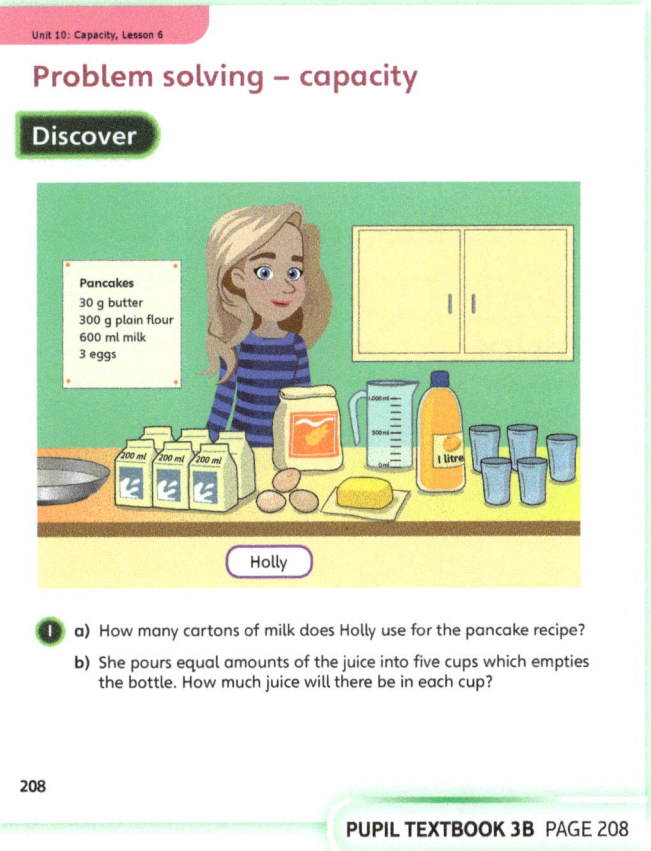

PUPIL TEXTBOOK 3B PAGE 208

Share

WAYS OF WORKING Whole class teacher led

ASK
- Question 1 a): *What addition facts did you use to find the solution?*
- Question 1 b): *What multiplication facts did you use to find the solution? Is there another way of finding the solution? How else could you share 1,000 ml equally between 2 cups? What about between 4 cups?*

IN FOCUS Known multiplication facts can be used to work out related division facts. These can be used to solve problems with much larger numbers. In question 1 b), draw attention to the relationship between 10 ÷ 5 = 2, and 100 ÷ 5 = 20 and therefore 1,000 ÷ 5 = 200. Discuss how to divide 100 between 2, 4 and 10, relating that to dividing 1,000 between 2, 4 and 10.

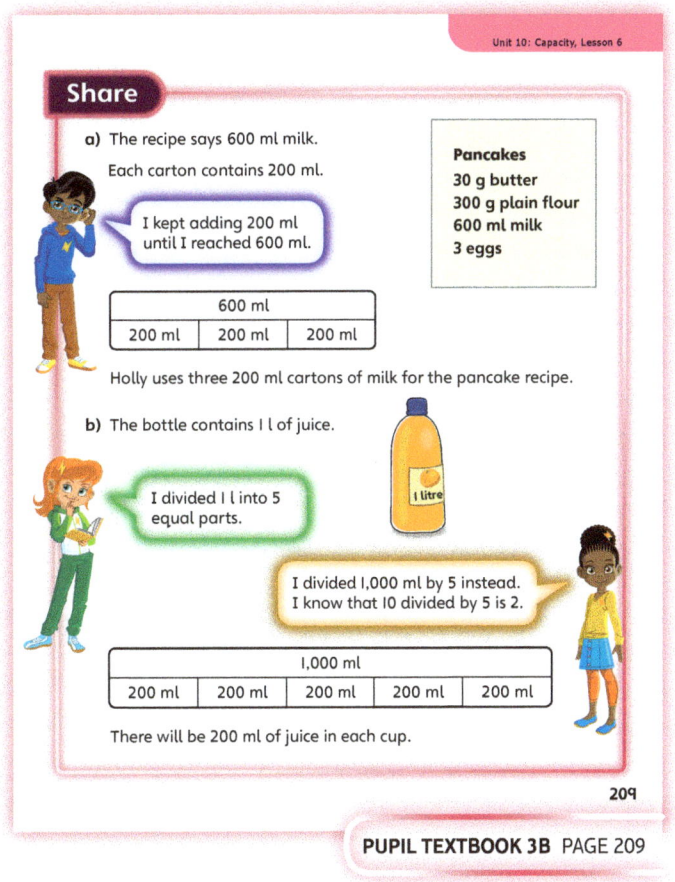

PUPIL TEXTBOOK 3B PAGE 209

Think together

WAYS OF WORKING Whole class teacher led (I do, We do, You do)

ASK

- Question ❶: *What addition or multiplication facts are going to help you?*
- Question ❸: *Have you solved it the same way as a partner? If not, which way do you think is easier?*
- Question ❸: *Can you draw a picture to help you solve the problem?*

IN FOCUS Question ❸ requires children to first convert from litres to millilitres and then carry out a division. Children may overlook that it is 2 l and only divide 1 l. They should remember that 250 ml is a quarter of 1,000 ml (from previous lessons), so they will get four cups from each litre. You could encourage children to draw a scale on a whiteboard for 2 l and to put in 250 ml divisions, so they can link the calculation to the learning they did on reading scales in previous lessons.

STRENGTHEN Ask children to draw pictures of the problems to help understand what is required. If possible, use equipment to replicate the problem, so children can see the calculations needed.

DEEPEN Can children adapt question ❸ so that the answer will be: 4 cups? 10 cups? 16 cups? 20 cups?

ASSESSMENT CHECKPOINT Can children solve word problems involving millilitres and litres, choosing the correct operation(s) and using their knowledge of the relationship between litres and millilitres?

ANSWERS

Question ❶ a): 180 ÷ 6 = 30 ml
Francesca uses 30 ml of shower gel each day.

Question ❶ b): 180 ml + 30 ml = 210 ml
Francesca does not have enough left for another day.

Question ❷: Francesca drinks 3 l 500 ml altogether in the two days.

Question ❸ a): Leon can fill 8 cups from the bottle.

Question ❸ b): If Leon fill 6 cups, 500 ml is left in the bottle.

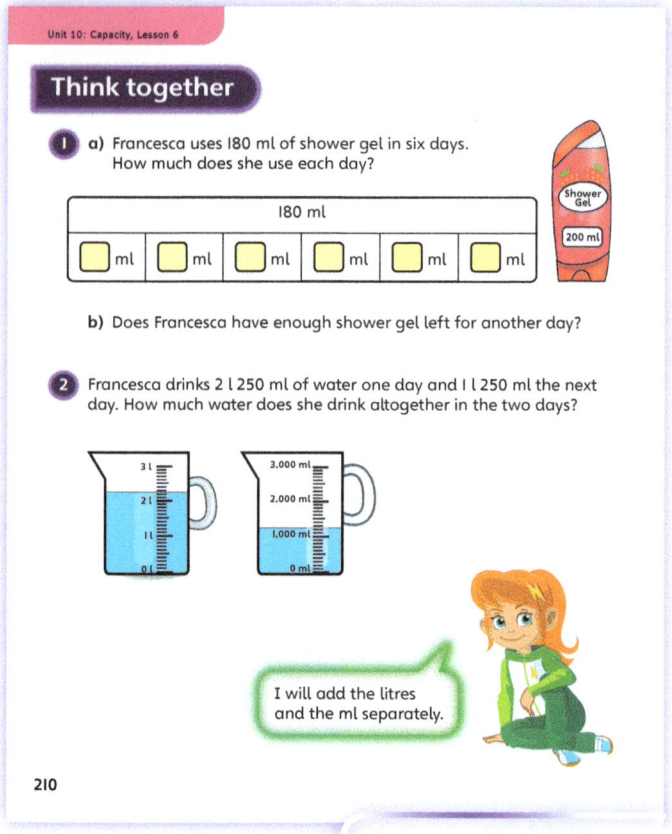

PUPIL TEXTBOOK 3B PAGE 210

PUPIL TEXTBOOK 3B PAGE 211

Unit 10: Capacity, Lesson 6

Practice

WAYS OF WORKING Independent thinking

IN FOCUS Questions from ③ onwards are two-step or multi-step problems with question ⑦ requiring a number of steps. Children may need support through diagrams. A common error is not working all the way through a question. For example, children may only do the first calculation or miss one of the calculations out. Encourage children to go back and read the problem after they have their answer, in order to check that they have answered it correctly.

STRENGTHEN Encourage children to talk through the problem with a partner. As they are talking it through, their partner can draw a picture of the problem. It may also help to read the problem to children and ask them to repeat it back in their own words. The use of bar models can aid children in identifying the calculations that are needed. You can adapt question ⑦ with fewer plants and/or fewer days.

DEEPEN Ask children to develop their own multi-step problems. You could challenge them further by specifying what operations children have to include in the problem. A good starting point is to write similar problems to those on these pages, with different contexts, values or operations.

THINK DIFFERENTLY Question ⑤ is a multi-step problem. Encourage children to look back at other questions to help them work out how many glasses Alfredo and Jen each need.

ASSESSMENT CHECKPOINT Can children solve a single-step capacity problem without crossing a litre boundary? Can children solve a one-step capacity problem crossing a litre boundary? Can children solve a two-step or multi-step capacity problem? Are children completing a multi-step problem or only partially completing it?

ANSWERS Answers to the **Practice** part of the lesson can be found in the *Power Maths* online subscription.

Reflect

WAYS OF WORKING Independent thinking

IN FOCUS This question encourages children to verbalise how they solved one of the problems on these pages. You may wish to specify the question they are to explain. Children should mention the steps they took in the correct order and which calculations they had to work out to answer the question fully.

ASSESSMENT CHECKPOINT Are children able to explain clearly how they solved a capacity problem with more than one step? Did children show the calculations needed in the correct order?

ANSWERS Answers to the **Reflect** part of the lesson can be found in the *Power Maths* online subscription.

After the lesson

- How confident were children in identifying the calculations within the problems?
- Did children apply any strategies that could be helpful to others?
- How successful were you and children in making the links between this lesson and previous lessons in this unit?

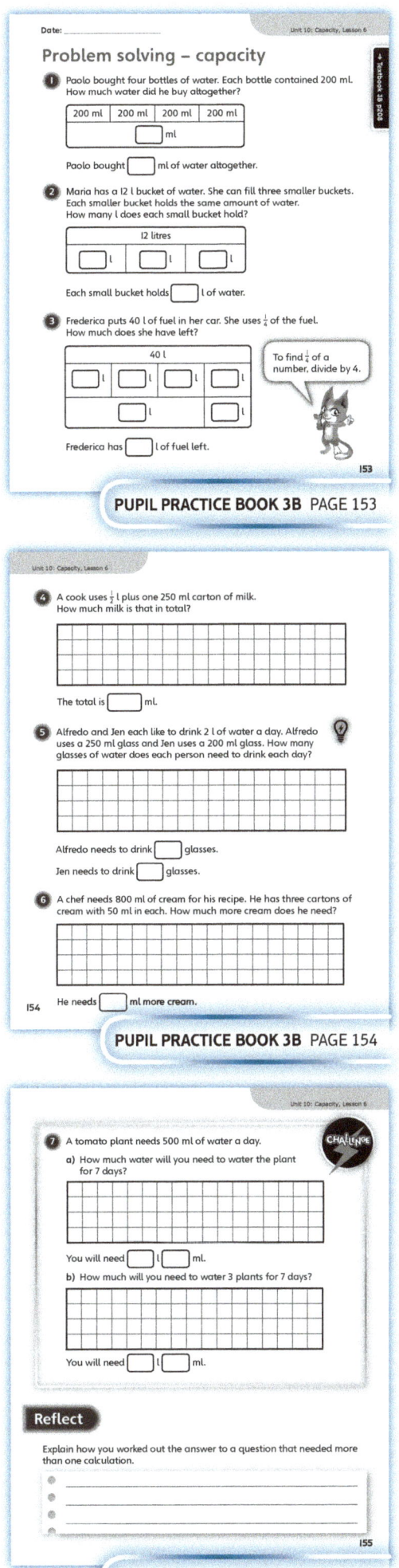

End of unit check

Don't forget the unit assessment grid in your *Power Maths* online subscription.

WAYS OF WORKING Group work adult led

IN FOCUS
- Question 1 assesses children's ability to read a range of scales involving millilitres.
- Question 2 assesses children's ability to read scales and convert from millilitres to mixed units.
- Question 3 assesses children's ability to convert from mixed units to millilitres.
- Question 4 assesses children's ability to compare capacities.
- Question 5 assesses children's ability to add capacities of mixed units without crossing the litre boundary.
- Question 6 assesses children's ability to solve multi-step capacity problems.

ANSWERS AND COMMENTARY

Children who have mastered the concepts in this unit will be able to read and measure amounts in litres and millilitres. They will be able to add, subtract and do simple conversions between the units and use this knowledge with the four operations in problem solving.

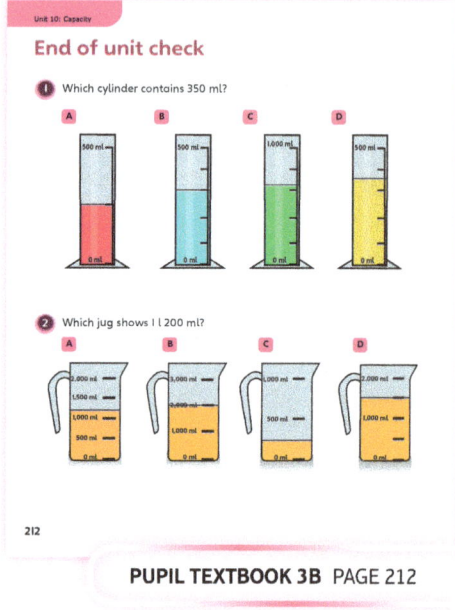

PUPIL TEXTBOOK 3B PAGE 212

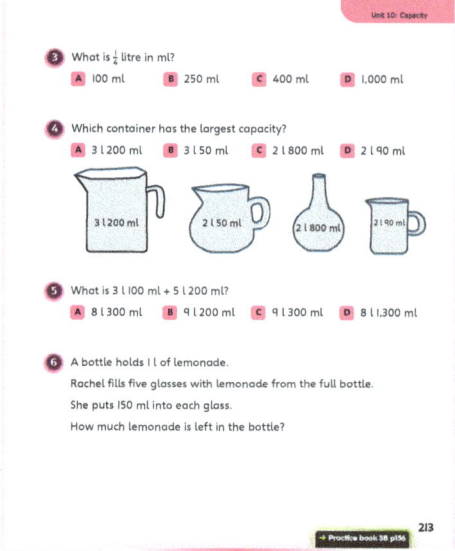

PUPIL TEXTBOOK 3B PAGE 213

Q	A	WRONG ANSWERS AND MISCONCEPTIONS	STRENGTHENING UNDERSTANDING
1	D	A suggests an error in halving 500 ml. B suggests misreading the target capacity as 300 ml. C suggests that the total capacity has been ignored when interpreting the scale.	Do not forget to make the links to understanding of number and calculating clear to children. The scale should be seen as a number line and children can use this to support their calculations. Providing opportunities for children to engage in practical exploration of capacity can help them in their understanding of the context.
2	A	C suggests children have looked at ml only.	
3	B	D suggests children have correctly converted 1 litre into millilitres but have not completed the next step of the problem.	
4	A	Any other answer suggests that children are not confident in comparing amounts given in litres and millilitres.	
5	A	C suggests children have added the millilitres correctly but the litres incorrectly.	
6	250 ml	An answer of 750 ml suggests children have completed the first part of the problem but they have forgotten to subtract to find the final answer.	

Unit 10: Capacity

My journal

WAYS OF WORKING Independent thinking

ANSWERS AND COMMENTARY

Children should be able to use the bar models provided to support their understanding of finding fractional amounts of litres in millilitres. Remind children that 1 l is the same as 1,000 ml. The use of base 10 equipment can support children's understanding. It is important that children understand that in part c), they are finding $\frac{4}{5}$ of $\frac{1}{2}$ litre rather than $\frac{4}{5}$ of 1 litre.

Question 1 a): 250 ml

Question 1 b): 750 ml

Question 1 c): 400 ml

Children should be confident in their understanding that there are 1,000 ml in 1 l.

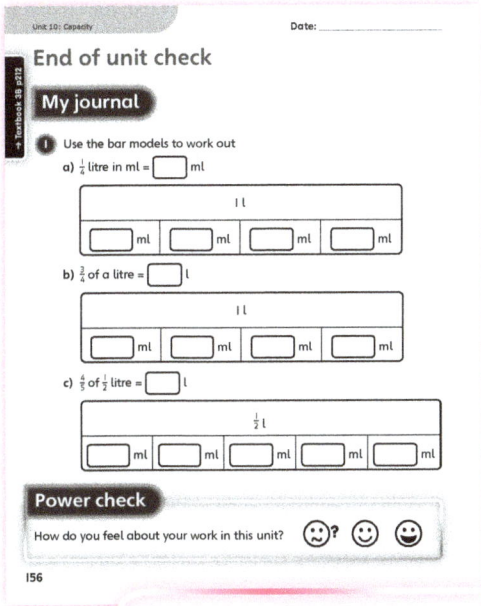

PUPIL PRACTICE BOOK 3B PAGE 156

Power check

WAYS OF WORKING Independent thinking

ASK

- How confident are you in converting between mixed litres and millilitres to millilitres?
- Do you think you can understand a range of scales to work out capacity?
- How happy are you in adding and subtracting capacities?

Power play

WAYS OF WORKING Pair work

IN FOCUS Use this **Power play** to determine how confident children are in reading scales as they add and subtract millilitres across the litre boundaries.

ANSWERS AND COMMENTARY The less confident of the pair could start by being the player who adds. They could use the scale as a number line to count on. Encourage children to try solving the calculation and finding the answer on the scale before moving their counter, recording their addition or subtraction in both ml and in mixed units.

For example, 2 litres (score of 4): 2 l + 400 ml = 2 l 400 ml = 2,400 ml, or 2 l – 400 ml = 1 l 600 ml = 1,600 ml.

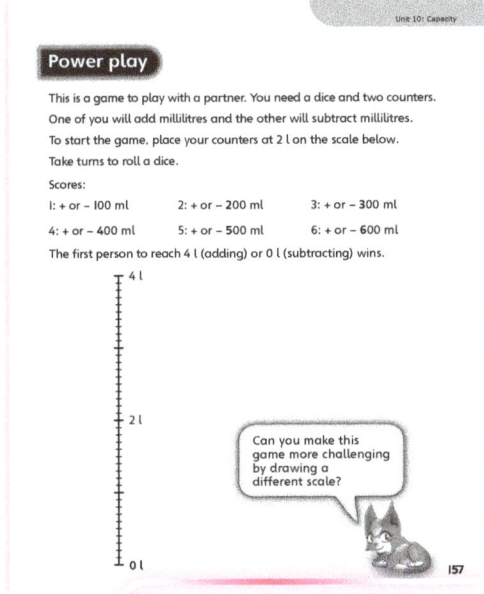

PUPIL PRACTICE BOOK 3B PAGE 157

After the unit

- The most obvious practical application for capacity is cooking. Could you provide an opportunity for children to apply their skills by following up with a cooking activity?
- The skills in this unit are transferable to other units of measure. Ask children what is the same or different about their learning in this unit and their learning in the length and mass units.

Strengthen and **Deepen** activities for this unit can be found in the *Power Maths* online subscription.

Published by Pearson Education Limited, 80 Strand, London, WC2R 0RL.

www.pearsonschools.co.uk

Text © Pearson Education Limited 2018, 2023
Edited by Pearson and Florence Production Ltd
First edition edited by Pearson, Little Grey Cells Publishing Services and Haremi Ltd
Designed and typeset by Pearson and PDQ Digital Media Solutions Ltd
First edition designed and typeset by Kamae Design
Original illustrations © Pearson Education Limited 2018, 2023
Illustrated by Laura Arias, Fran and David Brylewski, Nigel Dobbyn and Nadene Naude at Beehive Illustration; Kamae Design; and Florence Production Ltd
Images: The Royal Mint, 1971, 1990: 112, 113
Cover design by Pearson Education Ltd
Back cover illustration © Diego Diaz and Nadene Naude at Beehive Illustration

Series editor: Tony Staneff; Lead author: Josh Lury
Authors (first edition): Tony Staneff and Josh Lury
Consultants (first edition): Professor Liu Jian and Professor Zhang Dan

The rights of Tony Staneff and Josh Lury to be identified as authors of this work have been asserted by them in accordance with the Copyright, Designs and Patents Act 1988.

This publication is protected by copyright, and permission should be obtained from the publisher prior to any prohibited reproduction, storage in a retrieval system, or transmission in any form or by any means, electronic, mechanical, photocopying, recording, or otherwise. For information regarding permissions, request forms and the appropriate contacts, please visit https://www.pearson.com/us/contact-us/permissions.html Pearson Education Limited Rights and Permissions Department.

First published 2018
This edition first published 2023

27 26 25 24 23
10 9 8 7 6 5 4 3 2 1

British Library Cataloguing in Publication Data
A catalogue record for this book is available from the British Library

ISBN 978 1 292 45054 4

Copyright notice
All rights reserved. No part of this publication may be reproduced in any form or by any means (including photocopying or storing it in any medium by electronic means and whether or not transiently or incidentally to some other use of this publication) without the written permission of the copyright owner, except in accordance with the provisions of the Copyright, Designs and Patents Act 1988 or under the terms of a licence issued by the Copyright Licensing Agency, Barnards Inn, 86 Fetter Lane, London EC4A 1EN (http://www.cla.co.uk). Applications for the copyright owner's written permission should be addressed to the publisher.

Printed in the UK by Ashford Press Ltd

For Power Maths online resources, go to:
www.activelearnprimary.co.uk

Note from the publisher
Pearson has robust editorial processes, including answer and fact checks, to ensure the accuracy of the content in this publication, and every effort is made to ensure this publication is free of errors. We are, however, only human, and occasionally errors do occur. Pearson is not liable for any misunderstandings that arise as a result of errors in this publication, but it is our priority to ensure that the content is accurate. If you spot an error, please do contact us at resourcescorrections@pearson.com so we can make sure it is corrected.